Water Contamination Emergencies
Collective Responsibility

Water Contamination Emergencies
Collective Responsibility

Edited by

John Gray
Consultant

K. Clive Thompson
ALcontrol Laboratories, Rotherham, UK

RSCPublishing

The proceedings of the International Conference on Water Contamination Emergencies: Collective Responsibility held at the Royal Society of Medicine, London on 7–8 April 2008.

Special Publication No. 317

ISBN: 978-0-85404-172-5

A catalogue record for this book is available from the British Library

Published by The Royal Society of Chemistry,
Thomas Graham House, Science Park, Milton Road,
Cambridge CB4 0WF, UK

Registered Charity Number 207890

For further information see our web site at www.rsc.org

Preface

The third international conference in the generically entitled series "Water Contamination Emergencies" was the most ambitious. Two previous conferences asked the question "Can we cope?" and identified ways of "Enhancing our response". The main theme running through the two-day programme of the third conference was Collective Responsibility which developed the previous conferences by recognising the need for all those involved in responding to water contamination emergencies to share, collaborate and draw on lessons learnt. It was emphasised that achievement of the desired outcome required the co-ordination of significant effort in preparation and response. The conference provided the opportunity to understand others' roles, responsibilities and limitations, and positively encouraged the development of trust that is a vital ingredient of any emergency response team.

The book opens with an introductory chapter setting the scene from a public health perspective. The following 39 chapters were written by experts in their field, including medical and health professionals, environmental protection professionals, risk and business continuity managers, emergency planners, service and support providers, detection and equipment suppliers, disaster recovery specialists, water security experts, water distribution modellers and from water companies and regulators. Three broad themes of Operations; Information and Data Management; and Communications, and include chapters which cover a wide range of scenarios and follow the sequence of an incident, that is: the threat, the developing scenario, further response; and remediation. Chapters consider in detail the modelling of the threat; water safety plans and preparedness; security; incident management; contingency planning; remediation; vulnerability assessments; surveillance, early warning systems and detection; information management and databases; forensic investigations; interagency collaboration and integrated response; command and control; medical responses; mutual aid and exercises. The USEPA water security initiative; experience with vulnerability assessments; UK emergency planning; communicating with the public during water contamination emergencies; radioactivity monitoring in emergencies; and rapid analysis methods are also covered.

Responsibilities of particular post-holders and organisations are defined in legislation or other national requirements. Although individuals will have identified their own key responsibilities it is hoped that readers of this book will take further action to establish effective and efficient working relationships and knowledge-sharing networks with those with whom they will be required to work in an emergency - in other words, taking "Collective Responsibility".

John Gray
November 2008

Contents

Introduction: Themes and Objectives   1
J. Gray

Water Emergencies: Opening remarks   4
G. Nichols

Water is life: a view of organisational resilience in the Australian water industry   9
K.I. Gill

Online toxicity monitors and their use in distribution system and watershed early   22
warning systems
H. J. Allen, R. C. Haught and D. A. Macke

Water supply security issues and trends   36
B. Nguyen - presented by Hao-Nhiên Pham

Consequence management within the Environmental Protection Agency's water   44
security initiative
B.C. Pickard

Application of a risk based approach to security and integrity of assets – a   55
regulators view
M. J. Rink and S.A. Evans

Let's get real. Real world experiences with real-time on-line monitoring for   68
security and quality. Detecting and responding to events.
D. Kroll, K. King and G. Klein

The organisational culture of managing incidents and risks in the water sector   82
R. Bradshaw, S. J. T. Pollard, D. I. Jalba, J. W. Charrois, S. E. Hrudey and N. J.
Cromar

A simulation tool for contaminant warning system design and evaluation   117
W. Einfeld, S. A. McKenna and M. P. Wilson

CBRN modelling: application to water contamination   128
I.H. Griffiths

Planning, preparedness and security of the alternative water supply   139
K. Silcock

Procedures for the decontamination of building plumbing systems   155
S.J. Treado, M.A. Kedzierski, and V.J. Gallardo

Lessons learned from summer floods 2007. Phase 1 report – Emergency response 166
prepared by Water UK's Review Group on flooding
P. Mills

Risk assessment methodology for water utilities - RAM-W^{TM} – lessons learned 192
J. J. Danneels

Risk-based approaches to water quality management: integrating public health 198
metrics in water safety planning
G. Howard

How standards can assist the assessment of, recovery and prevention of future 207
emergencies
R. Greaves

The XX edition of the Torino Olympic Games experience: planning for and 218
responding to drinking water contamination threats
R. Binetti, P. Olivier, L. Meucci and L. Cappuccio

Sensitive, selective and simple UV-spectrometry for contaminant alarm systems 229
J. van den Broeke

Fully automated instrumentation for nucleic acid testing in the field 238
D.J.Squirrell, M.A. Lee and P. Wakeley

Optimisation of NMR methodology for non-targeted detection of water 245
contaminants
A. J. Charlton, J. A. Donarski, B. D. May and K. C. Thompson

Preventing water contamination – a co-ordinated response 252
H. Clay-Chapman

Potential sources of man-made radiochemical contamination of water resources 260
with special emphasis on the nuclear fuel cycle
N. R. Pacey and J. Cobb

Rapid methods 267
K.C. Thompson

Processing and databasing spectroscopic analyses and its use in the elucidation of 293
unkowns
I. Pierson

Handbooks to assist in the management of a radiological incident involving the 296
contamination of drinking water supplies
J. Brown, B.T. Wilkins and D. Hammond

Robust on-line total organic carbon (TOC) analyser for security monitoring 306
E. Milks

Water UK emergency planning 313
P. Fenton

The Scottish Waterborne Hazard Plan 316
M. McGuinness

Research related to water security 320
K. Fox

Early warning and reports 325
V. Murray

OK, we've got a problem, so who do we tell? Inter-agency communication – a 337
water company view
D. A. Woolloff

Review and evaluation of water concentration technologies for analysis by real- 342
time PCR
S.L. Cunningham and B.M. Dowling

Scientific and Technical Advisory Cell (STAC) – getting timely public health 353
advice to multi-agency frontline responders
R. Carr, S. Ibbotson and VSG Murray

Communicating with the public during water contamination events: addressing 360
vulnerable populations
P. A. Nsiah-Kumi

Medical preparedness for water contamination events 369
P. L. Meinhardt

Keeping the public on-side and maintaining reputation 383
J. B. Shaw

Sociological and psychological constraints to learning from failure 389
B.H. MacGillivray

Lessons learned from major contamination incidents – a discussion 397
M. Furness

Review of conference 401
J. Gray

List of posters 405

Subject Index 407

INTRODUCTION: THEMES AND OBJECTIVES

J. Gray

Consultant – Water safety and security. 1 Faraday Ride, Tonbridge, Kent, TN10 4RL, UK

1 INTRODUCTION

This is the third conference in the series "Water Contamination Emergencies" and is the most ambitious. The first conference in 2003 in Kenilworth, Warwickshire[1], challenged delegates with the question, 'Can We Cope?' The conference focussed on the more traditional water quality incidents that have recently occurred and considered the handling of seven major case studies of contamination events. Papers were presented by invited experts and, together with posters, reviewed lessons learned and identified ways to better cope in future.

It was concluded that there was particular key information needed to respond appropriately to incidents, including:

- the need to be aware it has happened;
- knowledge of the extent of problem;
- accurate and timely identification and quantification of contaminant; and
- identification of the origin of contaminant.

Other key features include consideration of:
- remediation issues;
- communication with consumers and users;
- liaison with health professionals and local authorities; and
- communication with media.

The conference concluded that, in general, water companies and key agencies respond well and were well positioned to deal with typical emergencies but more awareness and pre-planning would improve the response.

The second conference in 2005 in Manchester[2] addressed the issue of "Enhancing Our Response". It provided a forum for sharing new knowledge gained from two years experience following the first conference and considered the development of expertise and responses to contamination events, particularly in the areas of analysis and emergency planning. There was an international focus and particular consideration of possible malicious acts of contamination. New strategies and technologies were presented and again there was emphasis on communications with and between all parties, with the need to address consumer perceptions. A prime requirement for robust and rapid screening of samples was confirmed and areas of good practice were identified. The conference also considered the potential challenges associated with preparedness and heightened security in the CBRN context.

Again there was evidence of progress in preparedness, both in the UK and globally, and a desire to see greater interaction between key players in an incident to develop improved procedures and provide learning opportunities. This gave an immediate focus for this third conference.

2 THEMES AND OBJECTIVES

This conference has as its theme 'Collective Responsibility'. The organising committee believes that individuals should identify their own key responsibilities and take action to establish effective and efficient working relationships and knowledge-sharing networks with those with whom they will be required to work in an emergency – in other words, taking Collective Responsibility. A key aim of the conference is to provide a sharing forum for individuals in key organisations actively involved in any water contamination incident. Collective responsibility requires prior consideration of how to plan, respond, communicate and manage risk in an emergency and how to share, learn and develop after the event. We question whether individuals and teams have fully learned, shared best practice, established effective networks, collaborated, targeted research or exercise together. There appears to be no common set of values for evaluating risk and prioritising that risk; a constrained enthusiasm and desire to make a common approach work; and there seems to be little opportunity to modify systems. By encouraging international involvement, it is hoped that broader networks to promote these objectives can be established.

In order that all subject areas can be appropriately covered, three streams have been chosen for the delegates:

- Operations (systems, security and procedures);
- Data/Information (quality testing, analysis, on line monitoring, modelling and data interpretation); and
- Communication (emergency planning, public communication and inter agency working).

The two days of the conference are further divided into sessions which cover planning and preparedness, security and initial responses, incident management and the aftermath. The conference will end with three plenary sessions on day two.

This conference will deal with the full range of potential contaminations scenarios - chemical, biological and radiological. It will be of strategic, tactical and operational interest to organisations involved in civil contingency response, emergency planners, medical and health professionals, local authorities, service and support providers, detection equipment suppliers, disaster recovery specialists, remediation companies, research organisations and water companies and regulators. It will also provide a significant networking opportunity which all delegates are encouraged to utilise to the full.

References

1 Water Contamination Emergencies – Can we Cope?   Edited by J. Gray and K.C.Thompson, RSC Publications, ISBN 0-85404-628-3, 2004.

2 Water Contamination Emergencies – Enhancing our Response, Edited by K.C.Thompson and J. Gray and RSC Publications, ISBN 978-0-85404-658-4, 2006.

G. Nichols

Deputy Director, Environmental and Enteric Diseases Department, Health Protection Agency, Centre for Infections, 61 Colindale Avenue, London, NW9 5EQ, UK.

1 INTRODUCTION

Emergencies related to drinking water remain a cause of public concern. This conference will cover public health, waterborne disease and flooding looking at the planning and preparedness, security and initial responses, incident management and the examining the aftermath of such emergencies. It will look at operations, information and data management and communications and their importance in tackling these emergencies.

2 EXAMPLES OF OUTBREAKS

I have been involved in the investigation of outbreaks and incidents of food and waterborne disease for most of my career and in varying capacities. These range from a community outbreaks of *Shigella sonnei*, a hospital outbreak of *Clostridium perfringens*, early work on the role of *Cryptosporidium* in diarrhoeal disease, microsporidiosis in AIDS patients, outbreaks of cryptosporidiosis related to drinking water, badly pasteurized milk and swimming pools. There have been foodborne outbreaks such as *Salmonella* Enteritidis associated with imported eggs, *Salmonella* Schwarzengrund linked to chocolate coated nuts and *Salmonella* Barelli in sandwiches. I was involved in responses to flooding in 2000 and 2007, the foot and mouth outbreak in 2001 and the Yorkshire water drought in 1995. I have also investigated outbreaks of *Pseudomonas aeruginosa* folliculitis, Pontiac fever associated with a spa pool and an increase in deaths associated with *Clostridium novyi* in injecting drug users and have been involved in decisions about instituting and lifting boil water notices and withdrawals of bottled water that were subject to faecal contamination.

We have looked at 114 waterborne outbreaks involving 22,975 people associated with public and private drinking water supplies in England and Wales between 1910 and 1999[1]. Outbreaks during the first half of the 20th century that are recorded in the scientific literature were predominantly typhoid and paratyphoid fever. These organisms declined during the second half of the 20th century and in the last two decades the newly discovered organisms *Cryptosporidium* and *Campylobacter* were associated with the majority of outbreaks. These differences tell us that surveillance identifies real changes in disease (e.g. reduction in typhoid and paratyphoid), changes in laboratory detection (e.g. *Campylobacter* and *Cryptosporidium*) and changes in surveillance and outbreak detection. The outbreaks identified have a wide geographic spread and if you're in area hasn't had an outbreak there may be one within the next decade or two. There have been dramatic changes in pathogens over the last 100 years. Will we be looking back in the next century at similar changes and are there pathogens that we are currently missing?

3   THE IMPACT OF RAINFALL EVENTS

Of the outbreaks identified in this work, 89 had sufficient information about the time and place of occurrence to allow rainfall data to be collected for the 90 days before the onset of the outbreak. It was also possible to collect the rainfall for the same location and dates for the previous five years to use as a control.

There was a significant association between excess cumulative rainfall in the previous 7 days and outbreaks (p= 0.001). There was an excess of rainfall below 20mm for the three weeks previous to this in outbreak compared to control weeks (p= 0.002). These data imply that the attributable fraction of outbreaks associated with a sustained period of low rainfall is 20% compared to a period of heavy rainfall of 10%. Because the dataset used in this study are historical the results may not reflect current risks, because water companies may have adopted treatment strategies which limit the problems associated with heavy rainfall. Because 15% of the outbreaks were preceded by heavy rainfall in the week before the outbreak and 28% of outbreaks were preceded by a period of lower than average rainfall in the three weeks before the week of the outbreak it is important to consider weather when constructing water safety plans for public and private drinking water supplies.

4   WATERBORNE OUTBREAKS

From the outbreaks that have been published we can build evidence of what problems cause outbreaks related to drinking water[2]. These include the contamination of surface waters by animal waste and sewage, and problems of the streaming of waters on the surface of lakes. As I have indicated, weather can influence risk with groundwater vulnerability in drought, reduced dilution of sewage effluent into rivers in drought times, lower water table opening up new surface to ground routes, contamination of ground water, heavy rainfall and in some countries problems with ice melting quickly. There have been problems with the treatment operation and management of drinking water treatment works including filtration bypass, failure of slow sand filters, the performance of filtration in removing pathogens, the recycling of backwash water and turbidity control. In distribution there have been problems with cross contamination through incorrect plumbing backflow, post-treatment contamination, network repair and problems with aqueduct integrity. In addition there have been examples where several outbreaks have occurred at the same time that appear to be related as well as outbreaks occurring in the same location as previous outbreaks.

The introduction of new *Cryptosporidium* regulations in 1990, has been associated with improvements in the removal of cryptosporidiosis from drinking water treatment works, the removal of some poor supplies and the building of new water treatment facilities. A large drinking water outbreak in the North West of England in 1999 resulted in new drinking water treatment measures being introduced over the next few years. Surveillance data from the health authorities associated with this outbreak were compared to other health authorities in the area that were not affected. It appeared that unrecognized outbreaks had occurred in most of the previous years. Following the introduction of the *Cryptosporidium* regulations in 2001 the annual spring outbreaks have disappeared and the overall number of cases per year has declined substantially[3,4]. This works suggests that we can easily miss outbreaks. Problems occurring every year may be wrongly viewed as seasonal trends rather than regular outbreaks. Comparison between areas over time can be useful. There has been a significant burden of *Cryptosporidium* related disease associated with drinking water. This burden has substantially decreased but much of the burden has not been within identified outbreaks. The disease reduction has been predominantly with infection due to *Cryptosporidium parvum* and the reduction is mostly in the spring period. It is likely that a burden of illness associated with drinking water still remains.

5 FLOODS

In 2007 there were a number of problems relating to flooding. A bottled water product had to be withdrawn due to its contamination with *E. coli*. The contamination of the borehole occurred after heavy rain. An incident occurred in which surface flooding entered the final water tank in a water treatment works as a result of seals in the lid not being watertight. A boil water notice was introduced for a short period. The Mythe water treatment works supplying Tewkesbury was put out of use as a result of flooding of the treatment works. The population was supplied by bowser and bottled water supplies. A national emergency coordination centre (NECC), a national surveillance cell, regional emergency operation centres (EOC), health protection units and a command structure were set up to deal with the problem and there was a scientific and technical advisory cell (STAC) set up to provide scientific advice. These incidents raise the question of whether utilities are well enough prepared for the floods which can occasionally affect them. This was the first level 4 incident experienced by the HPA since the finalisation of its emergency plan. A coordinated approach based on emergency plans was established and the appropriate structures were created and their respective tasks undertaken. Limited resources were used efficiently and effectively. Various risks were identified and addressed using evidence and expert advice. Partner organisations were supported, advice to the public was produced and ultimately public health protected.

The summer floods in 2007 have been described as the country's largest peacetime emergency since World War II[5].The Mythe treatment works supplies water to around 350,000 people in Tewkesbury, Cheltenham and Gloucester[6]. Half a metre of flood water covered the site affecting buildings offices and equipment and preventing staff from returning to three days. Environment agency staff, fire and rescue services and other organizations quickly put up temporary barriers around the site and restored it to normal service as quickly as possible. The works was out of action for 17 days as a result of the flooding and 140,000 households were without water. More than 50 million litres of bottled water were provided to those affected. Following the floods more permanent defences were built around the site and extra pumping equipment was installed. The overall cost of the flooding at Mythe has been estimated at between 25 million and 35 million pounds.

The Walham electricity substation is built on the raised ground in the River Severn flood plane to the north of Gloucester. It provides power to half a million homes across Gloucestershire in South Wales. It was necessary to construct a 1000 m flood defence to protect the site. The response involved environment agency staff alongside the fire and rescue services, local authorities, utility companies and the military on 22nd July to work on temporary defences to protect the site from flooding. The work was conducted in extremely difficult conditions. It was dark and wet with floodwaters rising fast and it was potentially very dangerous with live high-voltage equipment within metres of where staff were working. After 10 hours the site was secured and the fire and rescue services began pumping water out of the critical area. Work was completed just in time narrowly averting a major shutdown of the site which would have left half a million homes without power. Power to 42,000 homes from nearby Castle Mead substation was cut temporarily whilst defences were put in place. More permanent defences have now been constructed around both Walham and Castle Mead substations and the electricity industry must make more effective long-term plans to protect the many other sites at risk from flooding.

What does this tell us? This was a severe flood. The flooding at both Mythe and Walham should not have been unexpected. Flood maps showed these and many other critical sites are vulnerable to flooding. Mythe flooded in 1947 and 2000 and narrowly escaped flooding in 1990 and 1998. The summer floods must now be a wake-up call for the water industry and other utilities to take action. A substantial percentage of public utilities may be at risk from flooding.

There are potential risks of infection following flooding. Flooded areas may act as breeding grounds for mosquitoes, there can be disturbance of rodent populations, waterborne outbreaks, contamination of water supplies and contamination of people and clothing of those in contact with floodwaters. Runoff from fields and storm drains can make floodwater contaminated at the start of a flood, but substantial dilution can reduce the importance of this. People whose houses are flooded may have problems with damp for months afterwards. Despite this, flooding in developed countries carries a relatively low risk of infectious disease outbreaks. Prevention can be attributed to effective public health systems and good surveillance. There can be psychological distress and illness as a result of the disruption associated with flooding, and there is a significant risk from carbon monoxide poisoning in people using pumping or drying equipment indoors.

5   REPORTING OF OUTBREAKS

Where a serious incident has occurred it is important for the outbreak to be reported by the investigating team, and where possible reported in the scientific literature. The reporting is an important opportunity for examining the successes and failures of the incident, including praise, censure, audit and learning. It can provide an opportunity for increasing awareness of new problems within the water industry and the health community. On occasions it can necessitate changes in water treatment paradigms and occasionally legislation. Legal action has on occasion delayed publication of scientific papers. In reviewing water emergency incidents there is an opportunity to examine the roles and perspectives, to raise questions and to examine shared objectives by the different organizations involved. It is important to ensure that arrangements are made to get across consistent messages to the public from the different organizations involved. There is an opportunity to examine what went well, what went badly, what should be done in the future and what we are already doing. There is a chance to clarify our roles and responsibilities and to share information with other professionals. There is an opportunity to influence policy to review research, to examine organizational resilience and to take any preventive action that may be deemed necessary.

6 CONCLUSION

In conclusion it is important for there to be regular contact between water companies and the health protection units and the HPA will be working with DWI to improve advice during incidents and to improve training.

References
1 G. Nichols, C. Lane, N. Asgari, V.Q. Verlander and A. Charlett, Rainfall and outbreaks of drinking water related disease and in England and Wales. 2008. J.Water Health, In press.
2 S.E. Hrudy and E.J. Hrudy, Safe Drinking water: Lessons from recent outbreaks in affluent nations, 2004, London: IWA.
3 W. Sopwith, K. Osborn, R. Chalmers and M. Regan, The changing epidemiology of cryptosporidiosis in North West England, 2005, Epidemiol Infect; **133,**(5),785-793.
4 G.L. Nichols, R.M. Chalmers, W. Sopwith, M. Regan, C.A. Hunter and P. Grenfell, Cryptosporidiosis: A report on the surveillance and epidemiology of Cryptosporidium infection in England and Wales, 2006, Drinking Water Directorate Contract Number DWI 70/2/201, Drinking Water Inspectorate.
5 Sir M. Pitt, The Pitt Review - Lessons learned from the 2007 summer floods, 2008, Cabinet Office, pp505.
6 Sir J. Harman, Review of 2007 summer floods, 2007, Environment Agency, pp60.

WATER IS LIFE: A VIEW OF ORGANISATIONAL RESILIENCE IN THE AUSTRALIAN WATER INDUSTRY

K.I. Gill

Department of Sustainability and Environment, Melbourne Victoria, Australia

"In all my experience at sea, I have never been in any accident of any sort worth speaking about. I have seen but one vessel in distress in all my years at sea. I never saw a wreck or have been wrecked nor was I ever in any predicament that threatened to end in disaster of any sort"

E.J. Smith Captain RMS Titanic

1 ABSTRACT

Most people are familiar with the term resilience. The question is often asked, why are some organisations more resilient than others? Is there a single common theme or are there a range of characteristics that contribute to a resilient organisation? This paper provides a view of organisational resilience in the water industry in Australia, particularly in the Victorian context of a risk and business continuity.

The objective of this paper is to further stimulate thought regarding issues associated with organisational resilience, with particular interest in indicators of resilient behaviour.

A key area for consideration is behaviours and characteristics of a resilient water business.

2 INTRODCUTION

Water is life.

Looking back over history, the supply of fresh clean safe drinking water has been the cornerstone of a well developed society. Water supply systems have sustained the growth of urbanised civilisation since ancient times[1] and today include large cities such as London and Melbourne.

Settlement in Australia, the driest inhabited continent in the world, has created many unique challenges for its water industry. The current drought, being one of the longest on record[2], has highlighted the scarcity of water in Australia. Drought over the last ten years in Victoria has left average stream flows 60-90% below average[3]. In 2006-07 annual rainfall and stream flow were lowest on record. This has resulted in reduced inflows of surface water into drinking water catchments. About 90% of all water used in Victoria is sourced from surface water and the rest from ground water.

The consequence of low water levels in many reservoirs means that 366 Victorian towns are currently on water restrictions[4]. This creates a sustained challenge for water businesses to manage water quality and quantity risks and combined with the current economic and security environment has created a new set of challenges[5] for a resilient water industry.

This paper draws on the latest study by the *National Resilience Working Group* conducted at Emergency Management Australia (EMA) attended by Australian delegates from across government and private sector business.[6]

3   WATER MANAGEMENT IN AUSTRALIA

Since 1901 Australia's seven states and territories have benefited from federation. The federal constitution[7] has delegated powers for water management to the states and territories. The states and territories also have constitutional responsibility for emergency management and control most of the functions essential for disaster prevention, response and recovery. These institutional arrangements provide a rich vein of water management strategies that vary from state and territory.

Within the state of Victoria the there are twenty independent water corporations[8] created under the *Water (Governance) Act* 2006. These organisations own and operate water, sewerage and dam infrastructure for a population of 5.2 million people. The *Emergency Management Act 1985* and the *Emergency Management Manual Victoria* provide the framework for emergency management in which the state and the water industry operate.

The Victorian state government Department of Sustainability and Environment has portfolio responsibly for water management that includes governance arrangements to ensure appropriate risk and emergency management arrangements are in place for water, sewerage and dam safety incidents.[9]

Governments and water businesses continue to investigate strategies in organisational resilience to minimise disruptions to water and sewerage services that have social, economic, political and environmental consequences.

4   WHAT IS RESILIENCE?

Resilience has been interpreted in a number of ways in recent times. The origin of the word resilience is derived from the Latin words *reiliens* and *resilire* – first recorded in 1626 – meaning "to rebound".[10]

The concept of resilience can be applied to individuals, organisations, sectors and communities. The abundance of definitions and the fact that this concept is shared by many different disciplines make it particularly difficult to define uniformly across the water industry. However, these two definitions are especially comprehensive in the context of risk and business continuity for the water industry:

- ...."the ability to face internal or external crisis and not only effectively resolve it but also to learn from it, be strengthened by it and emerge transformed by it, both individually and as a group."[11]

- "A resilient organisation is one that is not only able to survive, but also to thrive it the face of adversity."[12]

5   A PROPOSED ORGANISATIONAL RESILIENCE MODEL

A key concept introduced by the author in this paper is the Organisational Resilience Model (Figure 1). The model comprises three resilient themes:

   a.   Resilient Processes
   b.   Resilient Infrastructures
   c.   Resilient Behaviour

The underlying principle on which this model was founded is that for high levels of organisational resilience, a water business requires a capacity and capability in all three resilience themes.

There has been significant work in the area of resilient processes and this paper will expand on three initiatives; Critical Infrastructure Protection (CIP), the Security Vulnerability Risk Assessment Guideline (SV-RAG) and the Critical Infrastructure Modelling and Analysis Program (CIPMA).

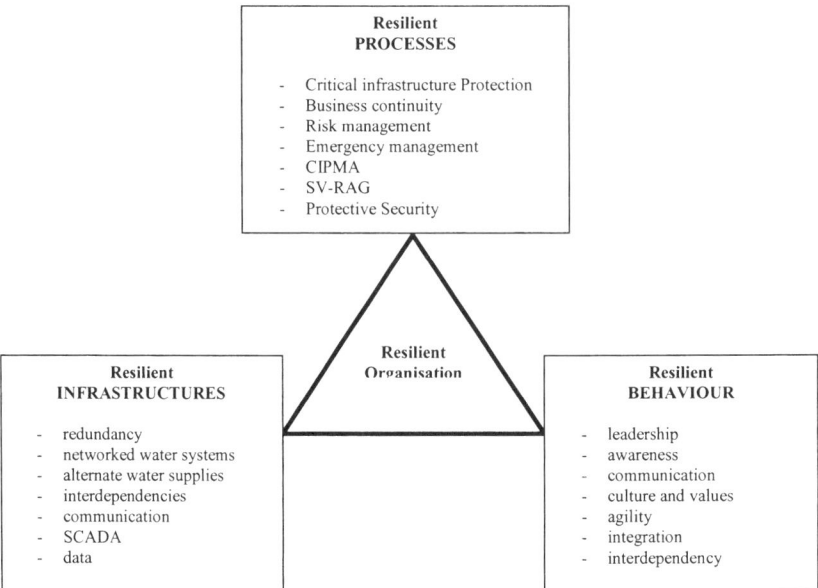

Figure 1 *Organisational Resilience Model*

5.1   Critical Infrastructure Protection (CIP)

Critical infrastructure protection[13] brings together a significant number of existing strategies, plans and procedures that deal with the prevention, preparedness, response and recovery arrangements for disasters and emergencies. The *National Guidelines for the Protection Critical Infrastructure from Terrorism* support owners and operators in this area. In the Victorian context, protection of designated critical infrastructure is now supported by legislation in the *Terrorism (Community Protection) Act* 2003, with operators of declared essential services having a requirement to prepare a risk management plan and undertake an annual exercise to test the plan.

5.2   Critical infrastructure Modelling and analysis Program (CIPMA)

The Critical infrastructure Modelling and analysis Program[14] (CIPMA) is an initiative of the Attorney General's Department to protect critical infrastructure through modelling and simulating dependency relationships of critical infrastructure systems. It provides an opportunity for owners and operators to understand interdependencies with other sectors and identify risks to the resilience of other infrastructures.

5.3   Security Vulnerability Risk Assessment Guideline (SV-RAG)

In 2003, the Victorian water businesses developed the Security Vulnerability Risk Assessment Guideline (SV- RAG). The SV-RAG was developed out of a need to provide a consistent methodology to identify system vulnerability and critical infrastructure using a risk management approach. While there are many other programs and manuals available in varying degrees of complexity, SV-RAG provides the water industry with a structured, logical process that has provided water businesses a decision support tool to identify security vulnerabilities, assess criticality of assets and identify critical infrastructure.

6   RESILIENT BEHAVIOURS – FURTHER EXAMINATION

Resilient behaviours seem to be less studied and require further examination.

The theme of resilience behaviours relates more to organisational culture. The capacity of resilience is found in a water businesses culture, attitudes and values. Dr Erica Seville of the University of Canterbury captures the essence of resilience *"Resilience is not something you do, it is something you are"*. These values complement the *"process"* and *"infrastructure"* elements of resilience. *"Businesses that acknowledge and embrace those elements which are non tangible truly make resilient organisations."*

Within sections of the water industry the context of organisational resilience has been described as a business outcome and a set of objectives. Therefore resilience is more than just a plan, process or a checklist. It is also about people and the resilient values of those people and how collectively they can transition those values to create a culture of organisational resilience.

The current research suggests that higher levels of resilient behaviour can complement the existing suite of resilient infrastructure and resilient processes and create a truly resilient organisation.

6.1   Resilient behaviour indicators

The Resilience Community of Interest[15] and Sydney Catchment Authority in New South Wales, Australia are considering the merits of resilience and working towards defining a possible approach to building resilience in the future via a Resilience Strategy. Table 1 shows an extract from the draft Resilience Maturity Model that is being developed and may be used to diagnose the level of resilience behaviours in an organisation[16]. This set of questions could also be used more broadly across the water industry to determine the levels of industry resilience.

Table 1 *Resilience Behaviours*

Behaviour	Rating					High Resilience
Agility						Builds active, flexible, agile, adaptive thinking and actions during response and recovery operations
						Solutions to problems encouraged at all levels in the organisation. Rapid adaptive behaviour
						Seeks opportunity during adversity
						Preparation for adversity is a priority
						Utilises teams with diverse skill sets
						Capitalises on incidents
						Retains lessons from past failures for future learning
Integration						Aligned risk management objectives across functional groups
						Resilience governance strongly integrated
						Governance outcome focused
Interdependency						Strong succession planning and redundancy
						Silos readily offer support to each other. One in all in approach. Silos unite to achieve objectives in times of adversity
						Supply chain vulnerability understood and planned for
						Mutual aid plans with industry peers
						Trusted relationships with key stakeholders
						Strong partnership with regulators
						Staff are willing and able to support the organisation in times of adversity.
Leadership						Empowered by crisis
						Clear direction in crisis
						Sense of hope and optimism in response and recovery
						Clear objectives and goals in response and recovery
						Partnerships - working with others
						Strategy to boost staff morale in times of adversity
						Empowered devolved decision making
Awareness						Anticipates and understands emerging threats
						Knows organisational vulnerabilities and breaking points
						Understands and mitigates staff vulnerability
						Understands community vulnerability
Communications						Diverse constant communications across silos
						Regular trusted communication with stakeholders
Culture and Values						Values are aligned, shared and believed
						Strong unity of purpose
						Enthusiasm for challenge
						Strong development of social capital
						One in all in. Pull together in adversity
						Disruptions are recognised as an opportunity for improvement and to build strengths

6.2   A proposal - What would a resilient organisation look like?

To further the discussion on organisational resilience the Federal Attorney General's Department of the Australian Government sponsored the *National Resilience Workgroup* in Australia.

This research to date suggests that resilience capability is strongest in an organisation that:

- Has good situational awareness and an ability to anticipate and understand emerging threats.
- Is able to understand the impact of threats on business, supply chain, the community in which they operate and upon employees' lives.
- Has supportive partnerships with governments and critical interdependencies,
- Respond to and recover from service disruptions as a unified whole-of-organisation team
- Has an adaptive capacity to disruptions and reacts flexibly to restore routine functions and strengthen the organisation
- Ensures staff are willing and able to support the organisation to achieve objectives in times if adversity.
- Leads with clear direction while enabling devolved problem solving.

7   CASE STUDY IN RESILIENCE – 2007 A YEAR OF CHALLENGES

Examination of a case study in the Gippsland region of Victoria, Australia will contextualise and illustrate the above aspects of organisational resilience.

Gippsland Water, East Gippsland Water and Southern Rural Water are three water businesses that have faced unprecedented extremes in weather conditions during 2007. While doing so they have been able to maintain water and sewerage services to stakeholder satisfaction.

Figure 2 depicts extreme bushfire conditions created after ten years of drought. The result - 69 days of bushfires that scorched 1.1 million hectares of private and public land including water catchments. Previously, in 2003, 87 bushfires burnt 1.3 million hectares over 59 days. (total area approx. 10% of England).   The provision of adequate water supplies to the communities and emergency services during these times challenged the resilience of water businesses given the direct threat of fire to water and sewerage infrastructures.

Figure 2 *Bushfire in Gippsland, Jan 2007*

Compounding the impacts of fire were two intense rain events that caused significant flooding in the region (figure 3). In the June 2007 incident, Bairnsdale airport received 322.6mm of rain, exceeding previous records.[17]

The floods impacted the Sale Water Treatment Plant and the Bairnsdale Sewerage Treatment Plant, with inundations of sewerage pipes and pumpstations and creating access problems to water and sewerage infrastructures.

Figure 3 *Gippsland floods, Jun 2007*

Apart from the direct impact of fire and flood, figure 4 shows an example of increased turbidity in runoff that was experienced in the burnt catchments that created raw water quality issues.

The challenge for the water businesses was to treat highly turbid surface water and create fresh clean safe drinking water.

Figure 4 *Mitchell River, Gippsland, Jul 2007*

Finally at the end of December 2007, a Blue-Green algal bloom of the genera *Synechococcus* began to spread across the entire Gippsland Lakes system[18]. This created a potential human health issues if the bloom turned toxic. Levels detected could cause skin irritation, mild respiratory and hay-fever like symptoms on contact. The Gippsland Lakes are of great environmental importance and contain a number of sites of international significance under the Ramsar Convention on wetlands (figure 5).

Figure 5 *Gippsland Lakes, BGA outbreak, Dec 2007*

An aggregation of the consequences from these weather events over a short period of time required high levels of organisational resilience, in particular resilient behaviours by staff and contractors. This was particular evident in the areas of leadership, agility and interdependencies in the water industry.

The township of Licola is a good example where Gippsland Water demonstrated these resilient behaviours. Staff responded to calls for assistance from the Lions's club who requested help to replace the townships water reticulation system despite Licola not being in Gippsland Water service area. Staff proactively assessed the situation, conducted tests and with high levels of adaptive capacity constructed a temporary water treatment system in the local swimming pool[19].

Similarity, with East Gippsland Water, resilient behaviours were required when dealing with poor water quality in the Mitchell river catchment. A strong unity of purpose from management, 40 consultants and contractors fast tracked the construction of emergency infrastructure at Woodglen, producing 18ML/day of drinking water for the towns of Bairnsdale and Lakes Entrance.[20]

Southern Rural Water's good situational awareness and ability to anticipate and understand emerging threats were demonstrated managing the impacts of torrential rain in the Macalister catchments and into Lake Glen Maggie[21] (Figure 6). High levels of adaptive capacity to disruptions allowed the organisation to quickly return to normal operations, while learning's from the response to fire and flood have strengthened the organisation.

Figure 6 *Lake Glen Maggie – intense rainfall event, Gippsland, Jun 2007*

Clearly, given these sustained, concurrent and compounding challenges these water businesses displayed high levels of organisational resilience.

8   THE NEXT BODY OF WORK

The United Kingdom is an ideal location for future collaborative work on resilience and to confirm assumptions from the recent body of work conducted by the Resilience Community of Interest in Australia. The two nations share a number of similarities, and risks. Both are modern western societies with like minded values, government structures and institutional arrangement in the water industry.

Resilient behaviours is an area that requires further research, particularly with refining the behaviours proposed in Table 1. There should also be an examination of methods that best allows us to retrospectively study resilient behaviours of organisations following a significant emergency.

The relationship between social resilience and organisational resilience requires further research. Some argue that organisational resilience relates directly to existing levels of individual and social resilience[22] and that resilient organisations should be fostering and involved in processes enhancing social resilience.[23] Similarly the underlying resilience at community level maybe a contributing factor at organisational level as mentioned Attorney-General Robert McClelland address at Hotel Realm, Canberra[24]. *"Community resilience is also an important part of achieving social inclusion. Helping your mates in times of crisis and supporting the community are key elements of the Australian way of life....It's a very Australian thing to do."*

9 CONCLUSION

Disasters, both natural and man made have happened before and will certainly happen again. While not all risks can be mitigated they can be managed through a range of organisational resilience strategies.

The underlying theme of this paper is that organisational resilience is a combination of three resilient themes – infrastructures, processes and behaviours.

Leading water businesses will be able to maintain higher levels of service delivery by continuing to develop capacity and capability in resilience, with particular focus on resilient behaviours.

With current world events and an ever changing risk profile now more than ever we need to manage this precious water resource well.

10 ACKNOWLDGEMENTS

The author fully acknowledges the efforts of the Victorian Water Security and Continuity Network (Water SCN) and the Water Services Information Assurance Advisory Group (WS-IAAG). The views expressed are those of the author alone.

Bibliography
David Templeman and Anthony Berigin, *Taking a punch: building a more resilient Australia* Strategic Insight, May 2008, Australian Strategic Policy Institute.

References
1   J. Fawell and G. Standield, *Pollution, Causes, Effects and Control*, 4th Edition, R.M. Harrison, RSC, 2001.

2   B.Timabal and B.A Jones, *Future projections of winter rainfall in southeast Australia using a statistical downscaling technique*, Springer Netherlands, 2007.

3   Department of Sustainability and Environment, *Annual Report 2007*, p10, 2007, Australia.
    http://www.dse.vic.gov.au/DSE/dsencor.nsf/LinkView/02076021149F07E1CA2571F80
    008065F9B8E64CAE7EA2CEACA256DAC002901AF

4   Department of Sustainability and Environment, *Monthly Water Report – March 2008*, Victorian State Government, 2008.
    http://www.dse.vic.gov.au/DSE/wcmn202.nsf/LinkView/01D1B3BC630EC312CA2574
    4B001068528C5DF181636FAC17CA25744B00175D62

5   K.I.Gill, *Water Matters*, 2006, Victorian Water Industry Association News, vol 5 issue 2, p1.
    www.vicwater.org.au/uploads/Water%20Matters/Water%20Matters%20Autumn.pdf

6   D. Parsons, *National organisational resilience Framework Workshop – The outcomes,* 2007, Emergency Management Australia.
    www.resorgs.org.nz/FINAL%20**Workshop**%20Report.PDF

7   Australian Government, Commonwealth of Australia Act 1900.

8   Department of Sustainability and Environment, *Our water Our Future - Your local Water Corporation*, Victorian State Government, 2008.
    http://www.ourwater.vic.gov.au/governance/water_corporations/your_local_water_corp
    oration

9   Department of Sustainability and Environment, *"Our water Our Future - Statement of Obligations"*, Victorian state Government, 2008.
    http://www.ourwater.vic.gov.au/governance/water_corporations/statement_of_obligatio
    ns

10 E. Oldfield, *Organisational Resilience* Organisational Resilience, 2008, pp 1-10

11 G. Brenson-Lazan, *Group and social Resilience Building,* 2003, URL:
    www.commmunityatwork.com/resilence/resilienceiaeng.pdf

12 J. Vargo, E. Seville, *Crisis Strategic Planning: Finding the Silver Lining*. 2008, World Conference on Disaster Management, Canada

13 Commonwealth, *Critical Infrastructure Protection,* 2003.
    http://www.ag.gov.au/www/agd/agd.nsf/Page/Nationalsecurity_CriticalInfrastructurePro
    tection

14 Commonwealth, *Critical infrastructure Modelling and analysis Program fact sheet*, 2007. http://www.tisn.gov.au/agd/WWW/TISNHome.nsf/Page/CIP_Projects

15 R. Oldfield, *Organisational Resilience*, Organisational resilience Community of Interest, 2007, Australia.
    http://www.thebci.org.au/NSW%20Resilience%20Mar%2008.pdf

16 S. Hancock, *Resilience Management Strategy 2008-2012*, Sydney Catchment Authority, NSW, Australia, 2008.

17 Gippsland recovery Committee, *Gippsland Emergency Recovery Program*, 2007,
    http://www.gippsrecovery.com/history.html

18 Department of Sustainability and Environment, *Blue Green Algae in Gippsland lakes*, Victorian state Government, 2008.
    http://www.dse.vic.gov.au/DSE/wcmn202.nsf/childdocs/-

90788670E97F8698CA257216007CA773-D655FF33C1395FD8CA2572C6001C9A27-77B382A162874CA7CA2573C6001739CB?open

19 Gippsland Water, *Annual Report Gippsland Water 2006-07*, p46, 2007, Australia. http://www.gippswater.com.au/Portals/0/Gippsland_Water_Annual_Report06-07_WEB.pdf

20 East Gippsland Water, *Annual Report East Gippsland Water 2006-07*, p4, 2007, Australia. http://www.egwater.vic.gov.au/Corporate/Annual_Report/East%20Gippsland%20Water%20Annual%20Report%202006%20-%202007.pdf

21 Southern Rural Water, *Southern Rural Water Annual Report 2006-2007*, p. 10, 2007 Australia. http://www.srw.com.au/annual_rep/SRWAnRep2007.pdf

22 George Mason University, *The CIP Report Vol6 no.12*, June 2008, p4

23 G. Sapirstein, Social Resilience : *The forgotten Element in Disaster Reduction,* 2008, Organisational Resilience International, Massachusetts, USA

24 Attorney -General R. McClelland, *Safeguarding Australia 2008*, Hotel Realm, Canberra, July 2008. http://www.attorneygeneral.gov.au/www/ministers/robertmc.nsf/Page/Speeches_2008_23July2008-SafeguardingAustralia2008

ONLINE TOXICITY MONITORS AND THEIR USE IN DISTRIBUTION SYSTEM AND WATERSHED EARLY WARNING SYSTEMS

H. J. Allen, R. C. Haught and D. A. Macke

U.S. Environmental Protection Agency, National Risk Management Research Laboratory, Cincinnati, OH 45269

1   INTRODUCTION

Interest in Early Warning Systems for watershed and distribution system water quality monitoring has increased in the wake of improving technical capabilities, particularly in the field of Online Toxicity Monitors (OTMs), and water quality data needs for ecosystem status and human health purposes[1,2,3]. This manuscript will describe an approach to Early Warning Systems (EWS) and efforts to develop tools and implement pilot scale watershed and distribution system EWSs.

The need for water quality EWSs can be illustrated by recent events. Source waters and distribution systems are vulnerable to unreported contamination. de Hoogh et al. describe an incident on the River Meuse in 2004[4]. An unknown contamination event was detected by an Online Toxicity Monitor (OTM) based on *Daphnia magna* swimming behavior at a drinking water intake. The utility responded by closing the intake. Samples were collected at the time of the OTM event detection and analyzed using HPLC-DAD and Q-TOF Mass Spectroscopy. The contaminant was determined to be a component of hydraulic fluid and toxicity was confirmed at observed concentrations.

In 2005, a disgruntled farmer dumped 10 L of Atrazine near the a drinking water intake on Lake Constance, Germany[5]. The utility had no chemical EWS in place at the time and only became aware of the contamination when they received a letter from the Farmer stating his demands. Subsequently, the utility decided to add an OTM at their intake.

A Methylene Chloride contamination event in the Ohio River was detected by the Ohio River Organics Detection System (ODS) administered by the Ohio River Sanitation Commission (ORSANCO) in July 2007[6,7]. This network of utilities samples Ohio River water daily and performs Flame Ionization Detector Gas Chromatograph analyses. The downstream drinking water utility responded to the event by adding activated carbon filtration. State environmental regulators were able to use the information to locate the source and issue a notice of violation.

Early detection of episodic contamination allows early responses by water utilities and regulatory/response agencies to minimize potential impacts and associated costs to the aquatic ecosystem, water supply, citizens, and industry that utilize the river. It is difficult to quantitatively measure all possible contaminants so a method to measure the effects of contaminants is needed to indicate their presence. Contaminants are of concern when they are found in toxic concentrations. Toxicity monitoring can be used as a water quality screen. However, there is no machine or analytical approach to measure toxicity. Only an organism in its own environment can integrate all factors that contribute to stress[8]. Measuring environmental toxicity using aquatic sentinel organisms can be described as the "Canary in the Coal Mine" approach.. Here we are using OTMs to measure water quality continuously and at a relevant time scale.

2   WATER QUALITY EARLY WARNING SYSTEM

The OTM/EWS model presented here may serve as a pattern for the site specific implementation of EWSs. Ideally water quality monitors (WQMs) would be strategically located within the system of concern to be protective of vulnerable locations, Figure 1. There are few absolutes when it comes to defining an EWS, however, the following elements must be included:

1. water quality monitoring tools,
2. data telemetry,
3. data analysis,
4. information distribution to decision makers, and
5. a response framework.

These five requisites and the underlying software/hardware that glues them all together compose the "system." The model presented here is one of a whole that is greater than the sum of its parts. The creation of the emergent system is an open area of research and is likely to occur in varied forms. The software linking the data collection, telemetry, data analysis, dissemination, and response is the major synthetic element. Of course, the human element is required to actually use the system and its importance cannot be overlooked. A dedicated team of managers is necessary to make the implementation of the system a success.

2.1 Water Quality Monitoring Tools

Water quality monitoring tools are the core sources for data driving the system. They may be analytical instruments capable of measuring physical, chemical, biological, radiological, or toxicological parameters. Ideally, some combination of these tools would be used to provide the most representative information regarding water quality possible. As the types of tools used and data collected increase, the uncertainty surrounding dynamic water quality decreases.

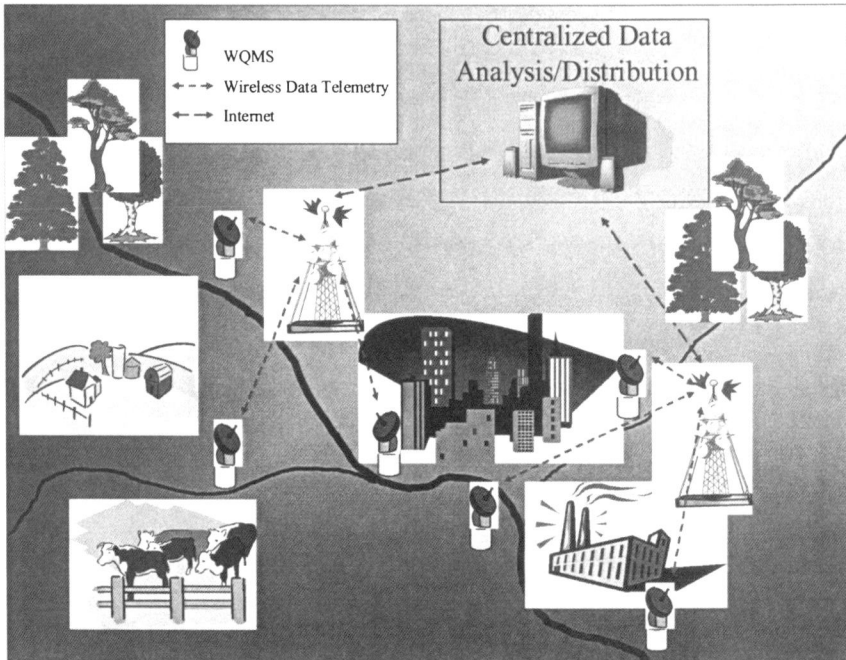

Figure 1: *Depiction of strategic placement of water quality monitoring stations to cover changes in land use and be protective of sensitive installations. A distribution system EWS would follow a similar approach.*

Usefulness of data collected by OTMs and other physical/chemical probes, collectively referred to as WQMs, is determined by how accurately it represents the true water quality condition, the relative temporal and analytical sensitivity, and robustness of the data stream. Data collected remotely is of little use if it does not provide an accurate representation of the water quality with little performance variability. Temporal sensitivity, the time the WQM takes to register a detection of the change event, must be on a time-relevant scale. In the context of a watershed or distribution system, the relevant time frame depends on the flow rates of water through the system, or the time required for water to flow from the data collection point to a point of concern e.g., a drinking water intake or point of use. In most cases, this time frame is on the order of minutes to hours. Also, the detection range of the WQM must match the potential range of the expected environmental and protective concentrations. Finally, the WQM must be offline for a minimal amount of time due to operational malfunction or maintenance. In source waters, fouling of instrumentation is a significant issue. Design of WQMs should include this consideration.

OTMs are a reasonable tool to monitor water for unknown contaminants in toxic concentrations. Several available OTMs have been used and evaluated by EPA, Figure 2.

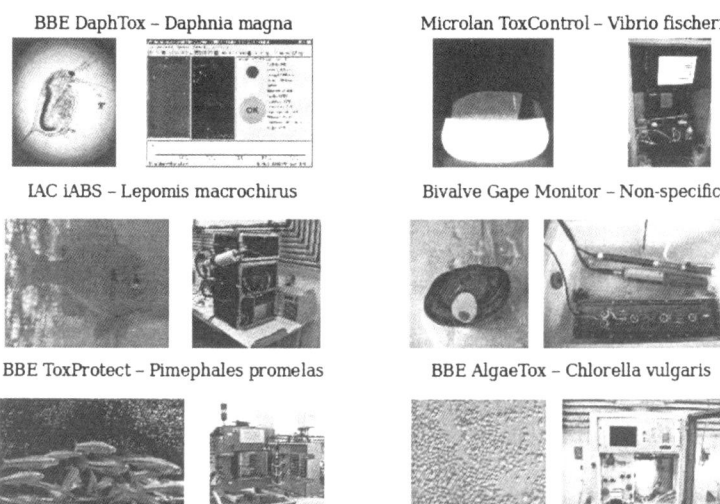

BBE DaphTox – Daphnia magna

Microlan ToxControl – Vibrio fischeri

IAC iABS – Lepomis macrochirus

Bivalve Gape Monitor – Non-specific

BBE ToxProtect – Pimephales promelas

BBE AlgaeTox – Chlorella vulgaris

Figure 2: *OTMs under evaluation by EPA and the respective organisms.*

These are all online, time-relevant, and either continuous or semi-continuous.Bivalve Gape Monitor. The bivalve biomonitoring system (BBS) analyzes gape behavior of bivalves in response to water quality [9]. When stressed, the bivalves will close their shells isolating their vulnerable tissues. Behavior of the individuals is analyzed using time series techniques eliminating the need for traditional controls.

2.1.1 *iABS Fish Monitor.* The Intelligent Aquatic Biomonitoring System (iABS), initially conceived at Virginia Tech University, was developed to its current form by the U.S. Army Center for Environmental Health Research and Intelligent Automation Inc., Poway CA[10,11]. The system continuously monitors muscle activity in fish (*Lepomis macrochirus*), particularly ventilatory activity, ie breathing. Changes in breathing can be indicative of the presence of waterborne toxins. In this monitor, eight fish are maintained in individual flow through chambers. Water of interest is pumped through the system and individual opercula and whole body movement responses are recorded. Fish "coughs" are episodes in which the individual reverses flow over the gills in an effort to clear the gills of mucus and other fouling material.

Data are analyzed using a neural-network built on historical data. Acclimation and baseline periods must be established in the source water before testing begins.

2.1.2 *BBE Daphnia Toximeter.* The Daphnia Toximeter[12,13] uses a video camera to monitor the activity of the test organisms, *Daphnia magna.* Computer software analyzes the images of the organisms while they are exposed to a continuous stream of sample water. The activity of each organism is monitored continuously to determine if changes occur in any or all of several parameters: average velocity, speed class distribution, average distance between organisms, average altitude within the chamber, the curvature of swimming patterns as determined by two fractal dimension equations, and the individual recognition rate. Mathematical reductions of these individual behaviors are used to create a Toxicity Index, a parameter that quantifies the overall status of the organisms. If changes occur suddenly or dramatically, the Toxicity Index will change and alarm when threshold conditions are met.

2.1.3 *BBE Algae Toximeter.* The Algae Toximeter is capable of measuring photosynthetic efficiency in algae (*Chlorella vulgaris*) using pulsed, amplitude modulated fluorometry. This instrument compares the effects of an unknown sample on the photosynthetic efficiency of an aliquot of algae with a control. In the case of significant deviations between the sample water and the reference water, an alarm is generated by the instrument to alert the operator.

2.1.4 *BBE ToxProtect.* The bbe-Moldaenke ToxProtect system uses the movement of fish, this case *Pimephales promelas,* to evaluate changes in water quality[14]. The movements are detected by a matrix of light emitting diode (LED) light barriers in the instrument. The toxicity alarm level is calculated using both the number of light barriers broken and their location within the matrix per unit time. The swimming pattern or movement of the fish in the instrument test chamber changes as the fish become stressed causing the changes in toxicity alarm level. A sodium thiosulfate solution is pumped into the inlet water before reaching the instrument test chamber to remove chlorine making this system suitable for use in a distribution system.

2.1.5 *Microlan ToxControl.* The Microlan ToxControl system is a completely automated OTM using the light-emitting estuarine bacteria *Vibrio fischeri* as a biological sensor[14]. The luminescence values for bacteria are measured simultaneously for reference water and sample water. These measurement values are compared to each other to mathematically compute a percent inhibition. The luminescent bacteria monitor is an automated version of the ISO 11348-3[15] allowing continuous monitoring of rivers, drinking water production or wastewater treatment plants. A 2% sodium chloride solution with 5mg/L sodium thiosulfate was used as a diluent to provide an isotonic environment and to dechlorinate the sample water respectively. Diluent control tests demonstrated no affect on *Vibrio fischeri* due to addition of sodium thiosulfate.

2.2 Data Telemetry

Data collected remotely is of little use if it is not made available in a time-relevant manner. As mentioned above, in the context of an EWS, this time frame is on the order of minutes to hours. Telemetry must also be reliable and secure, particularly during events when contamination is likely to occur such as heavy storm events. WQMs can be designed to maintain some degree of autonomy, reducing the critical nature of telemetry. However, notification of sample collection remains a link that cannot be severed. Samples must be retrieved for verification as rapidly as possible to maintain their integrity and reduce the time between the initial change detection and sample verification.

2.3 Data Analysis

Analysis of trends in a WQM parameter must be appropriate to the nature of the data. Many parameters are serial, continuous variables which should be analyzed using a statistical approach not affected by temporal dependence. Given the spatial heterogeneity of even small scale watersheds and distribution systems and their land use patterns, and the often patchy nature of weather events, traditional reference or control sites are not appropriate. The best way to analyze such data is to use each site as its own control and look for changes in the parameter(s) at each site.

Watersheds and distribution systems often cover large geographical areas. Flow within them may not be exactly understood, however, the spatial component of data collected is critical. Data can be spatially linked using a geographical information system (GIS). Dispersion characteristics can then be predicted using models to provide time of travel and concentration information, assuming concentration is known. Dilution is often the only means of remediation once a contaminant has entered a watershed and been detected. Flow information from gaging stations such as those maintained by the USGS can be used for modeling and predicting potential impacts downstream.

Seasonal trends can confound data interpretation. Factors such as changing temperatures, organic matter concentrations, and water quantity can have a dramatic effect on water quality but yet still remain within the range of tolerance established for the specific water body. An understanding of these changes and their affects on water quality monitors should be established and incorporated into any analytical decision framework to avoid false positives.

Alarm criteria should evolve from simple changes in individual water quality parameters to an algorithm considering all parameters and their spatial and temporal aspects. Certain events within a watershed will have a parameter signature. For example, a significant rainfall event is likely to result in decreasing temperature, conductivity, and pH , and increasing hydrograph and turbidity. These signatures must be identified and carefully assessed. On one hand, the EWS should be more than an expensive rain gage, yet the signature may mask relevant changes in contaminant concentrations. Establishment of alarm criteria will, at least initially, rely on site specific knowledge and professional judgment.

2.4 Information Distribution

Watershed and drinking water utility managers, and other decision makers are key players in an EWS information distribution network. Typically, as information moves up through organizational hierarchies, there must be a commensurate decrease in its level of detail. Information regarding system status must be packaged in a manner to concisely convey observed conditions without ambiguity or at least requiring minimal interpretation. This is a challenge when dealing with complex systems that play an important public health role. How much information does a decision maker need to turn off a drinking water intake, declare a water source off limits, or even a state of emergency? As information progresses through increasing levels of the decision hierarchy the confidence of the event and knowledge regarding specific details should increase.

2.5 Response Framework

Initial alarms must be investigated to determine their veracity and subsequently their cause(s). One model for this is a tiered response model found in Figure 1. The ultimate response framework directs many aspects of the first four items in the above stated EWS model. A tiered response model (TRM) is suggested here:

1. initial detection of dynamic water quality event and sample collection,
2. validation of water quality change,
3. biologically directed identification of causative agent(s),
4. appropriate public health, regulatory, and/or remedial response.

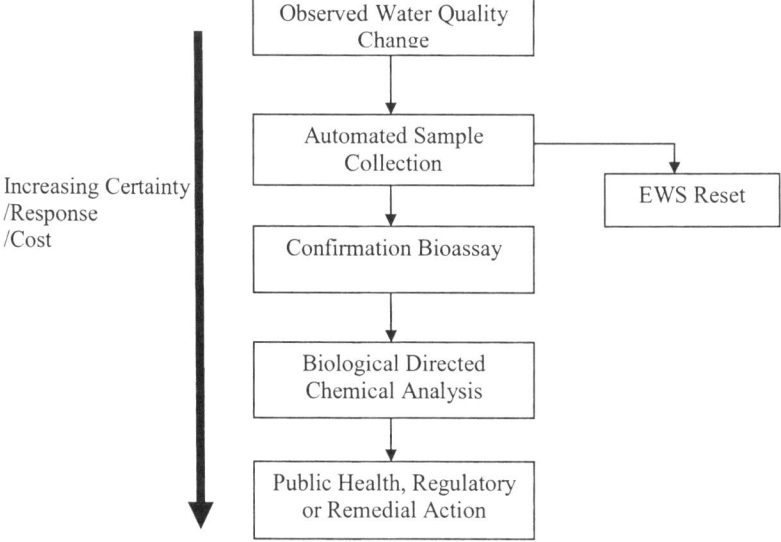

Figure 1 *Conceptual rendering of the tiered response*

With each increasing step in the TRM, uncertainty regarding the event decreases. At the same time, costs in terms of analysis and public response increase. One reason for this approach is the expected occurrence of false positives. Decision makers will be provided with the best information available in a timely manner with its associated uncertainty. This TRM requires swift responses at each step. Response time must be on the order of minutes to hours, rather than days.

2.5.1 *Initial detection of dynamic water quality event and sample collection.* Monitoring stations should be equipped with automated samplers that can be triggered by the monitoring hardware. This will provide for the sampling of the water causing any response in the biological or physical/chemical alarm algorithm. This is critical for any attempt to confirm biological response and identify the cause. An initial alarm may or may not be caused by an event of concern. The monitoring systems are intended as a screen of water quality, not a definitive diagnostic tool. Any information conveyed regarding an initial detection must be framed in the context of limited certainty.

2.5.2 *Validation of water quality change.* The first question to ask following an observed alarm is "So what?" Assuming the WQMs are functioning correctly, a change in water quality has occurred. However, the water quality change is, as yet, unconfirmed and the hazard of the change is unknown. The next step, upon collection of the sample, is to confirm the observed change. Ideally, this step would be rapid and increase the level of certainty that the observed water quality change is real and of significant concern to warrant further investigation or demonstrate that the change was of no consequence. Confirmation of the change should also be cost effective. If toxics are a concern, a bioassay should be used. Several rapid bioassay systems have become available recently[16]. Results must be obtained in a timely manner, on the order of minutes to an hour or two.

2.5.3 *Biologically directed identification of causative agent.* The number of potential contaminants is enormous. Some methodology must be employed to limit the scope of any analytical efforts while providing the greatest return of information for the level of effort expended. Toxicity Identification Evaluation (TIE) techniques described by the U.S.EPA [17] can be modified for this purpose. Conceptually, a TIE is performed when toxicity is observed in a bioassay caused by an unknown contaminant(s) with the goal of identifying the likely cause of toxicity. There are three TIE phases: contaminant class identification, specific causative analytical identification, and causative confirmation using spiking and mass balance approaches. Since time is critical in an EWS, effort must focus primarily on the first two stages. In phase one, an unknown sample is manipulated in an attempt to remove toxicity. This is done on a contaminant class basis. For example, Sodium EDTA is added to chelate metals, the sample is purged to drive off volatiles, and the sample is run through a C18 column to remove non-polar organics. If any of the manipulations remove toxicity, the class of the likely causative contaminant(s) can be identified. Phase two builds on information gained in the previous step for developing an analytical approach for the likely contaminant(s). Again, for example, if metals are identified in phase one, an ICP scan may be performed, if volatiles are likely, purge and trap GC methods may be used, and if C18 removes toxicity implicating non-polar organics, organic chromatography techniques may be used.

The TIE phase 1 methods use a modified acute bioassay lasting 24 hours. Of course, given the need for rapid turn around, the 24 hour time period is much too long to be of use in an EWS context. However, as mentioned previously, rapid bioassay methods are available with short exposure periods and appropriate sensitivities. Use of a rapid bioassay and streamlined manipulation methods would allow for class identification within hours of the initial event observation. First responders and decision-makers could proceed with the knowledge of at least the class of toxicant with which they are dealing. Phase 2 efforts could then follow with a focused effort to identify and quantify the contaminant while reducing costs by eliminating unnecessary analytical work.

2.6 Public health, regulatory, and/or remedial response

Appropriate authorities will make decisions depending on the outcome of the described process of identification of the contaminant. Procedures for this will likely be site and incident specific and are beyond the scope of this manuscript.

3 DISTRIBUTION SYSTEM RESEARCH

Drinking water distribution systems in the U.S. are vulnerable to both intentional and accidental episodic contamination. One step in the contamination threat management process is to understand various warning signs that might indicate contamination has occurred. Unusual water quality may serve as a warning of potential contamination if data are available. The U.S. EPA has initiated a program to investigate how changes in water quality parameters can be detected by using commercial off-the-shelf physical/chemical water quality sensors that measure the water quality in real or near real-time[18]. The available physical/chemical sensors utilize general water quality parameters, such as free chlorine, oxidation reduction potential (ORP), total organic carbon (TOC), turbidity, pH, dissolved oxygen, specific conductance, chloride, ammonia, nitrate to detect the contamination. Generally, one or more of these water quality parameters will change due to the injection of a contaminant. However, no single chemical sensor responds to all possible contaminants nor can they give any indication of the potential toxicity of complex mixtures.

Historically, monitoring of drinking water quality has generally relied on the collection of grab samples followed by extraction and laboratory-based instrumental analysis for both inorganic and organic pollutants. This approach provides a snapshot of the concentrations of analyzed chemicals at a single point in time and space, again giving little indication of the sample's toxicity due to single or multiple contaminants. However, research during the last two decades has shown that considerable limitations are associated with grab sampling approaches for determining total pollutant concentration[19]. To overcome the limitations of grab sampling techniques, online monitors are a topic of much interest for water security applications. The ToxControl and ToxProtect OTMs have integrated dechlorination and are being evaluated for use in distribution system water quality monitoring applications.

Based on the experimental results presented elsewhere[14] both MicroLAN Toxcontrol and bbe-Moldaenke ToxProtect are capable of detecting low levels of cyanide injected into a distribution system simulator (DSS). However, both OTMs were incapable of detecting considerably high levels of fluoroacetate in the DSS. The capability of OTMs to detect a particular toxicant in distribution systems depends on the physical/chemical activity of bioassay and nature of the toxicant. Interestingly, when the researchers reviewed the conventional online physical-chemical instrument monitoring data (e.g., pH, ORP, Chlorine, turbidity and TOC), it appears that the sensitivity of those parameters is inverse. The physical/chemical sensors were more sensitive to sodium fluoroacetate but less sensitive to cyanide. OTM responses to cyanide injections were much more sensitive to the lower cyanide concentrations (0.1 and 0.01 ppm nominal). Further pilot-scale DSS studies using OTMs are underway to establish toxicant-specific OTM performance. nother point of interest is the effect of the dechlorination agent, sodium thiosulfate, on toxicity of potential contaminants. Experiments with malathion indicate that the presence of sodium thiosulfate had a synergistic effect increasing malathion toxicity (Figure 3). This one example does not conclude the debate surrounding the use of dechlorination agents but it does show that not all contaminants will be masked.

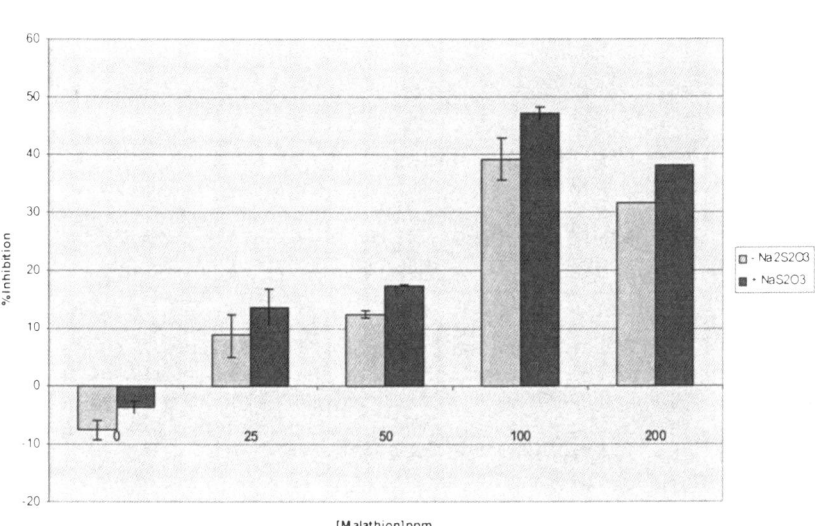

Figure 3 *Effect of Sodium Thiosulphate on Malathion toxicity to Vibrio fischeri.*

4   WATERSHED EARLY WARNING SYSTEM RESEARCH

Watershed EWS research has focused on two geographic areas: the Ohio River and its tributaries in the Cincinnati, OH area and the upper Mississippi River from St. Cloud, MN downstream to Muscatine, IA. Collaboration with various local, state, and federal entities was critical in the process of implementation of both EWSs (Text 1). No one group has the resources to develop an EWS at this scale. EPA developed the suite of monitoring equipment and fabricated the deployment housing. On-site infrastructure was provided by the hosts. Where possible, local universities were invited to participate for their unique perspectives and operation and maintenance capabilities.

Each water quality monitoring station   consists of a data collection and telemetry system capable of collecting data from a variety of serial interfaces and analog to digital (a/d) inputs (Figure 4). The system is capable of drawing power from the grid or a photovoltaic power supply. Data are telemetered using an appropriate technology based on site specific limitations. The biological component consists of a bivalve OTM.   Physical/chemical parameters (temperature, pH, dissolved oxygen, specific conductance, and turbidity) are collected using a commercially available water quality multiprobe (Yellow Springs Incorporated model 6820) and UV-Visible spectra are measured using a field Spectrometer (S:can carbolyser). Each WQMS is outfitted with an automated sampler which can be remotely triggered.

A SQL-compliant database has been developed with real-time Internet accessibility. The data include baseline water quality parameters, data telemetered from the remote field sites, and other relevant data. A web-based graphical interface for viewing collected data has also been developed.

When changes in water quality are indicated and a sample has been collected, the sample will be analyzed in the laboratory using short-term chronic ambient water quality bioassays using both *Ceriodaphnia dubia* and *Pimephales promelas* to confirm toxicity followed by Toxicity Identification Evaluation procedures as previously described.

4.1 Ohio/East Fork of the Little Miami Rivers

The East Fork of the Little Miami River (EFLMR), designated by the state of Ohio as an Exceptional Warm Water Habitat, is a primary source of drinking water for public water systems in Clermont County, OH. Historically, the watershed has been devoted to agricultural use but is rapidly transforming to suburban development and is experiencing one of the highest growth rates in Ohio.

Clermont County's Office of Environmental Quality (OEQ) has installed hardened stream-side structures for housing data collection equipment at critical sites within the watershed. OEQ has partnered with EPA to equip each site with a bivalve OTM, physical/chemical multiprobes, water sampling devices, and data telemetry at four locations on the EFLMR. This effort has integrated OTM data collection with the management needs of the watershed and ongoing work at the US EPA Experimental Stream Facility (ESF), also located in the EFLMR watershed. Physical/chemical and toxicological parameters as well as community structural and functional metrics provide representative information regarding water quality and its underlying processes. This research leverages the substantial investment the county has made in watershed environmental monitoring infrastructure and data collection including GIS land use and habitat/biota data.

One example of an observed event occurred in January of 2007. Snow was predicted for the Cincinnati area and local road crews salted the highways. Snow was followed by rain and warming temperatures resulting in the transport of the salt into surrounding catchments creek. Data presented in Figure 5 are from Hall Run, a tributary to the EFLMR. A strong relationship was observed between increasing conductivity and decreasing bivalve gape.

Table 1 EWS Collaborators

Upper Mississippi River Early Warning Network	**East Fork of the Little Miami River Pilot Early Warning Network**
1. Federal 　○ U.S. EPA ORD & Regions 5 and 7 　○ U.S. Army Corps of Engineers 2. State 　○ MN Pollution Control Agency 　○ MN Dept. of Nat. Res. 　○ Iowa Dept. of Nat. Res. 3. Regional 　○ Upper Miss. River Basin Assoc 4. Utilities 　○ Minneapolis Water Works 　○ St. Cloud, MN Water Works 　○ Moline, Il Water Works 　○ American Water 　○ Xcel Energy 5. Universities 　○ St. Cloud State University 　○ University of MN 　○ University of Iowa	1. Federal 　○ U.S. EPA ORD 2. Local 　○ Clermont County 3. Utilities 　○ Morehead, KY Water Utility 4. Universities 　○ Thomas More College 　○ Morehead State University

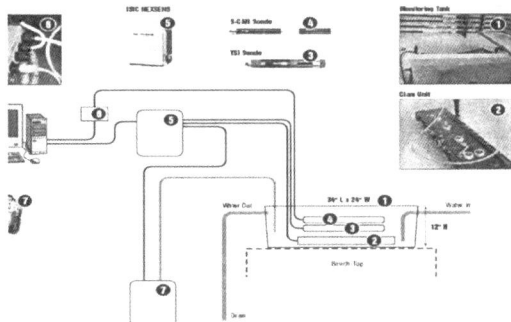

Figure 4 *Schematic of water quality monitoring sensor suite*

Upper Mississippi River EWS

The objective of this effort is the successful implementation of a water quality EWS for monitoring upper Mississippi river water at four sites. This system will increase the likelihood of detecting episodic contamination resulting from unintentional and intentional events. The first site implemented was located at the Minneapolis water works (MWW). Remaining sites are located at the St. Cloud Water Works, Excel Energy's Sherburne County Generation Plant, U.S. Army Corps of Engineers Lock and Dam 14, and University of Iowa's Lucille A. Carver Mississippi Riverside Environmental Research Station will follow during the summer of 2008. Key to this effort is the deployment of OTMs, based on the gape behavior of bivalves. Integration of time-relevant data (minutes scale) with current and historical contextual data in a temporal/spatial framework will provide water quality managers in the watershed with up to date information for drinking water process and ecological management decisions. A major outcome of this work is greater ability by local authorities to provide residents with clean water and the dual benefit of a greater understanding of water quality in the Mississippi river. The feasibility of implementing an Early Warning System and the level of data analysis and interpretation most appropriate for use by water quality managers in maintaining an effective monitoring program will be assessed.

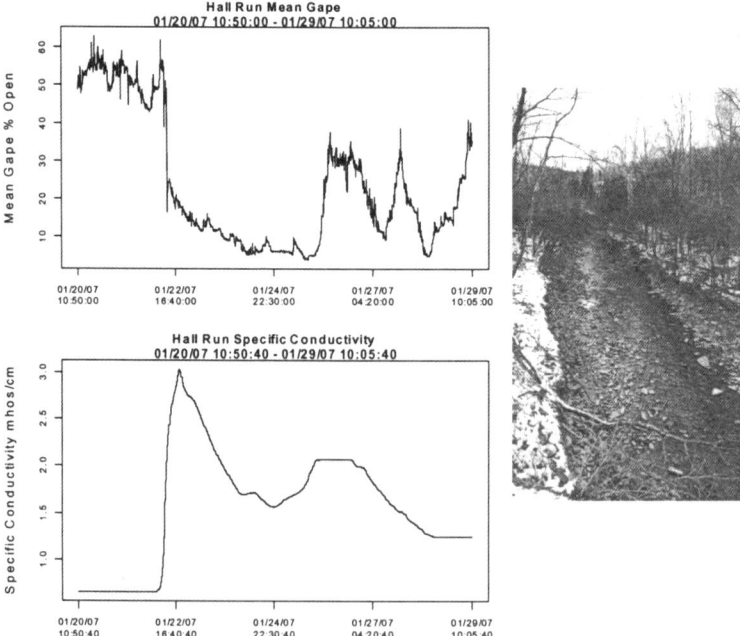

Figure 5 *Impact of a road salt runoff event on bivalve gape (top) and specific (bottom) in a small stream (right)*

5 CONCLUSIONS

In this manuscript a reasonable EWS/Tiered Response Model has been described to aid in the design and implementation of water quality EWSs for both watersheds and distribution systems. Laboratory results demonstrate OTMs are both technologically and, more importantly, toxicologically, appropriate for protection of distribution systems and source waters. To this end, pilot EWS efforts on the upper Mississippi River and Ohio and East Fork of the Little Miami Rivers have been described. These field deployments demonstrate the need for collaborative efforts to produce functional EWSs that can provide multiple benefits.

References
1. H.J. Allen, K.L. Dickson, H. Martin, K.A. Thuesen and W.T. Waller. *Journal of Urban Technology*, 2002, **9**, 1.
2. R.W. Gullick, L.J. Gaffney, C.S. Crockett, J. Schulte and A.J. Gavin. *Journal AWWA*, 2004, **96**, 68.
3. W.H. van der Schalie, T.R. Shedd, P.L. Knechtges and M.W. Widder. *Biosens Bioelectron*, 2001, **16**, 457.
4. C.J. De Hoogh, A.J. Wagenvoort, F. Jonker, J.A. Van Leerdam and A.C. Hogenboom. *Environ. Sci. Technol.*, 2006, **40**, 2678.
5. Associated Press, http://www.enn.com/top_stories/article/3044, 2005.
6. A. Fleck, http://www.eppc.ky.gov/press/press2007/july/7-30pregis.htm, 2007.
7. J.G. Schulte. Freshwater Spills Symposium, 2004, .
8. J. Cairns and D.I. Mount. *Environmental Science and Technology*, 1990, **24**, 154.
9. H.J. Allen, W.T. Waller, M.F. Acevedo, E.L. Morgan, K.L. Dickson and J.H. Kennedy. *Environmental Technology*, 1996, **17**, 501.
10. T.R. Shedd, W.H. van der Schalie, M.W. Widder, D.T. Burton and E.P. Burrows. *Bulletin of Environmental Contamination and Toxicology*, 2001, **66**, 392.
11. Y.B. Mikol, W.R. Richardson, W.H. Van der Schalie, T.R. Shedd and M.W. Widder. *Journal AWWA*, 2007, **99**, 107.
12. M. Lechelt, W. Blohm, B. Kirschneit, M. Pfeiffer, E. Gresens, J. Liley, R. Holz, C. Lüring and C. Moldaenke. *Environmental Toxicology*, 2000, **15**, 390.
13. U. Green, J.H. Kremer, M. Zillmer and C. Moldaenke. *Environ. Toxicol.*, 2003, **18**, 368.
14. H.J. Allen, S. Panguluri, N. Muhammad, D.A. Macke and G. Meiners. 2007 Awwa Water Quality Technology Conference And Exposition Proceedings, 2007, TBD.
15. International Organization for Standardization. 2007, Iso Is 11348-3:2007 Water Quality --Determination Of The Inhibitory Effect Of Water Samples On The Light Emission Of Vibrio Fischeri (Luminescent Bacteria Test) -- Part 3: Method Using Freeze-Dried Bacteria, International Organization for Standardization, Geneva, Switzerland.
16. R. James, A. Dindal, Z. Willenberg and K. Riggs. 2003, *Aqua Survey Inc. Iq Toxicity Test Rapid Toxicity Testing System*, http://epa.gov/etv/pubs/o1_vr_aqua_survey.pdf.
17. T.J. Norberg-King, D.I. Mount, E.J. Durhan, G.T. Ankley, L.P. Burkhard, J.R. Amato, M.T. Lukasewycz, M.K. Schubauer-Berigan and L. Anderson-Carnahan (Eds.). *Methods for aquatic toxicity identification evaluations phase i toxicity characterization procedures*, EPA/600/6-9 l/003, USEPA, Washington, DC, 1991.
18. J. Hall, J.G. Szabo and G. Meiners. *Journal of the AWWA*, 2007, **99**, 66.
19. I.J. Allan. *Talanta*, 2006, **69**, 302.

WATER SUPPLY SECURITY ISSUES AND TRENDS

B. Nguyen - presented by Hao-Nhiên Pham

Director of Operations, EAU DE PARIS, nguyen@eaudeparis.fr

1 INTRODUCTION

Water availability is a vital need for the human beings but only those who lack this drinking water in quantity and quality really know the value of it. This appears all the more true when people in developed countries find themselves confronted with unexpected interruptions of their water supply.

The apparition of wide access to safe water at the tap for any ordinary citizens which started like 150 years ago has led those citizens to rely on others, the local authority and by extension the water service operators, for ensuring their usual and basic needs for water.

The likelihood of being faced to a catastrophe and as a result suffering large and persistent interruption of the water supply, whatever the origins are, varies mainly according to the geographical situation, the availability of the resource and the growing needs. But any water utility should be aware of the potential risks of such events and be prepared for it.

The concern of the population with the water supply failure possibility is generally reversely proportioned to the level of service; and those who never questioned themselves about what lies behind the simple action of turning the tap also never imagined the dramatic consequences of such failures.

There probably lies the paradox where while the water infrastructure is generally very expensive compared to other infrastructure, like wireless telecommunications, most people consider that the acceptable cost for water supply should be very low.

The recent trend has shown the development of water supply security concern: the promotion of Water Safety Plans by WHO, the creation of a Specialist Group in the IWA and new ISO standards are just examples. On the other hand, governments also show their concern with new regulations while others, like the Swiss, already have done so during the last decade of the twentieth century.

The definition of water supply security is not worldwide accepted since it only really applies where actual water service is considered sufficiently satisfying. Indeed, maintaining the normal standard is the general objective, but then what happens when a crisis comes?

It is amazing to see how throughout the world supposed similar populations react differently when confronted to the same situation: as an example, boiling alerts issued by the utility or by the sanitary authority will be viewed somewhere as a simple incident, and elsewhere with suspicious by frightened consumers. The utilities have to take into account that culture and past experience strongly change the perception of the average consumer.

Whenever objectives in terms of minimum requirements are not fixed by the government or the local authorities, the water utilities are left on their own for defining their commitments. On the other hand, when these objectives exist, the utilities just have to comply with the imposed standard requirements.

Since September 11 2001, the concern about terrorism and its consequences has considerably raised in many countries. Like for other vital infrastructure, the water utilities were given new constraints more or less demanding for the protection of the populations. The potential risk of attacks on the drinking water supply is not to be diminished though it appears that other acts like bombing seem to have the preference of the terrorists.

In any cases, water utilities have to prepare the necessary organization in order to first reduce the risks of deterioration of the service, generally through a risk assessment methodology, and second mitigate the consequences of the failure what ever are the origins of this failure.

2 WATER SUPPLY SECURITY MANAGEMENT

Security Management is a subject in its own right and drinking water distribution operators are becoming professionalized in this field using tools and methods specific to the three important phases: upstream preparation, crisis management and the post- crisis period.

By definition, a crisis corresponds to a period of loss of control; its duration is not therefore generally known at the exact moment it begins. One thing is certain: it will come to an end when its effects are no longer felt.

During this complex period, management of the crisis aims to re-establish the equilibrium of a situation positioned in space and time just when all three are constantly changing.

Figure 1 below shows the positive effects coming with proper management applied to crisis situations; this approach concerns many services and not only for water supply.

3 STANDARDS

The risk management and quality assurance tools enable the operators to reflect on their practices and their vulnerabilities and to put in place organisations and resources that will make it possible for them to manage their risks in an optimised way. Tools such as audits and quality control allow them, moreover, to regularly check that the strategies chosen are actually deployed on the ground.

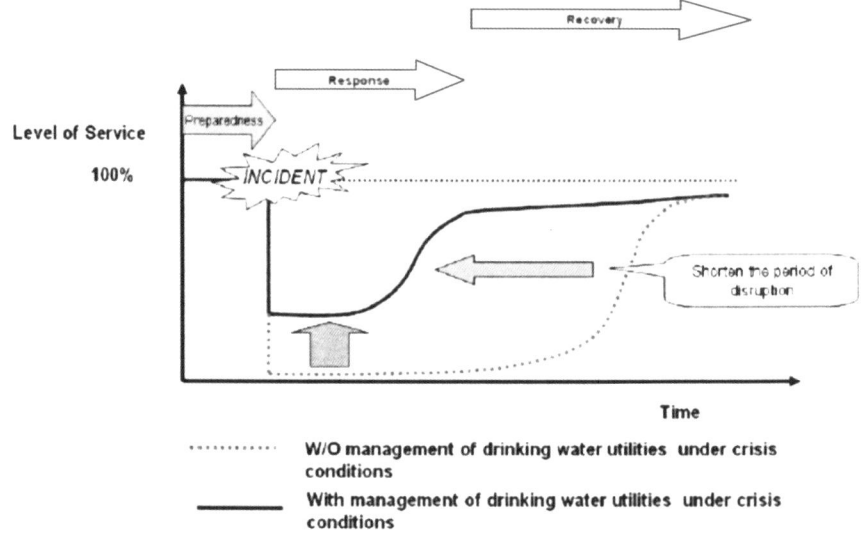

Scheme 1: Effectiveness of Crisis Management (*Source – ISO/TC 223 - ISO/PAS 22399*)

Figure 1 *Effectiveness of crisis management*

3.1 The HACCP method

HACCP (Hazard Analysis and Critical Control Points) is a standardised methodology applicable to all or part of a food products production or distribution system. It is based on an analysis of the dangers, the putting in place of measures to control these dangers and verification of the efficiency of these control measures. In accordance with this methodology, control of the dangers is essentially based on identification and management of critical points, which must be the subject of surveillance, and for which critical limits and corrective actions must be defined in the event these critical limits are exceeded.

3.2 The "Water Safety Plan"

The concept of a "Water Safety Plan" was developed by the World Health Organisation (WHO) specifically for drinking water and takes into consideration the safety of the consumer's tap water. The dangers under consideration in this case include then of necessity those relating to the resource, production, and distribution of the water. It makes no reference to the notion of critical point as regards control (the central stage in the HACCP approach), but stresses that of multi-barrier protection, the failure of a treatment stage being able to be offset by the following stage. For this reason, this concept is better adapted to drinking water than HACCP.

3.3 The ISO 22000 standard

ISO 22000 was developed in 2005 to enable food sector professionals to have their HACCP plans certified. It was drawn up with the aim of simplifying the necessary preliminary steps for these professionals, by including in one and the same certificate the requirements relating to ISO 9001, good hygiene practices and HACCP, thus bringing together under a single standard all the actions that contribute to the production and distribution of healthy, genuine-quality and saleable products.

This standard is consistent with the concept of the WHO « Water Safety Plan », and is therefore ideally suited to a drinking water application.

The incorporation of a HACCP plan or a « Water Safety Plan » in a ISO 9001-type quality management system is a plus for it guarantees in all cases its effective application. It offers what is more the possibility of a certification according to the ISO 22000 standard, a fact that bears witness to the high degree of safety as regards the water produced and distributed, as guaranteed by the operator.

3.4 Work of the ISO TC224

Under the umbrella of ISO/TC 224 – *"Service activities relating to drinking water supply systems and wastewater systems – Quality criteria of the service and performance indicators"*, a workgroup of experts (WG7) has started to work on the area of "Crisis Management of Water Utilities". This workgroup has merged with CEN/TC164, another group at the European level working on the same topic.

4 RISK MANAGEMENT LOOP

The risk management loop is designed for a continuous process through which the utility will use the return of experience that follows emergency situations to implement improvements.

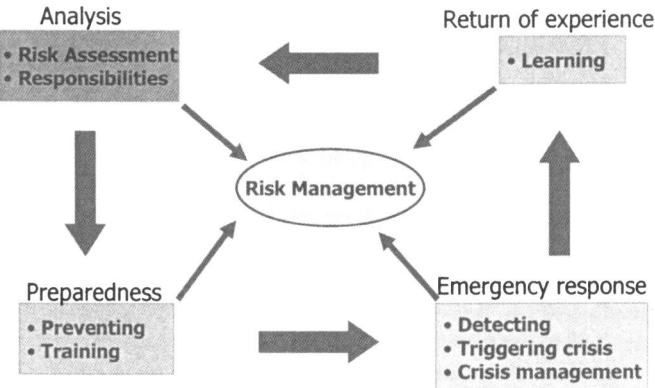

Figure 2 *The Risk Management Loop, source W-Smart*

5   REQUIREMENTS FOR THE OPERATION OF A WATER SUPPLY

To ensure the public water service, one needs:
- a resource in sufficient quantity and of minimum quality,
- operational treatment infrastructures,
- consumables (chemicals, vehicles…),
- « fluids » (energy, communication…),
- reservoirs,
- a distribution network (pipes, service pipes, pumping stations, hydraulic apparatus),
- qualified staff at the workplace and on call.

A sound water security management will assess the availability for each of these components, and their resilience as well.

Good operational principles must be put in place at installations level:
- redundancy of the critical installations,
- maintenance of security stocks (reagents, parts…),
- search for independence in regard to fluids such as energy, communications (note the risks of saturation of certain networks in times of crisis), transportation…
- optimised planning of limitations for maintenance or works (residual capacity potential and requirements to be satisfied).
- the distribution of strategic knowledge among multiple individuals.

6   THE WATER SUPPLY TARGETED BY MALICIOUS ACTS

Deliberate acts for deteriorating the level of service of the population's water supply are not a hypothetic vision of the mind. These are real possibilities, not necessary linked to terrorism but to vandalism or revenge against the utility: there are unfortunately several examples of such behaviour.

Here we are going beyond the usual grasp and know-how of the normal activity of the water supplier: the protection of the infrastructure requires new skills that generally don't exist within the water utilities. However, the development of new technologies will provide efficient tools at an affordable cost for the protection of our infrastructure.

The American Society of Civil Engineers (ASCE) and the American Waterworks Association (AWWA) have published in December 2006 the "Guidelines for the Physical Security of Water Utilities". Based on scenarios of deliberate action for the deterioration of the drinking water quality, the French Ministry of Health as issued in March 2007 a guideline for the self-assessment of vulnerabilities of drinking water supply systems by their operators (i.e. the water utilities).

Among many other publications, these guidelines help the utilities to establish precise objectives in their security Management scheme and prioritize their actions.

7   WHAT IS CRITICAL INFRASTRUCTURE

The selection of essential, nodal or major points or sites for the water service to the populations is a special exercise. In itself it is not fundamentally difficult for an operator in good control of his field of activity and his infrastructures.

On the other hand, difficulties emerge when one has to classify these sites in order to retain only the most important: how and using what criteria does one dispense with certain points while preserving others.

The "sensitive2 character of a site increases with:
- the proximity of the consumers, or the retention time of the water prior to use (lead times to put in place protective measures);
- release into atmospheric pressure of water (greater ease of pollution);
- the quantity of the volumes of water, or the number of consumers potentially affected;
- the unique non-redundant character of the site to carry out a vital function (no backup);
- concentration of several sensitive functions on the same site (for example pumping plant and reservoir located on the same plot).

The "sensitive" character of a site lessens with the existence of means of managing a crisis following the non-availability of the critical infrastructure. In this regard, one could make it clear that an alternative supply of bottled water to the population can be envisaged up to an urban density threshold corresponding approximately to a town of 100 000 inhabitants; beyond that the means generally available are no longer on a scale with the problem.

8 ACTIVE AND PASSIVE SECURITY

Security can be divided in two parts:

- Passive security concerns the infrastructure and its maintenance; it represents the basic level of security offered or guaranteed by the physical structure of the facilities. The passive security remains stable in time whenever due maintenance of the infrastructure is made. But passive security is made to respond to specific situations and will not go over its known limits. Passive security is generally expensive.

- Active security concerns the "human factor"; it is based on the training of people and the level of procedures as well as on the organization implemented. The active security has a natural tendency of decreasing with time if it is not kept at a good level through regular exercises and tests. Active security is generally less expensive than passive security, but it has the advantage of being adaptable to the situation. Active security will make the difference when confronted with unforeseen crisis situations.

9 EXERCISES

The exercises also constitute an indispensable and essential factor in the preparation for crises. They make it possible to judge and subsequently improve the scheduled operational response in the face of critical situations. To assist the operators in the United States, USEPA is putting online kits for the carrying out of crisis exercises. These kits each comprise a detailed scenario complete with the type and number of participants scheduled, the supports to gather reactions and the choice of players as well as all the necessary indications for the «game master» for the organisation of the exercise.

Heightened awareness may on occasion emerge during these exercises; these are opportunities to be grasped so as to improve the routine instructions regarding points, certain of which may a posteriori seem obvious (for example not to forget to fill up one's service vehicle before coming on duty call).

The exercises also lessen the emotional load and stress of the agents who will have rehearsed and experienced artificial crisis situations before being confronted with them in real life. While it is important not to underestimate the frailty of the human factor, one must also accord it the capacity for imagination and adaptation that only crisis situations can reveal.

10 NATIONAL AND INTERNATIONAL COOPERATION

Cooperation in Water Security Management has a potential for improving the preparedness and response of utilities. The benefits of such cooperation lies first through experience sharing rendered possible during conferences or working group meetings, and second in mutual aid agreements for emergency water supply or more generally for any cooperation between utilities.

Such cooperation exists and those who participate into these exchanges know very well how by defining best practices and standards they work for the benefit of many.

The W-Smart group (Water Security Management Assessment Research & Technology) created after the 9-11 attack aims at providing a unique platform between large utilities for studying and sharing experience, concern and know-how on man-made threats. One of the taskforces of the W-Smart group works on security drills and plans to hold an exercise in 2008 in different utilities based on the same scenario; observers from the group will help understand different approaches and try to identify best practices.

The IWA Specialist Group on Water Security and Safety Management (W2SM) is currently working on two main topics: the development of a database of experts from the association who will provide online support to local technicians working in post-disaster situations in affected areas.

The WOP's (Water Operator Partnerships) is another initiative supported by the IWA which brings together offers and request for operational cooperation between utilities mainly from North and South.

The decentralized cooperation is seen as a very effective mean of establishing partnership on the long run between cities or local governments and also a way to involve the politic level in the process for example with twinning.

11 CONCLUSION

Water supply security management covers a wide scope of topics and has seen growing concern by the utilities together with the customer demand for the quality of service. There exist a number of various tools that will help the water utility to optimize its management of supply security. Developing best practices, however as usual, is a very useful mean for implementing the continuous process of improvement. And the international community of water utilities can provide fruitful exchanges and experiences for the benefit of all.

CONSEQUENCE MANAGEMENT WITHIN THE ENVIRONMENTAL PROTECTION AGENCY'S WATER SECURITY INITIATIVE

B. C. Pickard

US EPA, Office of Ground Water and Drinking Water, Water Security Division; Washington, D.C. 20460

1 INTRODUCTION

Through the assessment of vulnerabilities to drinking water systems, water security experts have identified the distribution system as one of the most vulnerable components in a drinking water utility, with respect to contamination. To address this risk, the administration issued Homeland Security Presidential Directive 9, which required the U.S. Environmental Protection Agency (EPA) to develop and deploy a contamination warning system for drinking water. This directive gave rise to the EPA's Water Security initiative (formerly known as WaterSentinel).

The Water Security initiative contamination warning system detection strategy involves the use of multiple monitoring and surveillance components for timely detection of drinking water contamination in the distribution system. The monitoring and surveillance components include the following:

• Online water quality monitoring includes stations located throughout the distribution system that measure chlorine, total organic carbon, conductivity, and other parameters. Software analyzes the monitoring data to establish a water quality base state. "Possible" contamination is indicated when a significant, unexplained deviation from the base state occurs.

• Sampling and analysis is the collection of distribution system samples that are analyzed for various contaminant classes as well as specific contaminants. Sampling is both routine to establish a baseline and triggered to respond to an indication of "possible" contamination from another component. Analyses are conducted for chemicals, radionuclides, pathogens, and toxins using a laboratory network.

• Enhanced security monitoring includes the equipment and procedures that detect and respond to security breaches at distribution system facilities. Security equipment may include cameras, motion activated lighting, door contact alarms, ladder and window motion detectors, area motion detectors, and access hatch contact alarms.

• Consumer complaint surveillance enhances the collection and automates the analysis of calls by consumers for water quality problems indicative of "possible" contamination. Consumers may detect contaminants with characteristics that impart an odor, taste, or visual change to the drinking water.

• Public health surveillance involves the analysis of health-related data to identify disease events that may stem from drinking water contamination. Public health data may include over-the-counter drug sales, hospital admission reports, infectious disease surveillance, emergency medical service reports, 911 calls, and poison control center calls.

In addition to these monitoring and surveillance components, consequence management is a key aspect of a contamination warning system and consists of actions taken to plan for and respond to potential drinking water contamination incidents in the distribution system. These actions are meant to minimize response and recovery timelines through a pre-planned, coordinated effort.[1]

A consequence management plan (CMP) that successfully guides these actions is a cornerstone of an effective contamination warning system, and it is essential to have the plan in place and tested prior to operation of any contamination warning system components.

2 METHOD AND RESULTS

2.1 Overview of the Consequence Management Plan

Figure 1 provides a general overview of a CMP and how it relates to the monitoring and surveillance components of a contamination warning system. The upper portion of Figure 1 includes routine operation of the contamination warning system components, as governed by defined operating procedures and a process for initial trigger validation. EPA is developing a detailed guidance document, Water Security Initiative: Interim Guidance on Developing an Operational Strategy, for this aspect of the contamination warning system.

The middle and lower portions of Figure 1 identify planning and response actions during all phases of consequence management, including Credible Determination, Confirmed Determination, and Remediation and Recovery.[2] Each of the consequence management phases is described below.

2.1.1 Credible Determination. This initial stage of consequence management involves gathering additional information about the "possible" water contamination threat, as identified through routine monitoring and surveillance activities. The credible determination process includes further review of all contamination warning system components, execution of site characterization activities and use of other external resources when available and relevant. Some preliminary response actions may also be initiated during the credible determination process to limit or minimize impacts of suspected contamination. Based on additional information gathered, contamination is either ruled out and the system returns to routine monitoring and surveillance activities, or contamination is deemed credible, and additional confirmatory and response actions are initiated.

2.1.2 Confirmed Determination. In this stage of consequence management, additional information is gathered and assessed to confirm drinking water contamination. Response actions initiated during credible determination are expanded and additional response activities may be implemented. Notifications would likely be expanded to additional response partners or regulatory agencies. Confirmed determination also includes development or implementation of the public notification strategy (e.g., boil water, do not drink, do not use) and revisions to operational response plans once the contamination incident is considered confirmed.

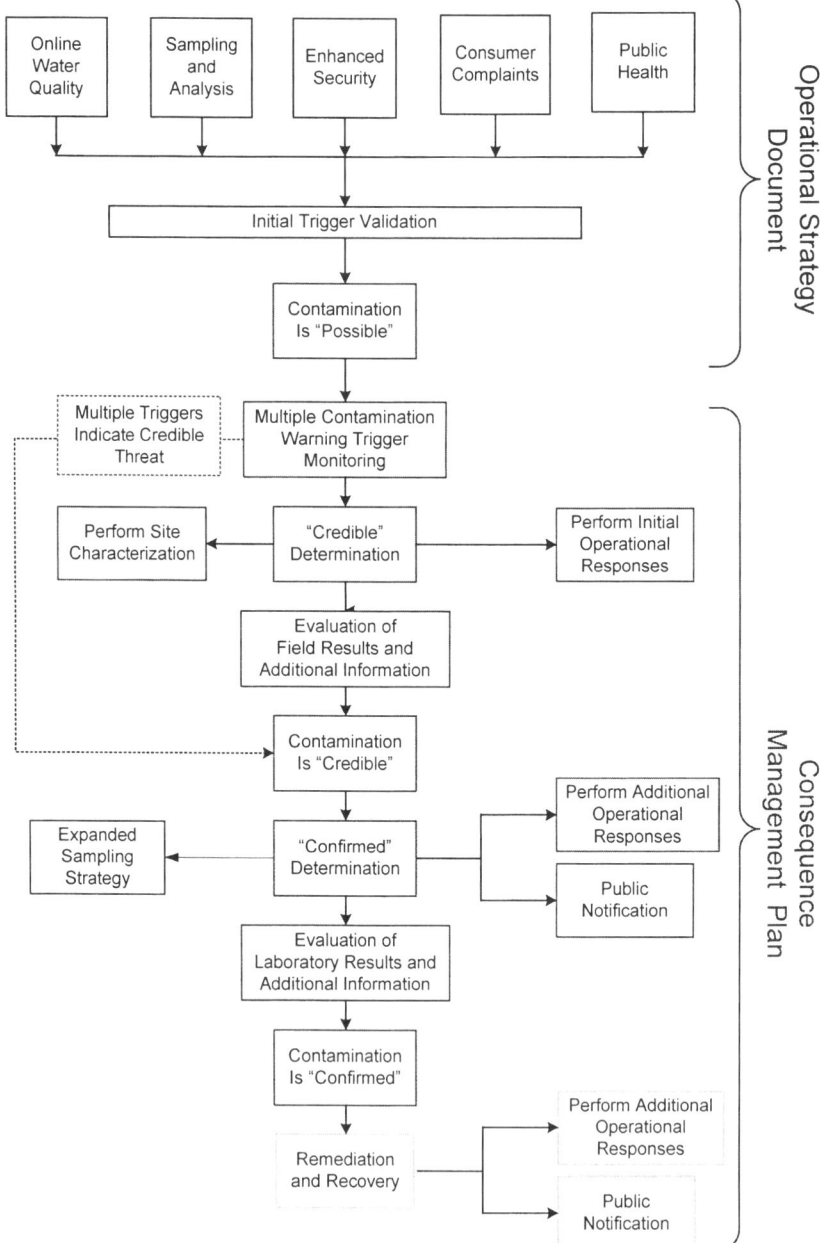

Figure 1 *Overview of the Consequence Management Plan within a Contamination Warning System*

2.1.3 Remediation and Recovery. Remediation and recovery occurs once contamination is "confirmed" and the immediate threat to the public and property has been mitigated. Remediation and recovery actions should be taken to quickly restore the drinking water utility to service, and include characterization of the contaminated area and identification of processes for treatment of contaminated water and remediation of distribution system infrastructure. Decisions regarding disposal of contaminated water and infrastructure must also be addressed.

2.2 Constructing the Consequence Management Plan

Construction of a comprehensive CMP can be categorized into the following steps:
1. Utility self-assessment and evaluation of existing emergency response plans;
2. Development of the CMP framework; and
3. Identification and engagement of key response partners and stakeholders.

2.2.1 Utility Self-Assessment. The first step in developing a CMP is to conduct a self assessment of the utility's existing emergency response plans and overall preparedness. A CMP developed in support of a contamination warning system should be a sub-set of the utility's existing emergency response plan, focusing specifically on the contamination threat to the distribution system.

The purpose of the self assessment is to identify and review existing plans and procedures outlined in the utility emergency response plan that may serve as a starting point for constructing a CMP. This will allow the utility to integrate existing material and related plans into the CMP. For example, utilities may have previously developed action plans and/or specific protocols and procedures within their emergency response plans for responding to the following:

• Water contamination, such as Cryptosporidium and Giardia, cross connections, chemical spills, intentional contamination, and "white powder" plans
  • Increased consumer complaint calls
  • Facility alarms, suspicious persons, or threats made to the system
  • Depressurization, power outage, adjusting water treatment parameters, or other operational problems
  • Severe weather
  • Mutual aid and assistance with other utilities
  • Need for water-use restrictions
  • Public Notification/Risk communication

In addition, the utility should assess their Incident Command System (ICS) training and National Incident Management System (NIMS) compliance. The CMP should contain provisions for the utility to implement an ICS to help manage a response to a contamination incident that may expand outside of its normal operations. The response partners engaged will most likely be well versed in ICS and NIMS, and should expect utility personnel to be as well. One of the first steps is to ensure that utility response personnel have basic NIMS and ICS training that is consistent with the Federal Emergency Management Agency NIMS Integration Center. The utility should consult their state emergency management and/or homeland security agency, as many states have NIMS requirements that are more stringent than the federal requirements.

2.2.2 Development of the CMP Framework. After assessing the utility's existing plans and overall preparedness, an initial CMP draft should be developed to allow the utility consequence management design team to begin conceptualizing the later stages of a response and determine when response partners should be engaged. Development of the CMP draft will also help identify key staff and/or utility divisions that should be involved in the development process. Development of the initial draft should occur before defining external agency involvement.

The initial draft of the CMP should include the development of decision trees for determining whether a contamination incident is credible and/or confirmed, and for remediation and recovery efforts. The decision trees should run through the time period up to and past the point where response agencies are contacted for assistance. The initial draft should identify the major steps, actions, decision points, communications points, and expected contributions by partners that are expected to occur. Decisions trees or other visuals will be helpful in clearly representing this information during investigation of a possible contamination incident. The following information provides high-level considerations during development of the credible, confirmed, and remediation and recovery portions of a utility-specific CMP.

Credible Determination Planning and Actions. Figure 2 provides a generic overview of the credible determination process. Credible determination begins when the utility person identified as the point of contact within the response team is notified that there is a possible contamination threat that has been initially validated. Credible determination activities include notification of internal and external parties, implementation of preliminary operational responses (e.g., limited system isolation), site characterization, development of a sampling strategy, coordination with internal and external laboratories, and review of other contamination warning triggers. It is important to note that credible determination activities may not occur in sequential order, and may start at different times, run concurrently, be iterative, or occur after credible determination.

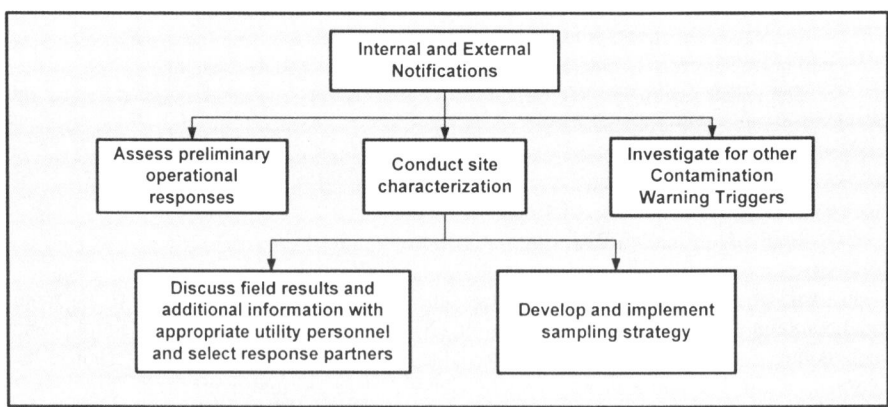

Figure 2 *Credible Determination Process Overview*

Each utility should develop its own process or plan for determining whether an incident is credible based on the activities presented in Figure 2. To accomplish this, a credible determination decision tree should be developed to clearly outline critical decision points and detail the factors that should be taken into account when making these decisions.

Notifications during the credible determination phase are critical for both internal utility personnel and external response agencies to investigate, control, and respond to a contamination threat. The utility may also want to consider implementing their ICS or portions of their ICS depending on the specific situation. Implementation of the ICS is at the discretion of the utility, and there is no standard timeframe established as to when it should be activated. It will depend on the utility's judgment based on circumstances surrounding an incident.

Preliminary operational responses should be considered during the credible determination phase, including how these responses will be implemented, and possible ramifications of implementing these response actions. Immediate operational response actions may decrease the urgency of the situation, but such actions may not resolve the incident. Operational response actions should specifically address whether the contamination can be isolated and the impact of isolation on water service and customer requirements. If it is determined that the impacts of isolation are negligible to customers and utility operators (e.g., minimal number of customers affected, pressure reduction compared to minimum pressure requirements, potential period of impact), then the utility should proceed with an isolation plan. On the other hand, if the impacts are not negligible, then the utility should consider other operational responses.

Site characterization is a critical step in the credible determination process and involves collecting information from an investigation site to support the evaluation of a drinking water contamination threat. This helps to characterize the incident once a threat, accidental or criminal, is suspected. Site characterization requires careful planning and execution, oftentimes with external agencies and response partners. A site characterization plan should be developed that describes the activities of the parties involved and highlights their roles and responsibilities. The plan should cover activities starting with performing a site hazard assessment when approaching a suspected contamination site(s) to collecting water samples and exiting the site.

After site characterization has been completed and samples have been collected, it is important to develop and implement a sampling strategy to coordinate the analysis of samples collected at the incident site. This includes coordinating with laboratories to ensure that roles, responsibilities and notification procedures are established. This effort should be coordinated in the same manner as with other utility response partners to ensure responsive, efficient and accurate handling of water samples collected from the site of a suspected incident. The utility sampling strategy should be well integrated with the plans and operations of the laboratories in the network.

Finally, field results from site characterization should be reviewed and discussed by the appropriate utility personnel and response partners. Field results include the information collected during site characterization (e.g., site hazard assessment, field safety screening, rapid field tests). This does not include the results of the laboratory analysis conducted on samples collected at the end of the site characterization process.

Confirmed Determination Planning and Actions. Figure 3 provides a generic overview of the confirmed determination process. Although presented in sequential order, confirmed determination activities may start at different times or run concurrently.

Each utility should develop its own process or plan for determining whether an incident is confirmed based on the steps in Figure 3. To accomplish this, a confirmed determination decision tree should be developed to clearly outline critical decision points and detail the factors that should be taken into account when making these decisions. Confirmed determination begins by evaluating field sample results (i.e., laboratory analysis of water samples collected during site characterization) and additional information with

Incident Command and other response agencies. The information is reviewed and response teams may implement or revise operational responses in order to isolate the contaminated area or mitigate the consequences of contamination. Next steps include the development or implementation of the public notification strategy (e.g., boil water, do not drink, do not use) and revisions to operational response plans once the contamination threat is considered confirmed. Confirmed determination activities also include additional field investigations; development of expanded sampling, health and safety, and risk communication plans; coordination of alternate water supplies; and use of any other tools that are required for the particular incident.

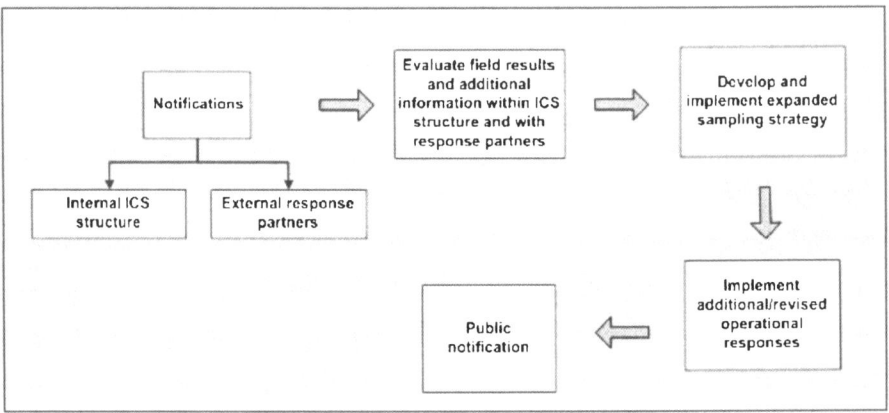

Figure 3 *Confirmed Determination Process Overview*

During the confirmed determination process, it is likely that the internal ICS has been established. If not, then this action should be considered at this point. If outside response partners have been involved, coordination with their ICS will also be necessary. The confirmed determination process should include close communication with and periodic update reports for appropriate response partners and laboratories. The ability of response partners to provide support relies heavily on these coordination efforts and timely information from the field.

Remediation and Recovery. Figure 4 provides a generic overview of the Remediation and Recovery process, which is performed when a contamination incident is confirmed. The goal of remediation and recovery is to return the drinking water utility to service as quickly as possible, while protecting public health and minimizing disruption to normal life. During the remediation and recovery stage, the immediate urgency of the situation has passed, and the magnitude of the remedial action requires careful planning and implementation. While rapid recovery of the system is crucial, it is equally important to follow a systematic process that establishes remedial goals acceptable to all stakeholders, implements the remedial process in an effective and responsible manner, and demonstrates that the remedial action was successful.

Each utility should develop its own process or plan for remediation and recovery based on the activities presented in Figure 4. To accomplish this, a remediation and recovery decision tree should be developed to aid in the connection of recovery steps from phase to phase.

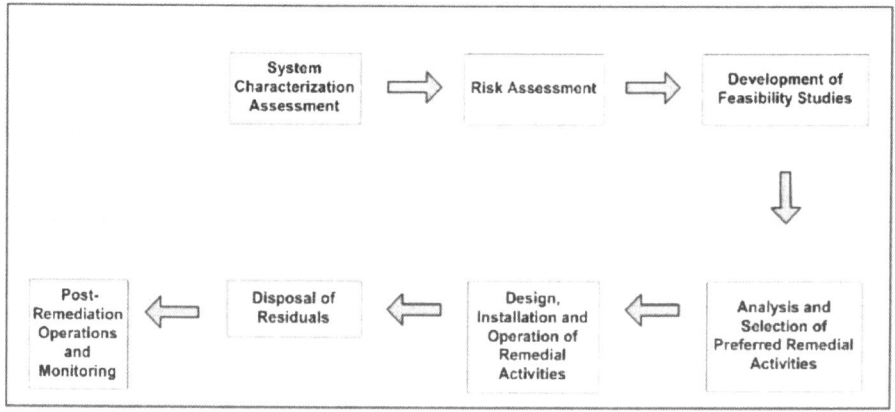

Figure 4 *Remediation and Recovery Process Overview*

2.2.3   Identification and Engagement of Key Response Partners and Stakeholders.
The utility should identify partners and stakeholders that may be involved in the
development of the CMP and corresponding response activities.   Utilization of and
teaming with partners and stakeholders should increase the success of response efforts and
allow all parties the opportunity to understand the standard processes and procedures used
during a drinking water utility emergency. It should also ensure that the CMP is integrated
and consistent with external emergency response plans. Figure 5 provides an overview of
potential partners involved in the development and implementation of a CMP.

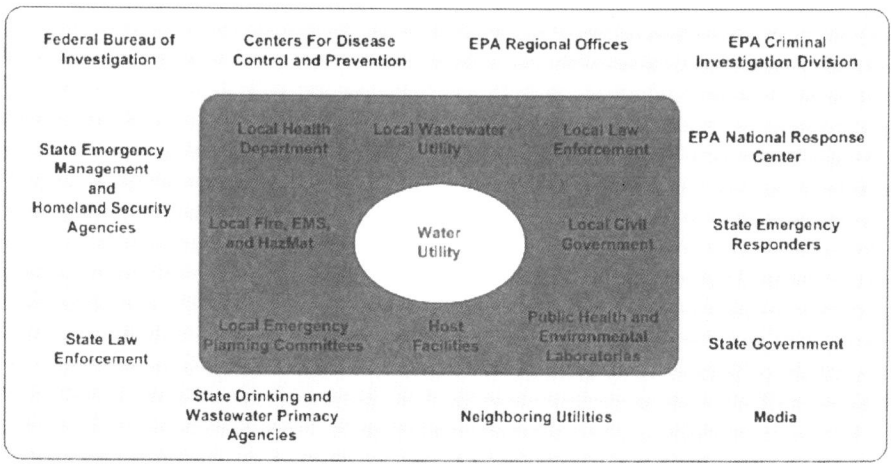

Figure 5 *Potential Contamination Warning System Partners*

Once identified, the utility should engage the response partners through coordinated
meetings to   determine   and   confirm   roles   and   responsibilities,   solidify   lines   of
communication, identify shared resources, and ensure that the draft CMP aligns with other
agency operational response plans. The utility should also confirm with each response

partner agency at what phase of the contamination incident they would be notified and that the appropriate points of contact are identified, including accurate, up-to-date contact information. Figure 6 illustrates the recommended approach for engaging partners.

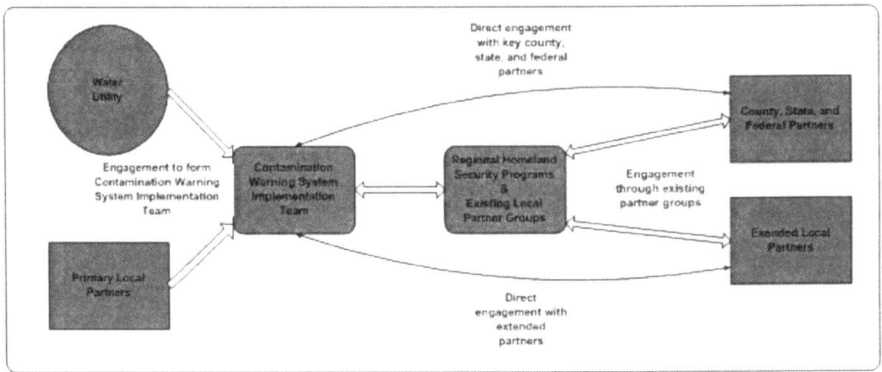

Figure 6 *Recommended Strategy for Engaging Consequence Management Plan Partners*

The recommended approach for engaging partners is to start with the primary local partners, followed by the county, state, and federal level agencies. Also during this process, the utility and partners should take any opportunity to leverage existing communication and response networks that may have been established by other agencies/programs, such as Local Emergency Planning Committees.

The outcome of these planning meetings with response partners should be a near complete draft of the CMP. Keep in mind that the CMP will never be "final," as it is always open to revision based on changing relationships, agency reorganization, etc. Additionally, as the implementation of the contamination warning system progresses, some adjustments to the CMP may be needed.

2.3 Implementing the Consequence Management Plan

Implementation of the CMP, once developed, involves several activities including developing communication plans, training, and maintenance.

 2.3.1 Communication Plans. Communication plans are an essential component of a contamination warning system, including the CMP. They prepare drinking water utilities for both routine and incident-specific communications with customers, response agencies (e.g., local, state and Federal government offices), the media, and the public at large. The overall communication strategy should include developing plans, defining roles, and identifying resources and equipment. Utilities can elect to develop communication plans as an integrated component of the CMP, or as a stand-alone plan.

 Communication plans include development of both general and risk communication plans. General communication plans describe lines of communication for internal utility personnel and with external response partners during a contamination incident. They should also specify the communication equipment (e.g., radios or cell phones) and shared emergency frequencies to be used (where appropriate). The risk communication plan should guide the utility and its partners on when and how to make public notifications, how to work with the media, how to define what the message will be and establishing delivery systems for the message (e.g., media, radio, television, reverse telephone systems).

   2.3.2   Training.   To have an effective consequence management program, training should be conducted to familiarize utility staff and response partners with their roles and responsibilities under the CMP to get them "up to speed" and "on the same page." In general, training should include how the CMP is organized (e.g., credible, confirmed, remediation and recovery), the corresponding steps associated with each response, a review of roles and responsibilities and working together under an ICS or a Unified Command.

   A comprehensive training strategy should include a suite of core courses in the ICS, augmented by a progression of exercises which are described by the Department of Homeland Security's Homeland Security Exercise and Evaluation Program (HSEEP).[3] The HSEEP describes "Discussion-Based" exercises, which include tabletops, seminars, and workshops to introduce and teach new concepts, followed by "Operations-Based" exercises including drills, functional exercises and full-scale exercises to test and evaluate program effectiveness. Additionally, training drills associated with specific CMP activities (e.g., field sampling, site characterization) may need to be conducted.

   2.3.3   Maintenance.   Since the CMP is a living and evolving guidance document, it must be properly maintained over time. To accomplish this, the utility should establish some maintenance guidelines. These guidelines should specify the actions needed for routine and non-routine updates to the CMP, the circumstances under which the updates will occur and the organizations responsible for the updates.

   A standing maintenance committee, comprised of utility management, field personnel, and support agency representatives, should be assembled to review and evaluate the operability of the CMP on at least an annual basis and after each scheduled training/exercise activity. This is important since operational and personnel changes within the utility and support agencies can occur frequently.

   The plan should also be reviewed for any potential changes following non-routine incidents such as:
   • After any personnel change which may alter management, field team, or ICS composition;
   • After any significant changes to sampling, analytical, or equipment procedures;
   • After any significant changes to the water treatment or distribution system;
   • After any significant changes to the monitoring system components, concept of operations, or procedures; and
   • After any off-normal occurrence that triggered the activation of the CMP.

3 CONCLUSION

The Water Security initiative contamination warning system detection strategy involves the use of multiple monitoring and surveillance components for timely detection of drinking water contamination in the distribution system. Consequence management represents the response mechanism of the contamination warning system, and consists of actions taken to plan for and respond to potential drinking water contamination incidents in the distribution system. Thus, a CMP that successfully guides these actions is a cornerstone of an effective contamination warning system.

   Construction of a comprehensive CMP includes a utility self-assessment and evaluation of existing emergency response plans, development of the CMP framework, and identification and engagement of key response partners and stakeholders. An initial self assessment allows the utility to identify existing procedures regarding planning, preparedness, and response that may serve as a starting point for constructing a CMP. The CMP framework can then be developed to begin conceptualizing the later stages of a response and determine when response partners should be engaged. This includes the development of decision trees that clearly outline response actions by the utility through each phase of the contamination incident. Finally, partners and stakeholders should be identified and engaged to increase the success of response efforts and allow all parties the opportunity to understand the official process and procedures used in the event of an emergency at a drinking water utility. It should also ensure that the CMP is integrated and consistent with external agency emergency response plans.

An effective consequence management program also includes communication plans, regular training and maintenance. Communication plans should be developed for both routine and incident-specific communications with customers, response agencies, the media, and the public at large. Training should be conducted to familiarize utility staff and response partners with CMP roles and responsibilities during a contamination incident. Maintenance guidelines should be established by the utility that specify the actions needed for routine and non-routine updates to the CMP, the circumstances under which the updates will occur and the organizations responsible for the updates.

EPA is currently developing a more detailed guidance document, Water Security Initiative: Interim Guidance on Developing a Consequence Management Plan, that expands on the concepts presented in this paper. The anticipated release date for this guidance document is summer/fall of 2008.

References

1   USEPA, *WaterSentinel System Architecture*, EPA 817-D-05-003, 2005.
2   USEPA, *Response Protocol Toolbox, Modules a-g,* EPA-817-D-03-007, 2004a-g.
3   HSEEP, *Volume I: HSEEP Overview and Exercise Program Management.*
     Department of Homeland Security, 2007. https://hseep.dhs.gov/.

Acknowledgements

The author would like to acknowledge the contributions of Ellery Savage and Kim Morgan from Computer Sciences Corporation and Bill Desing, Colm Kenny, and Todd Elliot from CH2M Hill. The author would like to acknowledge the extensive contributions of the following members of the US EPA Water Security Initiative implementation team: Steve Allgeier, David Harvey, Elizabeth Hedrick, Mike Henrie, Jessica Pulz, and Dan Schmelling. The author would also like to acknowledge the contributions of Kathy Clayton and Matthew Magnuson of the US EPA National Homeland Security Research Center. Finally, the author is grateful for the extensive cooperation, input and professionalism from staff at the Greater Cincinnati Water Works and local partners who were instrumental in establishing the first Water Security Initiative pilot.

Disclaimer

APPLICATION OF A RISK BASED APPROACH TO SECURITY AND INTEGRITY OF ASSETS – A REGULATORS VIEW.

M. J. Rink and S. A. Evans

Drinking Water Inspectorate, 55 Whitehall, London, SW1A 2EY

1 INTRODUCTION

Water Undertakers within England and Wales are directed by The Security and Emergency Measures Direction (SEMD)[1] to make, keep under review and revise plans to ensure the provision of essential water supply or sewerage services, at all times, including during a civil emergency or an event threatening national security. In doing this they have to have regard to additional guidance issued by Government. In the event of failure of the piped water supply the water undertaker is required to provide alternative supplies to domestic customers. This direction is underpinned in legislation by the Water Industry Act 1991 section 208[2]. Part of the requirement is protection of assets by exploration of assumptions, strategies, assessments and checks. This includes the physical security and the integrity of assets and the associated risks. When developing a risk assessment and the resultant appropriate measures, these should take into account the consequence and likelihood based upon defined criteria both current and on historical information of hazards and threats, whether they are internal or external and whether they affect actual assets or supporting systems. Furthermore, consideration should be given to the consequence and likelihood of occurrences which maybe less well understood such as analytical artefacts or changes in the integrity of the structure or environment which may lead to unexpected conclusions and actions. Companies in considering appropriate measures should also consider synergies which may benefit more than one objective.

2 OBJECTIVES OF SECURITY AND INTEGRITY

Water utilities will be motivated by a number of objectives directing the security and integrity of its assets. These often include the legal or regulatory drivers in the form of Acts, Regulations or Directions in England and Wales and the company's own financial drivers, the primary objective of any water utility must be to provide safe water reliably and in sufficient quantity to its customers. This objective should be driven at board level and throughout company operations paying due regard to the regulatory requirements.

There is often an emphasis on operational security towards general malicious attack which includes vandalism, extortion, terrorist and sabotage. Company operations, as an objective to providing safe water and therefore in doing so protect the public health, should operate in a manner that protects not only against intentional attack, but should protect against, detect and respond to threats both intentional, unintentional and natural from inside and outside the utility. As part of this strategy the company should form an objective which identifies critical assets and maintains those assets to reduce any impact should an asset be rendered inoperable. In doing so the company also benefits from secondary objectives including protection of the environment, its employees and the reputation and confidence in the water utilities.

3 REGULATORY DRIVERS

Legislation and guidance in this area is well established. The Security and Emergency Measures Direction 1998 (SEMD) requires water undertakers to make, keep under review and revise plans to ensure provision of essential supplies at all times and in constructing the plans to do this, to consider the worst possible case scenarios, with large populations and multiple sites.

The plans should be living documents and arrangements made to maintain as such. Each Water Undertaker must review the plans at least annually and send generic copies to Secretary of State (SoS) of Department of Environment, Food and Rural Affairs (DEFRA). The plans are then certified by independent auditors and checks made to ensure they make provision for adequate and appropriate personnel, stockpiles of equipment and materials, analytical services and dedicated communications.

The SEMD also specifies the requirement on water undertakers to provide a minimum volume of water during emergency situations in the form of alternative supplies (currently set at 10 litres/head/day).The quality of the water supplied as an alternative should meet the regulatory requirements, and special account taken for provision to the "vulnerable" consumers such as the sick, elderly, disabled, schools, nurseries etc and also the more sensitive users such as essential food industries.

The SEMD also requires water undertakers to notify the SoS (DEFRA) as soon as possible in accordance with 1(4)(b) of 1998 Direction and in turn DEFRA, notify other water companies using a well established cascade system, as well as other government departments and non government organisations.

In such emergency circumstances, the Drinking Water Inspectorate (DWI) have a specific remit; to act as technical advisors to ministers on matters regarding drinking water quality, to assist in briefings of DEFRA colleagues and to attend Cabinet Office Briefing Room (COBR) meetings as required.

DEFRA has emergency powers under S208 of the Water Industry Act 1991 and can issue general or specific guidance in respect of national security and/or mitigation effects for civil emergencies.

4 VULNERABILITY WITHIN WATER UTILITIES

Vulnerability is something which is liable or potentially exposed to attack. This might be a weakness of an asset either by design, management or its operation either physically, in procedures or through interdependent systems and networks, intentionally or unintentionally. For a water utility vulnerabilities may exist from source to tap and include the source itself, the works, storage and distribution including pumping stations and networks. Additionally, the reliance on SCADA, (Supervisory Control and Data Acquisition), systems can make these a target for exploitation by an aggressor intent on causing either physical damage, misinformation or the removal of data. Finally, as with all industries staff represent a significant vulnerability.

4.1 Source Water

The Drinking Water Inspectorate advocate the principle of the Water Safety Plan, (WSP), as an effective means of consistently ensuring the safety of drinking water. The WSP approach employs the use of a risk assessment from catchment to tap to identify hazards in a multiple barrier approach[3]. The first part of this is the source water which marks the beginning of the water treatment process and therefore in terms of vulnerability also marks the first point of potential attack. Source water for treatment is derived from either surface water or ground water or a combination of both. Surface water could be deemed difficult to contaminate due to the large volumes involved. Similarly, protected ground water as a consequence of its depth and protective geological layers such as clay also make this a difficult source to contaminate. However, in the case of the surface water, the transfer raw water main and in the case of the protected groundwater, the well head and the main are both vulnerabilities. Perhaps unprotected groundwater is a greater vulnerability in respect of the catchment water. Although scope exists in all cases for online surrogate monitoring, vigilance, good catchment management, and even physical barriers commensurate with the company's own WSP.

It is always of merit to review information from natural causes of contamination to inform company decisions of potential vulnerability. In England and Wales lessons can be learnt from cases of both surface and groundwater contamination that company's have been faced with and discovered through, good and proactive monitoring. Perhaps one of the best examples was the industrial contamination and subsequent exposure of the contaminated topsoil to the elements which resulted in a relatively rapid transmission of bromate and bromite to the aquifer used as a groundwater source by a number of treatment works. In this case, monitoring and the discovery of a rising trend permitted subsequent action by the water company to protect the drinking water supply. Contamination of groundwater and surface water is not unusual and company's must deal with this on a regular basis such as contamination by hydrocarbons or pesticides both which may challenge water treatment. Equally, there are the more unusual situations such as when Buncefield exploded on 11th December 2005 a water utility was challenged with the possibility that its source could be contaminated with multiple chemicals and not only hydrocarbons from the fuel but also Methyl tertiary butyl ether, (MTBE), Polyaromatic hydrocarbons, (PAHs) and chemicals used in the fire fighting itself such as Perfluorooctane sulphonate, (PFOS). PFOS incidentally also posed a problem for the sewage undertaker who collected the bunded firewater.

4.2 Treatment Works

Treatment works present a multiple stage approach and as such create a multiple barrier system to address hazards and in some cases flexibility can exist within the treatment process e.g. multiple streams which can enable an element of redundancy within the process. This perhaps creates a potentially robust base in terms of vulnerability and should also permit ongoing maintenance, training and the ability to shutdown in response to an emergency. Even with this, the vulnerability of staff, visitors, deliveries and deliberate external malice pose potential treats to electrical systems, plant, networks including SCADA systems, control of treatment processes, and chemicals, chemical storage and dosage. In its investigatory role, DWI assess incidents affecting the quality or sufficiency of drinking water supplies. Some recent examples highlighting the importance of risk management are cited. An example highlighting the vulnerabilities of electrical systems can be seen in one case where fluctuation the power supply resulted in the shutdown of pumps within the treatment works. This threatened the adequacy of supply as the clear water tanks progressively emptied and resolution to restart was fraught with difficulty due to damage to the process as a result. In this case redundancy of a stage was used to overcome the problem in the short term prior to full recovery. An example of the risk associated with inadequate control of dosing systems was illustrated in the instance of chlorine being dosed into a contact main being left "in manual-control" whilst the main was not pumping resulting in a significant rise in chlorine in the main before then going into supply as the main started pumping. There are also examples of coagulant chemicals over or under dosing due to failures in equipment and dosing lines. Finally, an example of improper chemical handling and delivery include the improper delivery of sodium hypochlorite into the aluminium sulphate tank resulting in a gas cloud which threatened workers in the near vicinity.

4.3 Storage and Distribution

4.3.1 Reservoirs

Reservoirs are a critical and necessary part of the water distribution system to balance fluctuating demand and pressure as well as providing a supply during shutdowns within the treatment or distribution process or just to provide a reserve to meet emergency demands. Their vulnerability can be simply due to them being unmanned and in remote locations and where access might not be discovered immediately. The construction may have access points such as doors and hatches, vents and intermediary stage dosing of disinfectant, all of which provide potential vulnerabilities. A number of measures can be taken to reduce the vulnerabilities including design of hatches, closure of vents and alarm systems. However, as with previous examples problems can and do occur which are unintentional. In one incident in 2006 the unintentional dosing of diesel into the water supply occurred at a service reservoir when a container marked as sodium hypochlorite (chloros) was used for dosing but in fact contained diesel. This resulted in four days where consumers were advised not to use their drinking water. This incident highlights the importance of control of chemicals not only at treatment works but also reservoirs and the ease at which an undesirable chemical can enter supply.

4.3.1 Pumping Stations

Pumping stations within the distributions system used for the augmentation of pressure or supply are also often unmanned and therefore can be vulnerable. The potential of exists for a pump to be damaged causing poor pressure or total loss of supply or as a possible point of introduction. In 2006 a booster pump failed due to a power dip. This directly affected over 20,00 properties causing a loss of supply for a proportion of those properties as the pressure in the system reduced and emptied and highlights the wide and significant effect such an incident can have.

4.3.1　Networks

The networks form the last part of the system by which drinking water is delivered to consumers. By their nature they can be wide and sprawling, have a number of attachments, such as hydrants, pressure reducing valves, break tanks, meters, valves as well as the connections to the domestic systems and the taps within them. Therefore, by virtue of necessary design, this part of the water supply system could be considered the most vulnerable. Multiple entry points exist for introduction of contaminants whether intentional or unintentional and these can be remote, difficult to be vigilant over, as well as being difficult to secure. The possibility of inappropriate cross-connections, inadequate back-flow prevention and lack of control of domestic systems or unauthorised use of the network connection points also create a significant problem to utilities. In 2006 there was an incident where up to 1000 consumers in a dockside town were supplied with sea water. In this incident a cross-connection occurred through an error in valve configuration from a ship taking on drinking water (illegally) whilst taking on sea water as ballast. The cross-connection resulted in sea-water entering the drinking water supply network. The ease at which such a gross contamination of the network can occur serves as a reminder as to the vulnerable nature of networks. Similarly, large distribution mains within the network should also be a consideration for vulnerabilities as companies will have experienced bursts and loss of supply due to other problems. In 2005 one water company experienced a valve which failed in a shut position on a 54" main causing loss of supply to in excess on 250,000 consumers for several hours followed by aeration on the return of supply.

4.4　Control Systems

Water utilities are becoming increasingly dependant upon Supervisory Control and Data Acquisition, (SCADA), systems within and throughout its works, reservoirs and networks. The increasing dependency, often due to the desire to reduce manning on individual sites, and the complexity to increase control and collection of data means that vital equipment are connected and controlled from a computer network. These can include the control of pumps and valves throughout the process and network, critical process controls, power, and equipment as well as information acquisition and alarms at every stage of process and distribution.. The convenience to have these systems connected to the internet for ease of information transfer and remote connection as well as the standardisation of the platforms add to the vulnerability of these systems to both internal and external attack. Often overlooked are those systems which are connected to dial-in modems for engineering purposes and have little or no security. The wide control of SCADA systems mean that an intruder who gains entry onto such a system could shut down or destroy plant, alter dosing of critical process chemicals or operate valves which either stop flow or inappropriately interconnect processes or systems or indeed disable critical alarms. All of these could be detrimental to the supply of safe drinking water and therefore have significant consequences for public health.

4.5　Staff

Staff are an essential part of a utilities operation and without them, their experience and intricate knowledge of the processes the function of the supply of water could not happen.

Equally, the resource to deal with an emergency rests with the staff themselves. However, staff are also a potential vulnerability because they have access to all parts of the process, systems, networks and computers that control and enable the supply of water as well as having knowledge of the vulnerabilities of the system. Utilities along with many other industries have to deal with rogue staff members who for whatever reason have not carried out their jobs appropriately and it is therefore difficult to entirely remove this vulnerability. Additional to this is the increasing use of contractors who enter sites to carry out work and whom a company may have less knowledge about their integrity. However, simple steps such as validation of background, training and use of identification badges, restrictions of access and information and clear policies reduce this vulnerability.

5 PRICIPLES OF PHYSICAL SECURITY

An effective and holistic security approach should employ equipment and systems which deter, detect, delay and if necessary deploy and then mitigate any consequences of an attack. In the case of a utility site a deterrent measure may be a visible impression that entering a site would be difficult or that the intruder would risk detection e.g. a high fence. For detection, this might be a system which should an intruder decide to attempt access they would be rapidly detected to enable an effective response. Delaying an intruder is often by physical barrier measures such as doors, hatches, locks, covers that will stop them for a period sufficient to permit successful deployment of a suitable response. Should an intrusion occur then it would be expected that a company would have pre-planned arrangements in place to permit rapid recovery, prevent, minimise or mitigate any injury that could happen whether that be intentional or unintentional. A utility should include, on and off site security for its assets where possible. Furthermore, when the company forms its plans, these should include loss of supply, introduction of compounds into the water, consideration of toxic chemicals on or off site including their, misuse of SCADA, pollution, staff , physical destruction and cyber and electronic attack. The recovery plan should encompass the incident management team, members of that team and their training, its communications and resources available to it including equipment, location, site information, design and layout, network connections and flow dynamics, consumer data and scientific and medical advice. All companies have such plans in England and Wales and a number practise such plans, however, in an emergency, plans do not always work as well as expected. In 2007 parts of England experienced extensive flooding resulting in the formation of incident management teams. Decision making particularly in the absence of sound technical and medical advice can result in the delay of an appropriate response.

6 LIKELIHOOD

In the last six years up until the end of 2007 there were just over 600 unauthorised access events notified to the Drinking Water Inspectorate in England and Wales. The majority of these (almost half) were site intrusions where there was no identified damage or theft. (Figure 1).

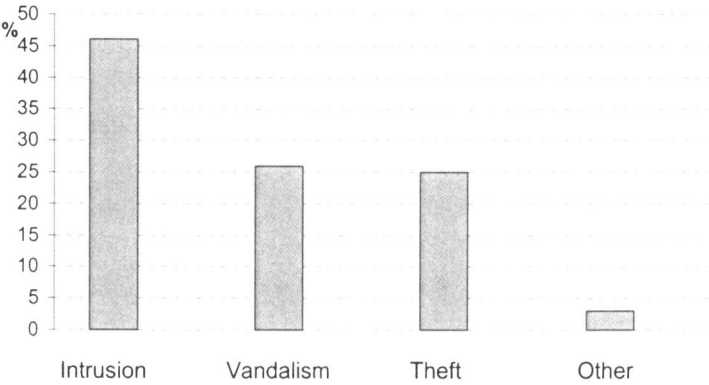

Figure 1 *Percentage distribution of unauthorised access to water company sites as reported to the DWI.*

Vandalism and theft constituted approximately 25% each and a smaller category listed as other were where sites were broken into and there was sufficient concern to the company such that they were required to take action in response to a perceived risk. For the vast majority listed as intrusions the reasons when known, were varied. Most probably intruders looking for something to steal but nothing of use was found, or they were unable to take anything and therefore left empty handed. More strangely were those reports of people camping on site, or those who were generally hanging around with nothing better to do and therefore decided to break into a utility site to have a barbeque and a few drinks, there were also instances where the site had been used as a race track or dumping ground, such as a place to leave a car, sometimes burnt out. Most unusually was the report of access to site for the erection of a pirate radio mast. The next main category was vandalism and these are purely damage, most commonly to doors, fences and locks, and graffiti for the sake of it.

Thefts are a major concern to British industry and with the rising cost of metals and fuels these are becoming an increasing target as well as the more traditional thefts of plant and hardware such as slates and tiles. The market value of metals has risen and appears that it will continue to rise. For example, copper has doubled in price over two years and its theft has affected all utility sectors. Metals theft is reported to currently cost £360 million a year to the UK industry since when stolen can result in the loss of service be it water, gas, electricity or other service in addition to the material loss[4]. Lead which is trading at a higher price than aluminium is often ripped from roofs. Cable theft emerged in 2005 and was increasing by early 2006. There is very little deterrent for these thefts, despite the fact that removing metals such as those in cabling often risks death. Instances of metal theft from railways, as reported by the British transport Police increased from 1,142 offences and 317 arrests in 2006 to 1,928 offences and 396 arrests in 2007[5].

When looking at the pattern of those intrusions on water utilities in England and Wales there appeared to be no preferences between treatment works and reservoirs as reported to the DWI. (Figure 2).

Figure 2 *Unauthorised access to site in England and Wales compared to site type.*

This would appear to be an unusual pattern since in terms of vulnerability, reservoirs are more remote and therefore they would expect to be less of a deterrent and so have a greater proportional intrusion. It must be considered however that these figures are dependant upon reporting, interpretation and geographical factors and reporting levels vary from company to company.

6 RISK BASED APPROACH TO SECURITY

The DWI itself has moved to a risk based approach for most of its activities including auditing and this is very much in line with the first recommendation made by Hampton in March 2005[7] in his report on reducing administrative burdens. Equally however, such an approach is and has been widely use in health and safety and other areas to become more effective and directed to resources. It therefore follows that using such an approach in security to direct resource could be labelled as working smarter. The objective for such a means of operating is to determine through a structured and consistent assessment, which sites are more likely to have problems and then to assign the most appropriate actions and timescales commensurate with the level of risk identified. At a conceptual level, a risk assessment is a tool that makes a distinction between an assessment of the risks and risk control. Ultimately this should inform better decisions on the level of security and resultant specifications required at each site. A concept developed by Deere *et al* in his publication of 2001[6] developed the concept of a consequence likelihood matrix adapted to risk and risk management of water related diseases but equally such a matrix is and can be applied to similar but unrelated circumstances (Figure 3). The matrix develops the 5 x 5 concept of likelihood verses consequence to create a mathematical value from which a categorisation can be made.

Increasing Consequence				
5	10	15	20	25
4	8	12	16	20
3	6	9	12	15
2	4	6	8	10
1	2	3	4	5

Figure 3 *Likelihood – Consequence matrix*

6.1 Consequence

Consequence is the impact on a supply and therefore the consumers following an intended or unintended action at a site such as a break-in. The level of consequence must be considered in respect of the consequence to the consumers (e.g. the lower end may represent loss of confidence in the company, whilst the upper end of the consequence represents a potential health impact). In addition to this the size of population as in few or many must be considered and who the site serves, such as a large user like a hospital or food producers.

Examples of levels of consequence could therefore be listed as follows going down the list in ascending order:

- loss of public confidence in company
- loss of supply for short time
- contamination or loss of supply for prolonged period/small-mid size population/key customer
- contamination or loss of supply for prolonged period/large population – potential for minor health Impact or rejection of supply
- contamination - potential for major or significant health impact

6.2 Likelihood

Likelihood is the likeliness or chance of a given event occurring or the state of being probable for an event, it is related to the consequence in only the possibility of a consequence happening. Therefore when assessing likelihood in respect of physical security then it should be supported by evidence of intrusion or intent. At the lower end of the scale it would be that there is no evidence of intrusion and therefore the likelihood is improbable and at the other end it would be that there is solid evidence of intent. This might be in the form of an actual received threat.

So like consequence, examples of levels of likelihood could therefore be listed as follows going down the list in ascending order:

- no evidence of Intrusion
- theft, Vandalism, Intrusion
- repeated Theft, Vandalism, Intrusion
- frequent Intrusion, Vandalism, Theft
- evidence of serious intent

6.3 Application of the Risk Assessment to site

Following on from the assessment of consequence and the assessment of risk the determination of assessment of risk can be made using the matrix shown in figure 3. So for instance where there is no evidence of intrusion at a site, or where the site is new or no data exists, there is no evidence in the opinion of the assessor that intrusion is likely, and should an incident occur, there would be at most a loss in confidence in the company since it is a small supply which has a redundancy measure or other reason, then the resultant score would be one Figure 5). Similarly, where a site has had definitive evidence of serious intent and this could cause a significant health impact, then the score would be 25.

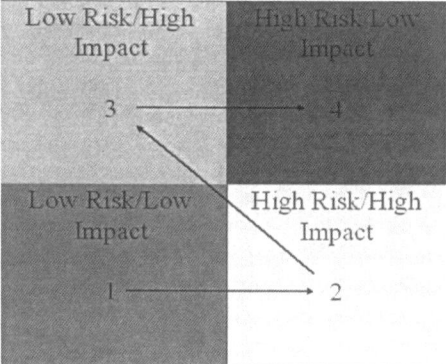

Figure 4 *Categorisation of site into risk levels*

Following an assessment, sites should be categorised as in figure 4. If the principles of deter, detect and delay are employed, each categorisation should carry a security design element appropriate for that level. Consideration of design elements should be made for fences, enclosures, detection system and ancillary devices, means of access and their control as well attendance and personnel on site and computer systems. A level 1 site would most likely have a basic level of security which may be a simple chain link fence, kiosk with an door and lock to a predefined approved standard and possibly an alarm. At each ascending level the standard of design element should increase and combinations of those elements used appropriate to the type and nature of site as well as the equipment and plant on it such that at level 4 the highest standard of equipment and any added barriers are employed. The ultimate aim being that appropriate resources are used in an appropriate way.

Companies should consider any synergies on site which might either complement the security arrangements or require a different level. In the instance therefore of a company having to build a wall for flood defence, then such a construction needs to inform the risk assessment. Similarly any such construction at a later date would require a re-evaluation of the risk assessment to update the new circumstance. This would also be the case for other synergies such as power sub-stations or mobile masts on site. Often such sites require access by non-company personnel and consideration must be given to how and who access is granted. However, it is of importance that companies review risk assessment periodically even there appears to be no change to those working on site to maintain objectivity and highlight a need to re-evaluate the categorisation. Figure 5 shows a conflict between a risk assessment for health and safety to ensure workers do not stand on a roof that could collapse causing the worker to fall into the water tank below. A security risk assessment may have highlighted that this roof is one of the only barriers to the treated water surface and that the roof is inappropriate for purpose and to notice points this out. The lack of synergy in this instance should be considered to provide a solution that meets both purposes.

Figure 5 *Photo of fragile roof notice on a roof directly above treated water.*

Companies must also understand the analytical procedures required for sampling and what possible outcomes there might be. A failure to consider and have an appropriate procedure in place could result in data that is incorrect or the company does not know how to react in response to an unexpected result. Whilst this is not part of a traditional risk assessment, a consideration of the consequence of actions should be coupled to the risk assessment as part of the deploy and mitigate strategy. Site categorisation could be altered if there is an inability to deploy or have an understanding for an appropriate strategy for a particular site.

6.4 Reliability of categorisation

When company's carry out a risk assessment, it is of importance to ensure that the assessment is objective. Risk assessments often err on the side of caution resulting in over categorisation. Assessors over categorise to be on the safe side and this is human nature to be cautious. However, over categorisation uses up time, equipment and resource on less serious sites creating unnecessary activities and work loads and ultimately decreases the effectiveness of a company. However, the consequence of under categorisation is such that a risk of a serious intrusion could be missed or the company could miss a critical widow of opportunity such as finance support or combining the upgrading of sites in a single job. Furthermore, such an assessment would decrease the effectiveness of the security measure in place or indeed any preparation for mitigation.

Figure 6 highlights a door which a risk assessment may have highlighted as inadequate or alternatively the risk assessment under categorised the site. In this instance the lock on the door was cut and the door opened, however, the door was weak in a number of other points. It is therefore imperative that company's set out both the risk assessment strategy and the standards to which they must work to in order to avoid such a situation arising.

Figure 6 *Door to access treated water at a service reservoir.*

7 CONCLUSION

In the application of a risk based approach to security and integrity of assets from a regulators view, fulfilment of the regulatory duties can be achieved by an appropriate risk assessment resulting in an appropriate categorisation balanced with the specified application of security measures. Whilst this may not stop a determined intrusion, taking reasonable an proportionate action in respect of and to reduce likelihood and consequence, delivers a structured and consequently more effective approach.

References

1 Security and Emergency Measures (Water and Sewerage Undertakers) Direction, 1998
2 Water Industry Act, 1991.
3 A. Davison, D. Deere, M. Stevens, G. Howard and J. Bartram, Water Safety Plan Manual, Framework for Safe Drinking Water, 2006, World Health Organization.
4 Rochdale Online News 22 April 2008.
5 "BTP cracks down on cable theft in Warwickshire", British Transport Police Media briefing 5 March 2008 16:00 hrs.
6 D. Deere, M. Stevens, A. Davison, G. Helm, and A. Dufour, 2001, Management Strategies. In Water Quality: Guidelines, Standards and Health – Assessment of risk and risk management for water-related infectious disease. (Editors. J. Bartram and L. Fewtrell) pp. 257-288, World Health Organization, IWA Publishing, London, UK.
7 P. Hampton, Reducing administrative burdens: effective inspection and enforcement 2005.
8 AWWA Security Guidance for Water Utilities, 2007, American Water Works Association.
9 The Water Supply (Water Quality) Regulations 2000, (Amendment) Regulations 2007 The Stationery Office Limited.
10 Drinking water in England and Wales, CIR 2004 – 6, Drinking Water Inspectorate.
11 Health and Safety Executive, Use of Risk Assessment within Government Departments Guidance, 2007.
12 Water Quality: Guidelines, Standards and Health. Chapter 12. Edited by L. Fewtrell and J. Bartram. 2001 World Health Organization, Published by IWA Publishing, London.
13 Guidelines for Drinking-water Quality, Third Edition, Volume 1, Recommendations, Chapter 4 Water Safety Plans, WHO, Geneva, 2004.

LET'S GET REAL. REAL WORLD EXPERIENCES WITH REAL-TIME ON-LINE MONITORING FOR SECURITY AND QUALITY. DETECTING AND RESPONDING TO EVENTS.

D. Kroll[1], K. King[1] and G. Klein[1]

[1]Hach Homeland Security Technologies, Hach World Headquarters, 5600 Lindbergh   Drive, Loveland, Colorado, 80539 Phone 970-443-2436. email: DKROLL@hach.com

1   INTRODUCTION

The distribution system represents the last analytical frontier in the water quality industry. The monitoring of source water and treatment plant processes has progressed to a level at which we can be confident that we are providing good quality water from the plant to the distribution system. Once the water reaches our aging distribution systems, however; our knowledge as to its continued integrity is limited by the quality and amount of available data. Most monitoring in the distribution system is relegated to the occasional snapshot provided by grab sampling for a few limited parameters or the infrequent regulatory testing required by mandates such as the Total Coliform Rule in the US. The development of water security monitoring in the years since 9/11 has the potential to change this paradigm. Since 9/11 numerous communities have installed multi-parameter monitoring stations in various locations through out the distribution network as early warning systems to detect potential water security threats. These continuous on-line systems have recorded large streams of data (at some sites for a number of years) relevant to water quality in the distribution systems in which they have been deployed.

For the past several years, scientists at Hach Homeland Security Technologies (HST) have been actively engaged in the development and testing of an early warning system for detecting water quality problems including those related to an intentional terrorist attack. This paper gives a brief summary of how the developed system operates and presents results form some of the recent field-deployment efforts that have been undertaken.

In this study, data streams from a number of communities (both small and large) are analyzed for pertinent information as to the health and operation of the distribution system. Changes in water quality are correlated with known causes attributable to day-to-day operational changes and also anomalous events.   Case study information concerning what, if any, action was taken to ameliorate the problem will also be reported for the identified events.

This sort of information is critical in understanding and improving the operation of our distribution systems and can also be valuable as we consider regulations that affect that operation.   Any future efforts to regulate the distribution system will need to consider databases and monitoring techniques such as the one described here before we can determine the best course of action to ensure our public water supplies meet acceptable levels of quality and safety from source to tap.

2 THE HACH METHOD AND APPROACH

Monitoring in the distribution system is a difficult proposition. The shear number and diversity of potential threat agents that could be utilized in an attack against the system makes monitoring for them on an individual basis an effort that is doomed to failure. To counter and detect the unprecedented number and types of compounds that may be encountered, what is needed is a broad-spectrum analyzer that can respond to any likely threat and even unknown or unanticipated events.

Rather than attempting to develop individual sensors to detect contaminants, the Hach HST approach was to utilize a sensor suite of commonly available off-the-shelf water quality monitors such as pH, electrolytic conductivity, turbidity, chlorine residual and total organic carbon (TOC) linked together in an intelligent network. The logic behind this approach is that these are tried and true technologies that have been extensively deployed in the water supply industry for a number of years and have proven to be stable in such situations. One of the difficulties encountered when designing such a device is that the normal fluctuations in these parameters found within the water can be quite pronounced.

The problem then becomes, can we differentiate between the changes that are seen as a result of the introduction of a contaminant or anomalous condition and those that are a result of normal everyday system perturbation? The secret to success, in a situation such as this, is to have a robust and workable baseline estimator. Extracting the deviation signals in the presence of noise is absolutely necessary for good sensitivity. Several methods of baseline estimation were investigated. Finally, a proprietary, patented, non-classical method was derived and found to be effective.

In the system as it is designed, signals from 5 separate orthogonal measurements of water quality (pH, Conductivity, Turbidity, Chlorine Residual, TOC) are processed from a 5-paramater measure into a single scalar trigger signal in an event monitor computer system that contains the algorithms. The signal then goes through a crucial proprietary baseline estimator. A deviation of the signal from the established baseline is then derived. Then a gain matrix is applied that weights the various parameters based on experimental data for a wide variety of possible threat agents. The magnitude of the deviation signal is then compared to a preset threshold level. If the signal exceeds the threshold, the trigger is activated.

The deviation vector that is derived from the trigger algorithm is then used for further classification of the cause of the trigger. The direction of the deviation vector relates to the agents characteristics. Seeing that this is the case, laboratory agent data can be used to build a threat agent library of deviation vectors. A deviation vector from the monitor can be compared to agent vectors in the threat agent library to see if there is a match within a tolerance. This system can be used to classify what caused the trigger event. This system can also be very useful in developing a heuristic system for classifying normal operational events that may be significant enough in magnitude to activate the trigger. When such an event occurs the profile for the vector causing it is stored in a plant library that is named and categorized by the system operator. When the event trigger is set off the library search begins.

The agent library is given priority and is searched first. If no match is found, the plant library is searched and, the event is identified if it matches one of the vectors in the plant library. If no match is found, the event is classified as an unknown and can be named if an investigation determines its cause. This is very significant because no profile for a given event need be present in the libraries for the system to trigger. This gives the system the unique ability to trigger on unknown threats and events. Also, the existence of the plant library with its heuristic ability to learn plant events results in a substantial and rapid decrease in unknown alarms over time and offers a tool for system optimization. The developed system has been subjected to strenuous testing in both laboratory and field scenarios. This has resulted in this approach being the only water quality monitoring system listed by the US Department of Homeland Security as a designated and certified antiterrorist device under the US SAFETY Act.

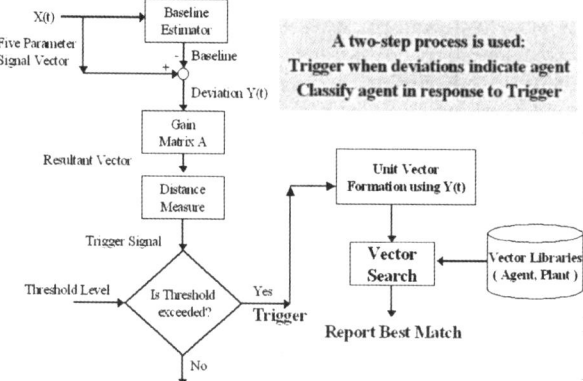

Figure 1 *The use of intelligent algorithms with standard bulk parameter monitoring equipment allows for a robust system that is capable of triggering on and classifying a wide diversity of threat agents including unknown events.*

Figure 2 *A real world deployment site showing the system in operation.*

3 FIELD TESTING OF THE DEVELOPED SYSTEM

Prior to the onset of this project, there was a definite lack of data concerning conditions in the distribution system. Very few utilities carried out data collection in the distribution system other than periodic grab samples. Those that did have some on-line continuous data were usually limited to only one or two parameters. Since the initiation of this program over, 500,000 hours of real time data has been collected across a wide variety of different distribution systems exhibiting different water matrix profiles revealing many interesting attributes of the distribution systems. These systems represent a variety of water quality conditions and operational situations. The site locations need to remain anonymous due to security considerations and non-disclosure agreements, but they represent a wide diversity of system sizes and geographic locations throughout the United States. The deployments are at both civilian and military sites. At the time of the preparation of this article, all deployments are within the United States. Hach HST has only recently received International Traffic in Arms Regulation (ITAR) clearance from the US Department of State to deploy the systems on an international basis. The following are a few examples of incidents that have been recorded during these real world deployments. These incidents help to demonstrate the systems ability to learn and to become a useful tool not just for security but also for every day operational improvements.

3.1 Caustic Overfeed Event

In this deployment scenario, the plant uses caustic feed to control water pH. The system experienced a trigger that when investigated was identified as an operational problem that resulted in the feed of excess caustic. The result was that the overfeed affected the pH and the conductivity of the water, causing the alarm. See figure 3.

The reason behind this was that the vendor from which the casuistic was being purchased had delivered the wrong concentration of the solution. No one had checked to see if the concentration was correct before feeding in the material. New procedures were put in place to verify incoming raw materials. The Event Monitor learned this Plant Event and can identify a recurrence of the event in the future if there is another failure in the system and it is repeated.

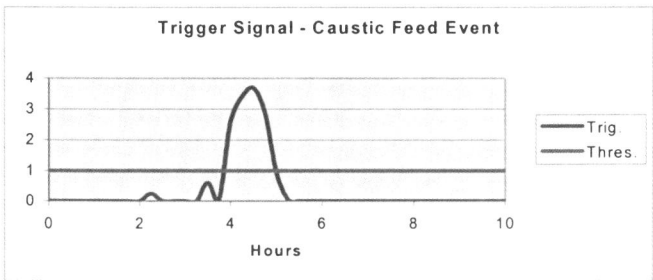

Figure 3 *Caustic Event.*

3.2 Road Work Event

In this event, roadwork (jackhammers) near a distribution line dislodged biomass and other particulate matter from the lining of the pipe. There was a massive increase in turbidity, which not only showed up on the turbidimeter, but also showed up as an interference in the chlorine measurement (optical). As expected, the conductivity and pH also showed minor changes. The increase in biomass in the water was indicated by the TOC analyzer. See figure 4. This event illustrates the ability of the Event Monitor to detect and alarm on unanticipated events. This event also provides a signature for the materials adhering to the walls of the pipes in this location and should recognize any future excursions of this type.

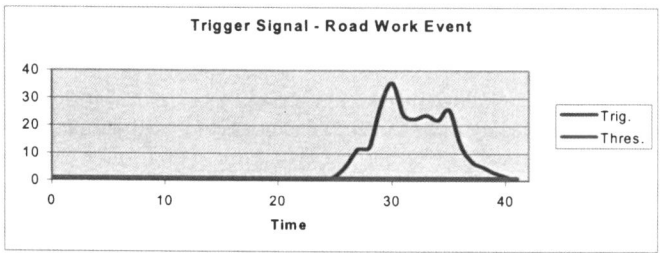

Figure 4 *Road work event.*

3.3 Pressure Event

In this scenario, the system was located in a building, which experiences a daily variation in water pressure. The sample variation is associated with a turbidity increase that causes a Trigger. See figure 5. There is also a small pH decrease at that time, possibly because of increased solubility of CO_2 in the water, dropping the pH slightly. After recognition of the cause and proper naming of this pattern, it is recognized by the Event Monitor as a "Normal" event, rather than an alarm condition, and appropriately classified and named as such.

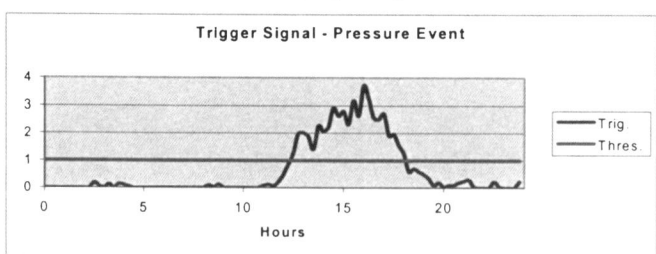

3.4 Rain Events

Large amounts of rain fell in the area of a reservoir, raising turbidity and affecting other water quality parameters. These events were large enough to cause a Trigger. See figure 6. The system was able to store this pattern and recognize it upon recurrence.

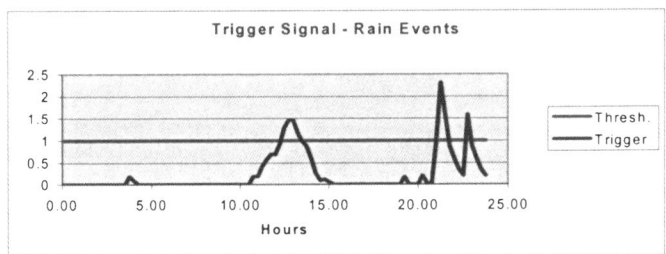

Figure 6 *Rain Events.*

3.5 Effect of Variable Demand

In this deployment, daily events influencing turbidity, chlorine, pH and conductivity are not completely understood but are suspected to be caused by water demand fluctuations in the area. This may indicate a need for more routine flushing of the areas pipes and the instillation of a chlorine booster station.

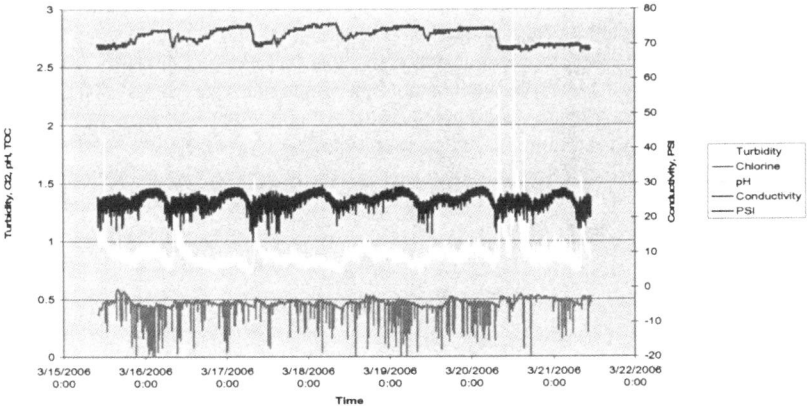

Figure 7 *Variable Demand.*

3.6 Ammonia Overfeed Events

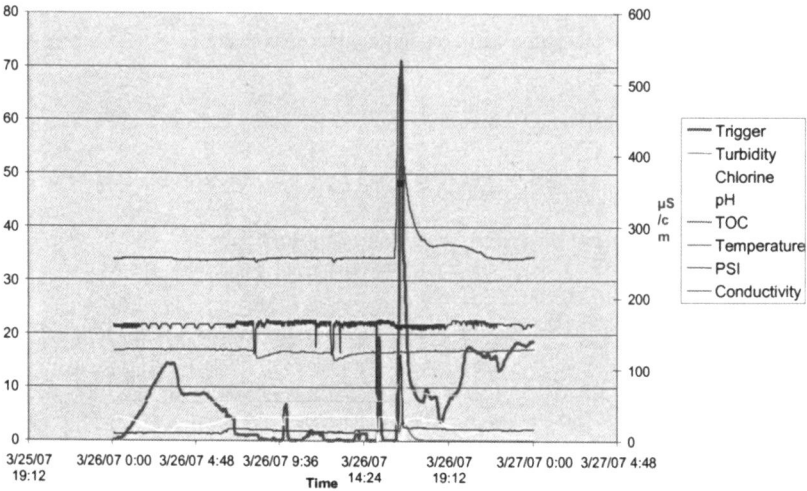

Figure 8 *March Ammonia Overfeed Event.*

On March 26th, 2007, maintenance was performed at the plant supplying water to the distribution system being monitored. After the maintenance was completed, the plant was restarted and the system that feeds ammonia to create monochloramine as a residual disinfectant overfed the chemical. The person in charge of the on-line monitors immediately noticed the increase in pH and notified plant operations. Operations reported a problem with ammonia feed pumps. The problem was temporarily fixed, but a slug of ammonia was sent into the distribution system. Several customers called, complaining about an ammonia smell and taste coming from the tap. The exact amount of ammonia released was unknown, but was believed to be less than 10 ppm. The facility continued operations but temporarily switched to free chlorine as a disinfectant until July 2nd. See figure 8.

On October 3rd, the same treatment plant experienced a brief ammonia overfeed. In this case, a pump was turned on and not switched off at the proper interval. There was a drop in chlorine and a decrease in pH. See figure 9.

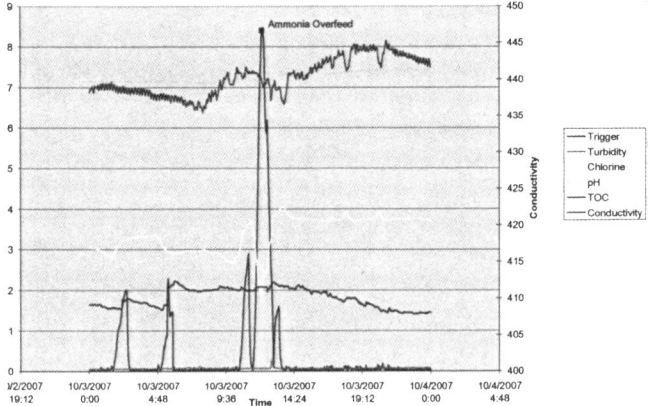

Figure 9 *Second Oct 3rd Ammonia Incident.*

Between August 22 and August 23, 2007, three similar events occurred. Increases in turbidity pH and possibly TOC with drops in chlorine caused triggers. These changes resemble the large ammonia event that occurred in March. The operator believes that these could be ammonia feed events, but could not confirm it or find the fault. See figure 10.

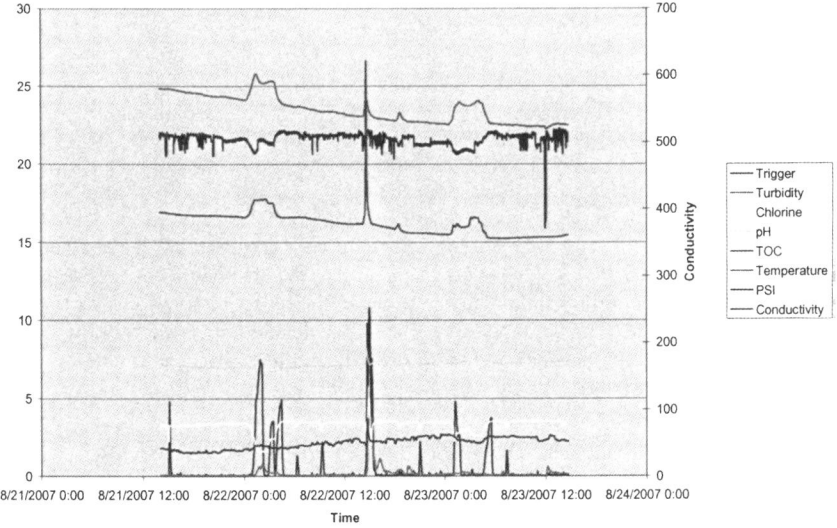

Figure 10 *Possible August Ammonia Events.*

3.7 Possible Chlorine Feed Event

An April 3rd, 2007 there was a turbidity and pH increase and a decrease in chlorine and conductivity. The operator suspects that there was a problem with the chlorine feed at the plant. However this cannot be confirmed. The plant was using free chlorine at the time which rules out the possibility of an ammonia feed problem.

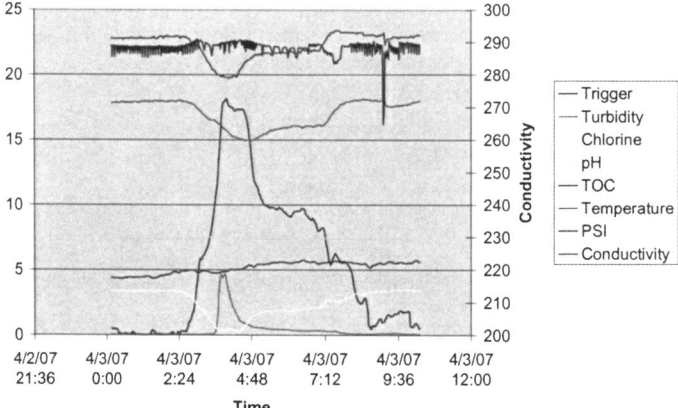

Figure 11 *Potential Chlorine Feed Event.*

3.8 Air Bubble Event

In one Northern Midwestern city's distribution system, every Friday, the deployed sensors would behave extremely erratically resulting in multiple alarm signals being generated. Investigation led to the discovery of extreme amounts of entrained air bubbles being present in the system's water on Friday afternoons and evenings.

Further investigation into the root cause of the air bubbles being present revealed that school buildings in the area that were to be vacant over the weekend had a policy of using air to blow out their water lines to prevent freezing so that the heat could be turned off over the weekend and holidays. A faulty check valve at one of the schools allowed the air to bleed into the distribution system. The valve was replaced, thus closing a possible backflow route into the system. After this maintenance was performed, the erratic readings ceased.

3.9 Distribution Flushing Case Study

In this case, the water utility customers were experiencing poor water quality in certain areas of the distribution system. The systems operations team could not pinpoint the problem. A series of distribution system monitoring devices were installed at various locations throughout the distribution network. The devices were able to help in locating a number of transient dead ends and low flow areas based on the monitoring results. Using this data the team revised the normal flushing schedule for these areas. The results were improved water quality and fewer customer complaints.

3.10   Grab Sample Versus On-line Case Study

In this situation, the local water utility had in place an extensive system of water monitoring through grab samples. All indications were that the water quality was good. After installation of several water monitoring panels in the distribution system they found that the turbidity spiked to levels as high as 20 NTU during the night and early morning hours when typically no grab samples would have been collected. They also found extremely high variability in chlorine levels during these time periods.

A series of changes to their treatment plant operations and distribution system procedures allowed the system to regain control of the water quality in the distribution system. They were able to lower the turbidity spikes to around 1.5 NTU at night and maintain more consistency in chlorine residual levels resulting in better water quality and consistency for the end consumers.

3.11   The Strange Case of the Chlorine Spikes

In one field deployment scenario, the system was very quiescent and rarely came anywhere close to causing a trigger alarm to go off. Except, that every night at around midnight, the chlorine level would spike dramatically and cause an alarm. This was deemed very strange and extensive trouble shooting of the instruments and power supply revealed no abnormal conditions that could be causing the problem.

After a thorough investigation, the night operator for the treatment plant was queried about the strange chlorine response. His reply was that of course the system's chlorine level spikes every night at midnight that is when he super chlorinates the system just like he had been told to do. It appears that several years ago, when there was a pipe rupture in the system that may have allowed contamination to seep in, the night operator had been told to super chlorinate the system. Unfortunately, the operator was new at the time and the instructions were not explicit that the super chlorination should take place that night only. The operator had continued to perform the operation every night for years resulting in a huge unnecessary cost in chlorine. This situation was remedied and should result in substantial chemical cost savings in the future.

3.11 Main Break Events.

In this situation the system had only just been installed a few days previously. Hach HST personnel were informed that the instruments were behaving abnormally and were giving strange readings. An investigation of the sensors found no problems. A short time later a major main ruptured in a catastrophic mode. The system was able to detect the perturbation in water quality parameters that were precursors to the main break and trigger upon them. Unfortunately, the system was newly installed and the event was not recognized until it was too late.

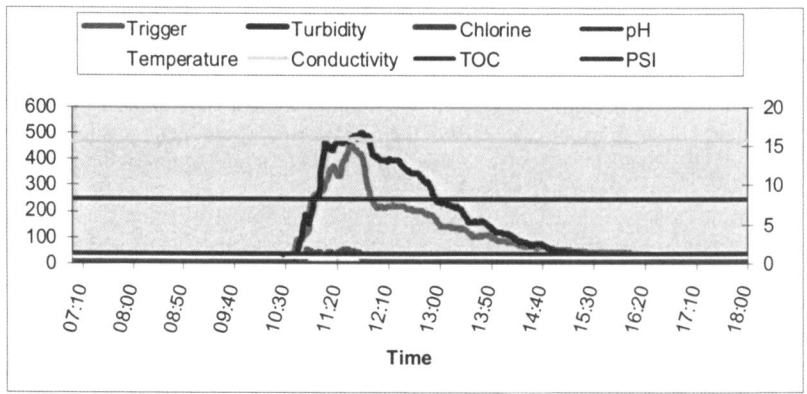

Figure 12 *Erratic readings soon after installation.*

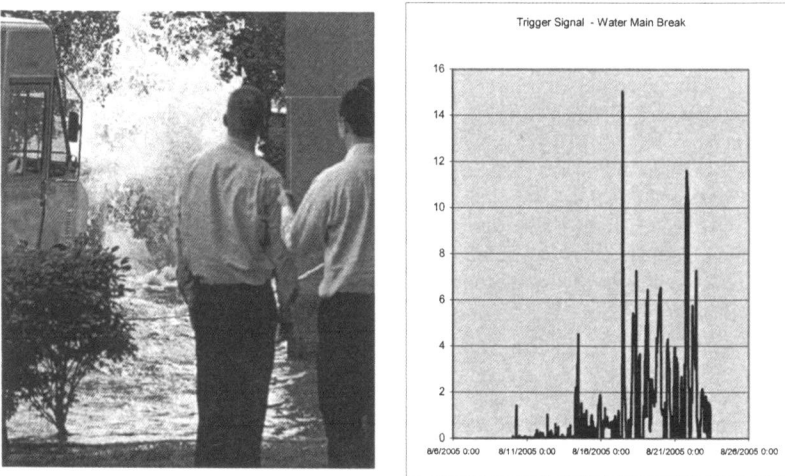

Figure 13 *Photo of main break event (left). Trigger signal for main break event (right).*

Another incident occurred on June 29th, 2006 the pipe was an 8-inch line and was located 1.8 miles down stream from the monitoring sensors. There was an increase in turbidity and chlorine. The turbidity appears to have increased the day prior to the break. See figure 14.

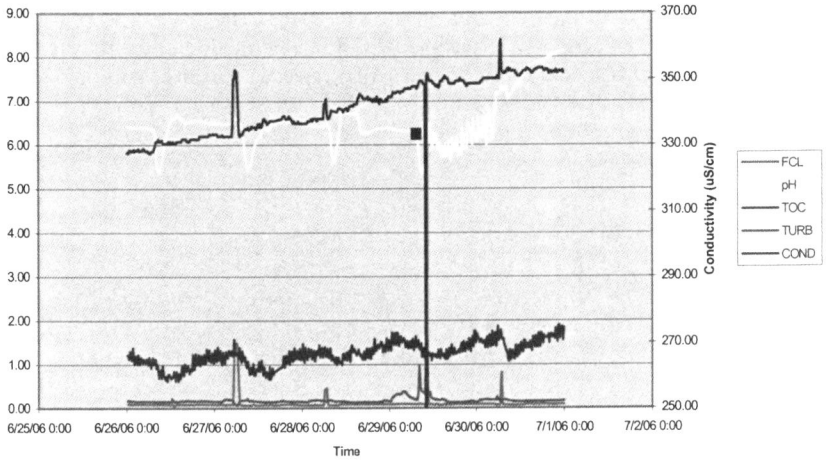

Figure 14 *8" Pipe Break.*

A third break occurred on July 30th, 2006. The line was a 36" water main located 1 mile from the monitor. Conductivity, turbidity, and chlorine spiked. There appears to be two water flow interruptions to the WDMP the night before but it's unclear if they are related to the break on the 30th. See figure 15.

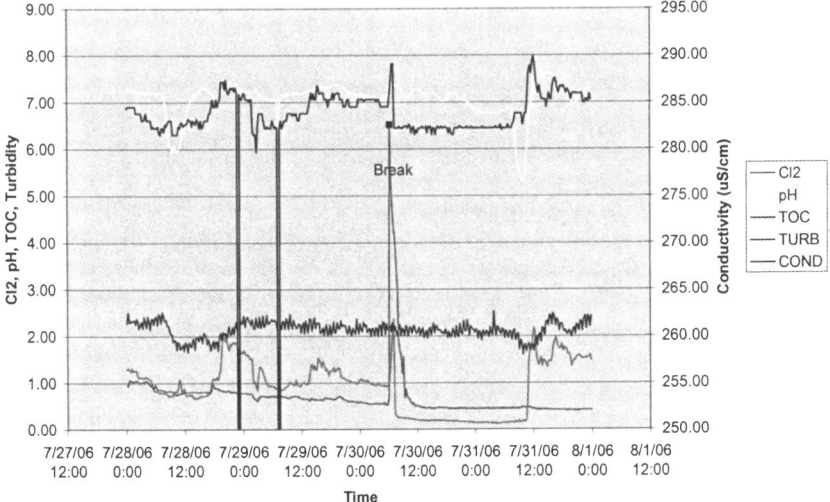

Figure 15 *36" Main Break.*

This break occurred at night on September 20th, 2006. The exact time of the break is currently unknown, although there appears to be a flow interruption to the monitoring panel the previous morning. The line involved was a 12" water main, which is 0.4 miles from the panel.

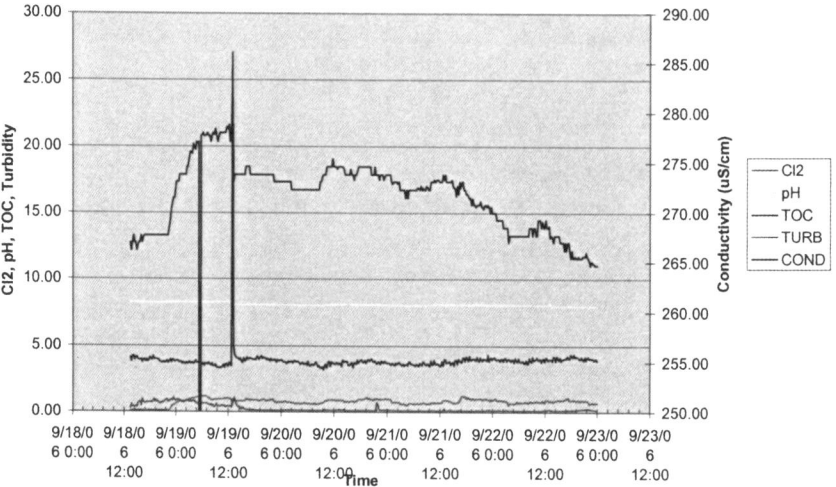

Figure 16 *12" Main Break.*

4 CONCLUSION

Extensive laboratory and pipe loop testing that is detailed in a separate papers[1,2,] indicate that these types of monitoring systems appear to be a good choice for detecting water quality excursions that could be linked to water security events. There are a number of advantages to using such systems. The chief advantage is that these instruments are not new. They are based on common everyday parameters with which the average industry worker is quite familiar, thus, adding a degree of comfort in operations not afforded by other new technology. As existing technologies, these instruments have been proven to be robust and dependable in prior field deployments. They represent measurements that would be of interest and use to water utility personnel above and beyond their role as water security devices. The deployment incidents detailed in this paper confirm this and also demonstrate the applicability of utilizing these everyday parameters by linking them with advanced algorithms. The field deployment studies not only demonstrate robustness in the field and the ability to recognize a wide variety of events, but these studies also demonstrate such system's ability to learn. It is foreseeable that these devices will become much more than a system that is capable of detecting terrorist events. They could easily become a critical tool for improving everyday operations.

For example, through many years of experience, the best old hands at treatment plant operations have developed "a sense" for knowing something in the treatment system is amiss. It can be a smell, color, clarity (or lack there of), sound or just tingling in the nape of the neck.

One gains this sense only by extensive experience in a particular facility. These senses do not exist in distribution systems because there has typically been little measurement done upon which to gain these "senses" and, therefore; "Bulk Parameter Monitoring in the Distribution System with Interpretive Algorithms" has the potential to become the artificial "sense" able to quickly "learn" the quirks of the distribution system and have those quirks labeled by those with extensive experience so that less experienced employees have the benefit of that knowledge without having to wait 5, 10 or more years. A good phrase to describe this knowledge base would be "institutional intuition."[3]

With the aging of the work force and rapid employee turnover "institutional intuition" has the chance of quickly dying out. Above and beyond their obvious security benefits, algorithms could be a way to circumvent this loss of knowledge and to build a knowledge base where none has previously existed. This in turn could allow improvements is system operation that may result in cost savings and definitely will result in a higher quality product being delivered to the consumer.

References

1 D. Kroll, and K. King, "Safeguarding the Distribution System: On-Line Monitoring for Security and Enhancing Operational Performance", 2006, Journal of the New England Water Works Association, June.
2 D. Kroll, "Operational and Laboratory Verification Testing of a Heuristic On-Line Water Monitoring System for Security", 2007, International Journal of High Speed Electronics and Systems. 17, No. 4
3 D. Kroll, 2006, "Securing Our Water Supply: Protecting a Vulnerable Resource, PennWell Publishers, Tulsa, OK.

THE ORGANISATIONAL CULTURE OF MANAGING INCIDENTS AND RISKS IN THE WATER SECTOR

R. Bradshaw[1], S. J. T. Pollard[1]*, D. I. Jalba[4], J. W. Charrois[2], S. E. Hrudey[3] and N. J. Cromar[4]

[1]*Cranfield University, Centre for Water Science, School of Applied Sciences, Cranfield, Bedfordshire, United Kingdom*
[2]*Environmental Management and Health Laboratories, Alberta Research Council, Canada*
[3]*School of Public Health, Faculty of Medicine and Dentistry, University of Alberta, Canada*
[4]*Department of Environmental Health, School of Medicine, Faculty of Health Sciences, Flinders University, Australia*

1 INTRODUCTION

In this paper we report on an investigation into the technical and organisational reliability of a regional water supply system. The study was conducted in a Water Utility in the United Kingdom. We introduce a High Reliability Organisations (HRO) framework for a water utility, conceptualised from a review of the literature on other industry sectors. 'High reliability' reflects the desire for an absence of failure and for high performance management during the "trying conditions" of incidents and emergencies. In this paper, two perspectives are investigated: firstly, the requirement for organisational reliability during an incident or an emergency situation; and, secondly, the requirement for technical reliability within the system and its delivery through risk-based asset investment and maintenance planning.

The HRO framework was conceptualised to consist of the following elements:
- Organisational culture of reliability
- Continuous learning and intensive training
- Effective and varied patterns of communication
- Dynamic decision making and flexible organisational structures
- System flexibility and redundancy
- Precise procedures in managing technology
- Human resource management practices that support reliability

Three aspects are introduced and discussed: (i) the nature of incidents and their impact on customers; (ii) the need for an organisational capability to manage incidents and its role in maintaining a resilient water supply system that minimises the impact of incidents on customers; and (iii) risk-based asset management strategies that secure technical reliability of the water supply system. For the latter, the opportunity to learn from previous incidents

* to whom correspondence may be addressed
(s.pollard@cranfield.ac.uk) SJT Pollard; Tel: +44 (0)1234 754101

to enhance asset risk assessments is investigated. The relationship of these three themes is conceptualised in Figure 1.

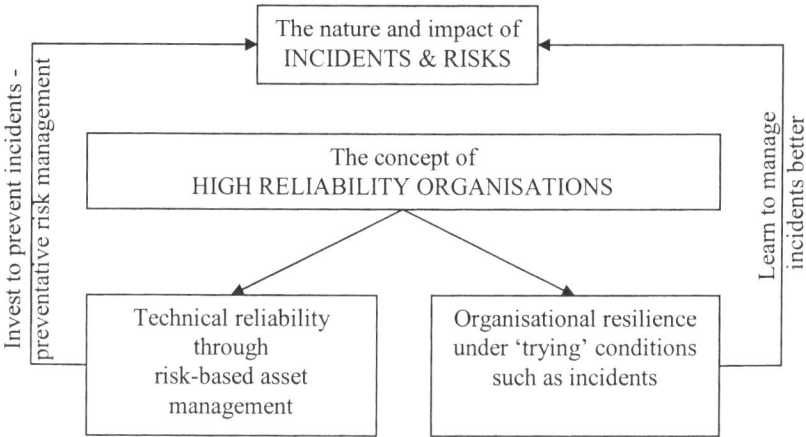

Figure 1 A h*igh reliability organisations' perspective on technical reliability and organisational resilience*

2 HIGH RELIABILITY ORGANISATIONS IN THE LITERATURE

High reliability organisations (HROs) operate under trying conditions and yet have outstanding safety records[1]. They have been described for nuclear power plants, aircraft carrier flight decks and air traffic control operations[2,3,4], for offshore oil platforms[5], nuclear powered aircraft carriers[6] and certain parts of the energy sector[7]. The organisational attributes of an HRO are only partially tested however, in that they have not been challenged under conditions of major failure to evidence that their absence contributed to that failure.

2.1 Organisational culture of reliability

A strong organisational culture of reliability is required as a bulwark against failure resulting in catastrophic consequences. Staff need to have a strong sense of the primary mission of the organisation and share a common system of beliefs and perceptions.[8] With the development of such a 'mindful' (vigilant) culture the formal system can be monitored, understood and likely failure events foreseen.[9] Members of staff require a highly developed understanding of their contribution and role in the system, acting in a collaborative manner to deliver 'collective intelligent' interactions.[10]

Constant vigilance and concern for reliability dictate behaviour[11], and alertness, attentiveness and care[10] can prevent cascading error and the subsequent escalation into system failure. Employees are encouraged to take responsibility, in particular where problems are identified and immediate corrective actions are required.[12] Errors are regarded as system faults and employees are encouraged to report mistakes without fear of punishment. On the other hand, individual behaviours that deliberately jeopardise the primary mission of reliability are not tolerated.

Within HROs, the commitment of senior management to the reliability of the organisation is communicated to all levels in the organisation and demonstrated with investments in technology, processes and personnel.[13] There is a strong sense of collective needs and goals. Individuals "monitor, advise, criticize and support" one another, in particular in situations where mistakes are more likely to occur.[12]

2.2   Continuous learning and intensive training

In order to facilitate continuous learning and intensive training, HROs constantly review their processes and standard operating procedures (SOPs). Staff training is extensive and focuses on the requirements for maintaining a safe system embedded in formal rules, generalised guidelines and standardised frameworks[14] prior to performing tasks under "live" conditions. The emphasis is not only on adherence to SOPs, but also on identifying potential sources of failure and actions to stop faults from escalating. Staff maintains a commitment to continuous learning and seek the acquisition and improvement of skills.

HROs also learn by studying the failures, near failures and mistakes by others. They use these as a means to study the failure susceptibility of their own organisation.[4] Even minor errors and incidents provide a source for learning[4], which might be assessed using root cause analysis.[12] In this way, the organisation develops a collective memory for failures, incidents and their root causes that helps the organisation anticipate future problems.[12] Much research on HROs has been undertaken in 'high hazard – low probability' environments where, because of the high consequence of failure, trial and error is not a realistic learning method. Offline methods of learning are required, consisting of realistic drills, simulations and exercises to replicate potential scenarios[1].

2.3   Effective and varied patterns of communication

Effective communication allows a complex system to become more understandable, predictable and controllable.[8] HROs create information-rich environments. Processes are measured and understood, with data made transparent and available to all. Within an HRO, information is a public good and staff are encouraged to share their experiences relating to the reliability of the system. Communication is designed as both bottom-up and top-down to ensure rapid flow of information through the hierarchy of the system. Rapid dissemination of information helps the organisation respond to mistakes, with corrective action aiming to prevent the escalation of error into failure.[12]

Communicating information allows staff to shape and share the 'big picture' of the HROs vision, mission and responsibility of individuals towards reliability.[11] HROs use multiple channels to transmit different types of information – direct and complementary. Indirect information enhances information reliability and provides a form of redundancy[6]. Multiple signals from a variety of sources provide information that allows individual signals to be scrutinised for fitting into the whole pattern. Abnormal signals are treated as an indication of latent errors about to unfold into failures.[2,6] Where possible, communications are formalised in a brief, precise, unambiguous, impersonal and efficient manner. This does not allow individuals to complicate or distort and ensures clarity of message.[12]

2.4   Adaptable decision-making dynamics and flexible organisational structures

Perrow[15] argues that complex and tightly-coupled systems can only prevent accidents with a high level of centralisation, because low level decision makers have insufficient understanding of the inter-relationship between their actions and consequences on other elements of the system.[14] HRO research on the other hand, has demonstrated that decentralisation is required to respond rapidly to unfolding failures. Centralisation is essential in tightly coupled technical systems where interdependency is high. Where systems can be de-coupled, decentralisation provides for action at the point of need.[12,16,14,4] HROs can therefore be described as 'holistic' or 'decomposable'. In emergency conditions, a holistic HRO needs to be centrally maintained in order to maintain an overview of the entire system. In a decomposable organisation, emergencies can be confined to one sub-unit which is then isolated from the entire system. Control over such an emergency is decentralised to this sub-unit until cleared.

HROs therefore enforce stringent adherence to SOPs aiming for a repeatability of actions and routines. Such formal rules and procedures identify and preventatively mitigate risk.[17] Activities based on decisions that are not defined in SOPs are taken at the most senior levels, for these individuals should have the best overall knowledge of the system[18]. Effective HROs build slack into their decision-making processes[19,1] in order to assess and challenge decisions, and so avoiding faulty decisions escalating into failures.

2.5   System and human redundancy

HROs maintain reserve capacity in their organisations and systems that includes back-up functions, overlapping tasks and responsibilities.[14,16] This said, it is important to recognise that designing redundancy for a system can be counterproductive, as back-up functions can increase technical complexity, conceal errors and lead individuals into not performing their required tasks under the assumptions that someone else takes care.[20] This 'diffusion of responsibility'[21] may itself be a significant cause of system error.

2.6   Precise procedures in managing technology

An HRO does not necessarily require 'state of the art' equipment. Such technology can add unnecessary complexity.[12] HROs aim to simplify complex technical systems and avoid unnecessary automation.[12] New technology acquisition is only justified if existing equipment does not perform to required specifications.[14] On the other hand, existing technology is maintained to exceptionally high standards with zero tolerance of defective, substandard or malfunctioning equipment.[16] Maintenance activity and protocols, as well as performance data, are used to monitor the healthy operation of the system.[13]

2.7   Human resource management practices that support reliability.

According to Weick[1], "humans who operate and manage complex systems are themselves not sufficiently complex to sense and anticipate the problems generated by those systems". In recruitment and selection, HROs try to recruit and select suitable and skilled candidates aiming to match, as closely as possible, the complexity of the environment with appropriate people skills and competences. Having recruited, it is vital to align reward and control systems, remunerating reliability with incentives, recognition and career opportunities. Job rotation can increase networking between teams and help the organisation transfer and disseminate knowledge and lessons learnt.[14]

Taken together, all these dimensions contribute to a design template for HROs. Individually, each dimension is important, but it is when acting together as a coherent configuration that failure susceptibility might be reduced and "trying conditions" managed effectively. In the next section, sector specific "trying conditions" are investigated in the context of incidents affecting customers in a regional water supply system between 1997 and 2006.

3   THE NATURE AND IMPACT OF INCIDENTS ON DRINKING WATER CUSTOMERS

3.1   The causes and effects of incidents

In this section, a series of incidents that occurred between 1997 and 2006 are investigated to understand their nature of unfolding, and to quantify their impact on customers using an incident impact metric. Our review of unfolding incidents used a number of models to conceptualize incidents, including:

- 'Hazard Analysis and Critical Control Points' (HACCP) methodology, from catchment to tap;[22]
- Failure modes and the analysis of their effects (FMEA); and
- An asset system model that investigates asset types (e.g. physical, information and human assets) involved during an incident.[23]

In Table 1, a histogram of the primary incident causes and effects is presented for 420 incidents in the years 1997 to 2006 in the partnering water utility. The causes and effects are presented in descending order.

The top 50% of primary incident causes were identified as 'failed water mains (31.67%)', 'IT failures (11.19%)' and 'maintenance work (10.71%)'. The remainder is distributed over 21 further categories that reflect the diversity of primary incident causes resulting in water quality incidents.

Similarly, the top 50% of primary incident effects were identified as 'interruption to drinking water supply (29%)' and 'discolouration / aesthetics (27%)'. The remainder is distributed over 22 further categories that reflect the prevailing diversity of primary incident effects governing the incident management response.

345 of the 420 incidents had a direct impact on customers between 1997 and 2006. Their cause-effect relationships are presented in Figure 2. In 23% of all incidents between 1997 and 2006, a burst main has primarily resulted in the loss of supply to customers. In 10% of all incidents a burst main has primarily resulted in aesthetical problems related to the drinking water supplied to customers. 8% of incidents were attributed to maintenance work that subsequently led to aesthetical problems with the drinking water supplied to customers. 5% of incidents were attributed to chlorination systems failure that potentially could have led to pathogens being present in the drinking water for customers. All other interrelationships between causes and effects in the matrix fields constitute less than 5% of incidents.

In a detailed analysis, the 145 incidents between 2004 and 2006 were coded using a structured incident assessment template. A more detailed description of the incident enabled a precise identification of the failed equipment and components. In Figure 3, the asset, equipment and component types where incidents occurred are shown for the years 2004 to 2006. The majority of incidents occur in the distribution network: 33% occurred due to the failure of water mains. Another 14% occur as a result of trunk main failures. The second largest asset type causing an incident is equipment for chemical treatment in water treatment works (14%). This is followed by incidents due to power failures (10%).

Table 1   *Primary incident causes and effects between 1997 and 2006 in the Region*

Primary incident cause	10 year histogram	in %	Primary incident effect	10 year histogram	in %
Burst main	133	31.67%	Interruption to supply	120	28.57%
IT failure	47	11.19%	Discolouration	115	27.38%
Maintenance work	45	10.71%	loss of M & C	42	10.00%
Asset failure	41	9.76%	Potential biological pathogens present	24	5.71%
Power failure	25	5.95%	Chemicals present above guidelines	23	5.48%
Operational intervention	23	5.48%	Biological pathogens present, health effects envisaged	18	4.29%
3rd party	19	4.52%	Potential biological pathogens present, health effects envisaged	13	3.10%
Chlorination failure	18	4.29%	Biological pathogens present	12	2.86%
Asset contamination	15	3.57%	Empty Service Reservoir	11	2.62%
Unknown	13	3.10%	Loss of asset	8	1.90%
Treatment failure	12	2.86%	Damage to asset	5	1.19%
Raw water quality	8	1.90%	3rd party impact (Gas)	4	0.95%
Asset damage	4	0.95%	Aesthetics above guidelines	4	0.95%
M & C failure	3	0.71%	Environmental	4	0.95%
Severe weather	3	0.71%	Chemicals present above guidelines, health effects envisaged	3	0.71%
High Demand	2	0.48%	low pressure	3	0.71%
Security	2	0.48%	3rd part damage	2	0.48%
Adverse weather	1	0.24%	Disruption To Normal Processing Of Work	2	0.48%
Chemical spillage	1	0.24%	Risk of cross contamination	2	0.48%
Chemical supply contamination	1	0.24%	3rd party accident	1	0.24%
Illegal connection	1	0.24%	Statutory monitoring failure	1	0.24%
Water quality	1	0.24%	Treatment failure	1	0.24%
Total	420	100.00%	Total	420	100.00%

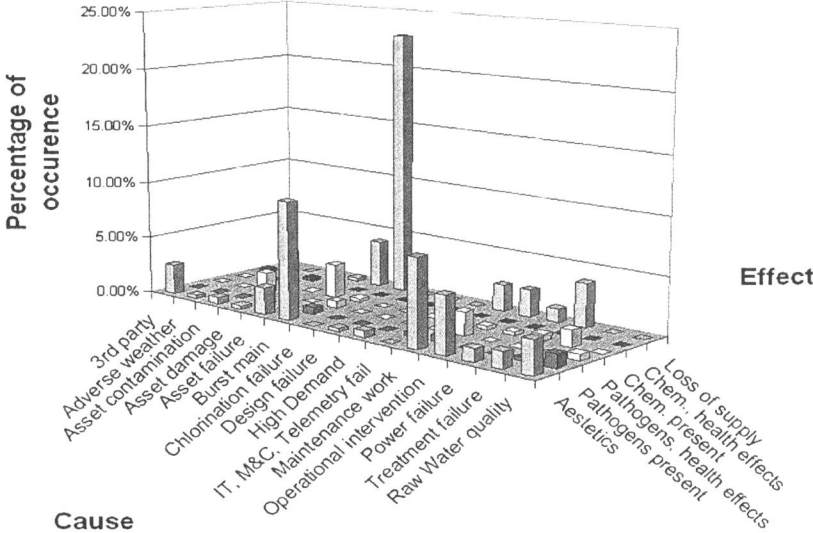

Figure 2   *Percentage of incident occurrence for primary cause and effect relationships for incidents between 1997 and 2006 in the Region*

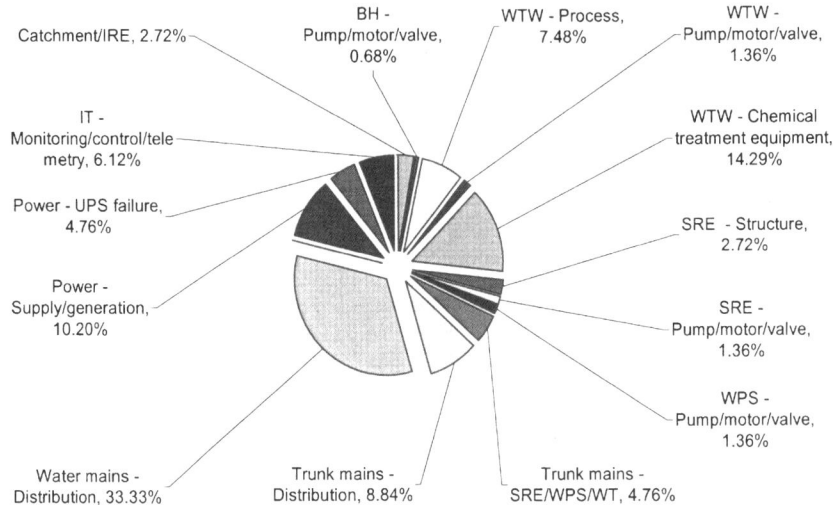

Figure 3   *Asset type causing an incident between 2004 and 2006*

In Figure 4, the immediate availability of redundancy for the failed asset is identified. The majority of incidents between 2004 and 2006 that occurred on assets had no immediate redundancy or stand-by available. These duty-only assets were largely water and trunk mains. The lack of standby or redundancy is one reason why the incident had an impact on customers and, it could be argued, that the incident impact could have been avoided, if immediate redundancy had been available. This, of course, would have significant cost implications and duplication of the entire distribution system is, of course, impracticable. We also observe that 30% of incidents had immediate standby assets or that failed in the course of the incident unfolding. This was attributed to duty-standby failures *e.g.* due to a common cause failure.

In Figure 5, the catchment to tap model was used to identify intermediate assets between the asset where the incident occurred, and the customer. In this analysis, all intermediate assets were identified even if they had no reducing effect on the impact on customers. For example, a failure of a water treatment works could have had a significant impact on customers. Fortunately, in 15% of the incidents, the system design allowed for a service reservoir between the water treatment works and the customer, reducing or eliminating any impact on customers.

It can be seen that in 68% of the incidents, an asset failure had an immediate impact on customers without any asset between the incident origin and the customer. This 68% contained the majority of "burst water and trunk main" incidents. In 8% of the incidents, the rezoning capability of the distribution network significantly reduced the impact on customers. In 15% of the incidents, a service reservoir mitigated against the full impact of an incident. As in the example above, the impact of a failure at a water treatment works was largely reduced due to the availability of drinking water in the service reservoir for supply to customers.

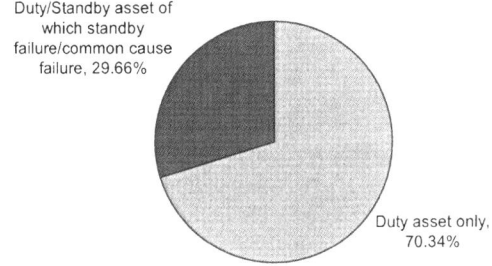

Figure 4 *Immediate redundancy of assets that failed during incidents between 2004 and 2006*

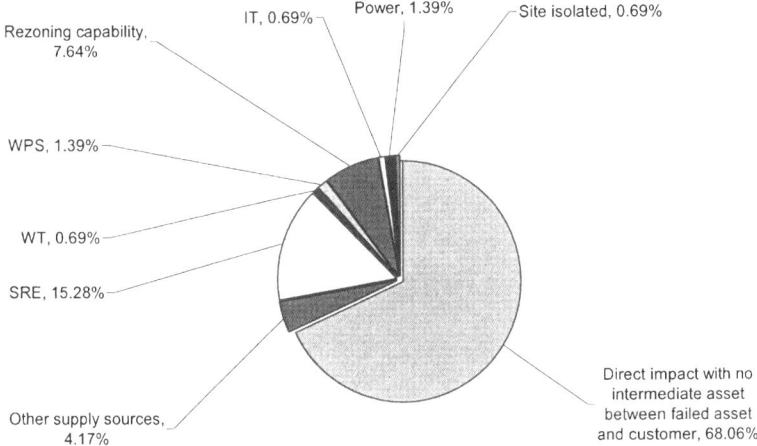

Figure 5 *Alternative redundancy in the water supply system between the failed asset and customers for incidents between 2004 and 2006*

In Figure 6, we assessed the beneficial use of the intermediate asset to reduce the impact of an incident on customers. As identified above, the majority of incidents had no immediate asset between the failed asset and the customer that could have reduced the impact. In 10% of the incidents an intermediate asset was available. However, it was not designed to, or it failed to reduce the impact of the incident on customers. In 4% of the incidents, the intermediate asset made a substantive contribution to reducing the incident impact so that the residual impact was minimal. The remainder of incidents had intermediate assets between the failed asset and the customer. Their effectiveness in reducing the impact ranged from very low to high.

In Figure 7, the number of attributed asset-related incident causes is shown. In total, 316 asset-related causes were identified for the 145 incidents. The categories of incident causes presented in this figure were identified as such in the incident documentation. These categories are not necessarily mutually exclusive and the analysis allowed multiple categories for the assessment of individual incidents. For example, a burst water main could be attributed to water mains failure due to material fatigue and corrosion (if this was so identified in the incident documentation for a particular incident). 'Corrosion', 'Material fatigue', 'Wear & Tear', 'Age' and 'Poor condition' are often the underlying nominal factors for different types of asset-related failures. 'Asset failure' denotes any failure of an asset which has not been explicitly recorded as 'Mechanical failure', 'Electrical failure', 'Civil failure' or 'Water mains failure'. Within the 145 incidents, the largest number of incident causes was recorded as 'water main failures'; this is followed by 'material fatigue', '3rd party impact' on assets and 'corrosion'.

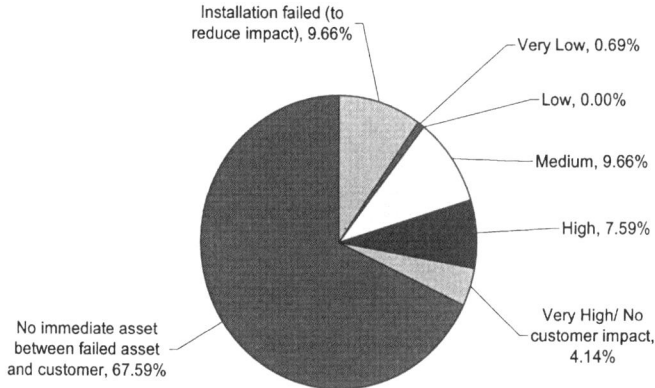

Figure 6   *Beneficial use of systems redundancy during incidents between 2004 and 2006*

Incidents do not only result from asset-related failures. In Figure 8, the number of attributed process-related incident causes is shown. In total 110 process-related causes were identified for the 145 incident between 2004 and 2006. Again, the categories are not mutually exclusive and the analysis allowed multiple categories for the assessment of individual incidents (if so identified in the incident documentation). Process-related incidents may relate to process engineering issues, impact from the environment but also operational issues affecting the human-machine interface. The three largest process-related incident causes were recorded as 'water main scouring', 'treatment process failure' and 'ingress of contaminants'. Water main scouring denotes changes in flow patterns, velocities and pressures in distribution systems that re-suspended deposited solids to cause discolouration incidents. Hydraulic effects, too, are changes in flow patterns, velocities and pressures in a distribution system which may disrupt the continued water supply to customers or dissolves air in the drinking water. 'Operating environment' denotes an adverse operating environment for assets.

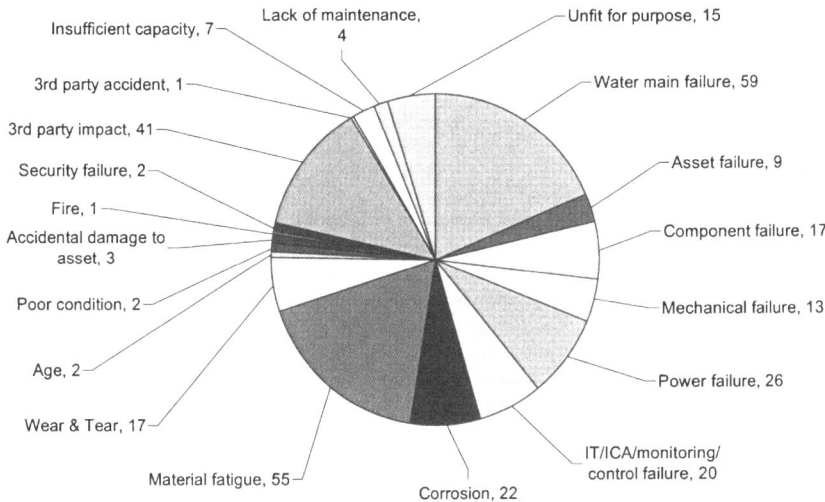

Figure 7 *Asset related causes for incidents between 2004 and 2006*

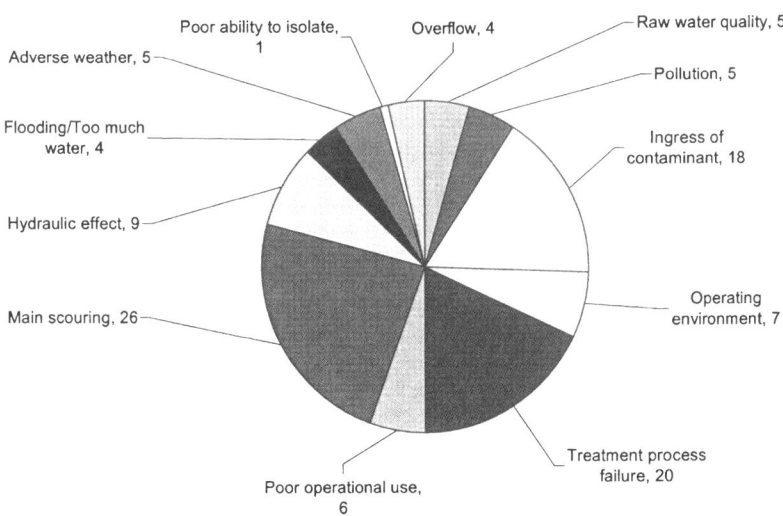

Figure 8 *Process-related incident causes for incidents between 2004 and 2006*

So far, 426 asset- and process- related incident causes were identified for the 145 incidents. On average, 3 asset- and process-related incident causes were attributed per incident. This confirms that incidents can have multiple causes and contributing factors leading to failure.

Beyond asset- and process-related failures, human factors may also cause incidents. In 127 instances, a causal relationship between human factors/errors and an incident could be attributed. This represents an average of 0.9 human factors contributing per incident. Although this was not necessarily the root cause of the incident, the human factor was seen as a contributing factor. In Figure 9, the numbers of occurrences in the respective categories of human factors are presented. A number of these categories are arguable: 'unanticipated effect' and 'acted in good faith' do not describe an insufficiency in the decision making of an operator, rather denote a reflection with hindsight. Had the person involved in the incident known about the potential effect, s/he would probably not have pursued this course of action. Other categories are self-explanatory. The largest attributed human factor was identified as 'poor design' and 'unanticipated effect'. 'Lack of information' on the potential consequences of action and 'poor outage planning' represent the third and fourth largest group in human factors contributing to incidents.

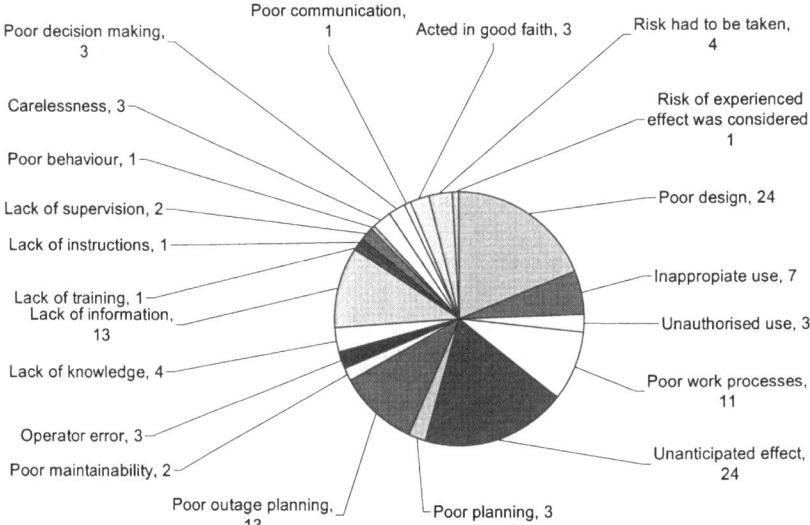

Figure 9   *Human factor-related incident causes for incidents between 2004 and 2006*

In total, 553 asset-, process- and human factor- related incident causes were identified for the 145 incidents. This represents an average of 4 causes and contributing factors for each incident.

We now return to the effects of these incidents and considers the explicitly stated, multiple effects an incident can have. In Figure 10, the effects of the 145 incidents between 2004 and 2006 are shown. In total, 170 incident effects were recorded which equate to 1.2 incident effects per incidents. The percentages shown in figure 10 are based on 145 incidents. For example, out of the 145 incidents, 59 resulted in a "loss of supply" to customers. This equates to 41% of incidents resulting in "loss of supply". Out of the 145 incidents, 107 were recorded to have a no impact or only one impact category affecting customers. Thirty-seven were recorded explicitly stating two distinct incident effects affecting customers. One incident was recorded explicitly stating three different impact categories affecting customers. In Figure 11 the percentage of incidents are shown with no impact on customers, a single impact on customers and the percentage of incidents with double and triple impact on customers.

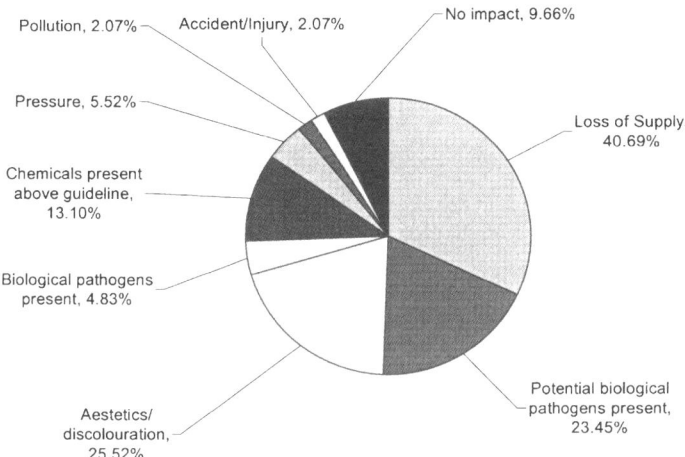

Figure 10 *Effects of incidents between 2004 and 2006*

We now turn to the incidents with two or more incident impacts on customers. In Figure 12, the percentage of incidents with multiple impacts on customers is shown. Ten per cent of the 145 incidents caused a 'loss of supply' to customers followed by aesthetical problems – mainly due to discolouration – on resuming normal operations. Six per cent of the incidents involved potential pathogens present in the drinking water in combination with an exceeding chemical parameters above guidelines. Five percent of the incidents were recorded as 'loss of supply' for customers and subsequently or simultaneously 'low pressures'.

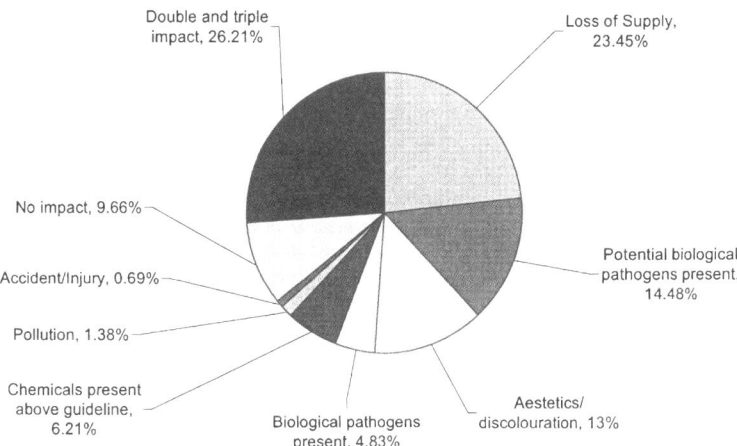

Figure 11 *Percentage of single, double and triple effect incidents between 2004 and 2006*

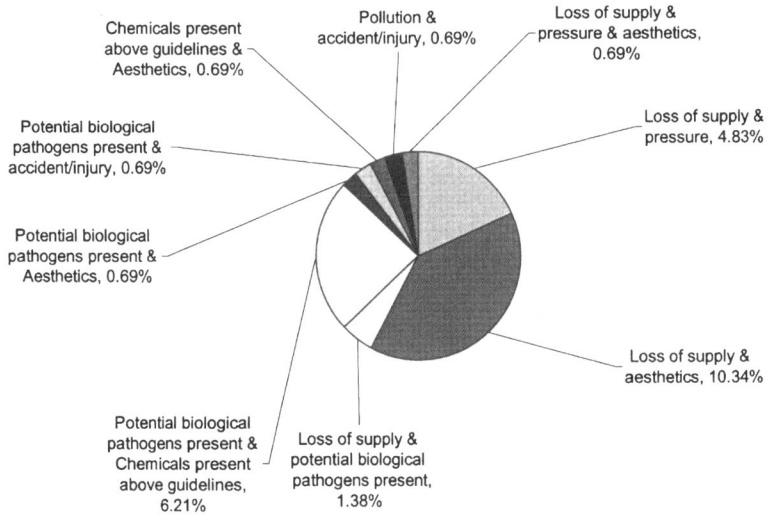

Figure 12 *Percentage of incidents between 2004 and 2006 with a multiple incident effect*

3.2   The impact of incidents

Based on a methodology advanced by Fewtell and Bartram[24], a comparative measure for incident impacts was used that considers the hazard type, affected population, and the duration of hazard exposure. This methodology is used to assess and compare the impact of incidents on drinking water customers. The methodology is conceptualised in Figure 13. The metric for incident impact is calculated as

$$\text{Incident score} = [(H \times 0.33) + (D \times 0.33) + (P \times 0.33)] \qquad (1)$$

Table 2 is used to derive a severity score for the hazard, the affected population and the duration of an incident.

In the water supply region investigated in this study, a trend of increasing frequency for incidents can be identified between 1997 and 2006. Whereas the number of incidents has gradually increased, the average incident impact has marginally reduced (Figure 14). For the incidents recorded between 1997 and 2006, an increase in the number of incidents in our partnering water utility could be observed. We offer three possible reasons. Firstly, the threshold for defining an incident may have changed during that time period. Collecting incident data corresponds with a need to report incidents to the water quality regulator in England and Wales, the Drinking Water Inspectorate. Since 2004, the current definition of an incident is governed by the Water Undertakers (Information) Direction 2004[25] and the Guidance on the Notification of events.[26] If the threshold in the definition of an incident had been reduced, it would reflect in the number of incidents in the incident catalogue. Secondly, the water utility may commit more resources to document incidents. This, turn, may contribute to an increased availability of incident documents that would, otherwise, not be available. Thirdly, incidents may actually occur more frequently due to increased vulnerability of the water supply system, lack of maintenance, increased 3rd party impacts, etc.

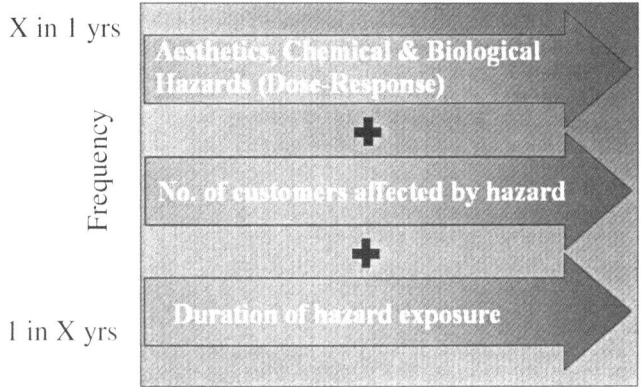

Figure 13 *The incident assessment matrix (adapted from Fewtell & Bartram[21])*

Table 2 *The incident scoring matrix (adapted from Fewtell & Bartram[21])*

Estimate frequency of hazard	Estimate magnitude of hazard		Estimate duration of hazard		Estimate no. of customers affected by hazard	
Score (F)	Hazard type	Score (H)	Duration in days	Score (D)	Customers	Score (P)
1 in X yrs	Aesthetics above guidelines	32	< 0.5	2	0 – 7,500	2
	Unwholesome, potential health effects	48	0.5 – 1	4	7,500 – 15,000	4
	Chemicals present above guidelines	8	1 – 2	8	15,000 – 30,000	8
	Chemicals present above guidelines, health effects envisaged	32	2 – 4	16	30,000 – 60,000	16
	Potential biological pathogens present	6	4 – 8	32	60,000– 120,000	32
	Potential biological pathogens present, health effects envisaged	48	8 – 16	64	120,000– 250,000	64
	Biological pathogens present	8	16 – 32	125	250,000 – 500,000	125
	Biological pathogens present, health effects envisaged	64	32 - 64	250	500,000 – 1,000,000	250
	Loss of supply, potential contaminant ingress	16	64 – 128	500	> 1,000,000	500
X in 1 yr			> 128	1000		

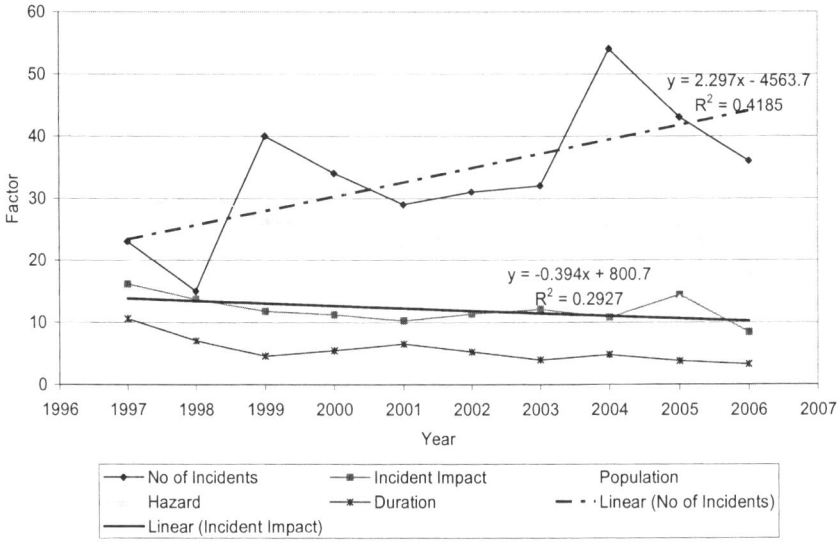

Figure 14   *Frequency and impact of incidents between 1997 and 2006*

Over that 10 year time period, the annual average impact of incidents reduced marginally. The annual average incident impact is derived from three components, namely, the average factor for the hazard type, the average size of the population affected during an incident and the average duration of an incident. The number of occurrences for specific hazard categories directly relates to the failure proneness of the system that facilitates hazard related incidents. The average size of the population affected by an incident not only indicates the size of a population supplied by individual supply system arrangements but also indicates the ability of the organisation to reduce the affected population during an incident. The average duration of an incident indicates the speed of the incident management team to identify the hazard source and the speed of re-instating normal operations.

In Figure 15, the annual, average size of populations affected during incidents is presented. It can be seen that the annual average fluctuated between 1,000 and 10,000 customers affected by incidents. In the year 2005, the annual, average peaked at above 100,000 customers due to one extreme event that caused an impact on customers in the entire water supply region. In 2006, the average number of customers has again fallen below the 10,000 customer threshold.

Figure 16 shows the average duration of incidents in hours of exposure to hazards. Over the years, the incident duration has reduced by more than 50%.

So far, we analysed and determined the incident impact for individual incidents and calculated the annual average incident impact. This was used to compare the annual, average incident impact with the impact from incidents in subsequent years. This analysis was used to monitor the trend of annual incident impacts, however, it does not tell anything about the nature of these incidents.

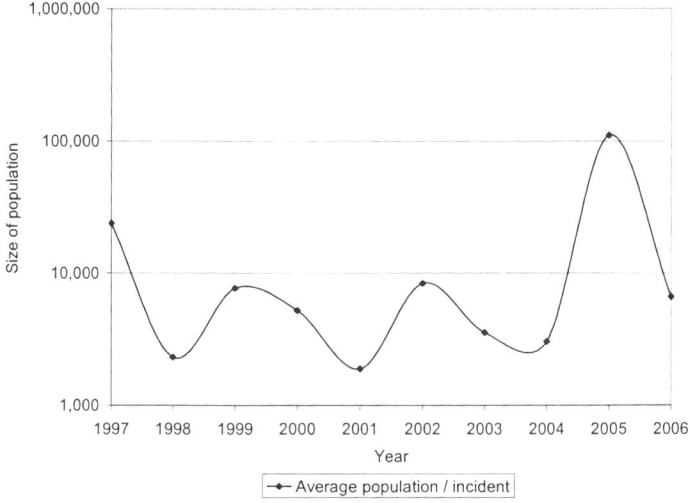

Figure 15 *Annual, average population size affected by incidents between 1997 and 2006*

In the following figures, the incidents are investigated with a specific focus on the different hazard types affecting customers. The overall annual customer impact from incidents in their respective hazard categories are presented in Figure 17. Annual incident impact on customers is calculated as (frequency of incident for hazard category)*(average incident impact for hazard category).

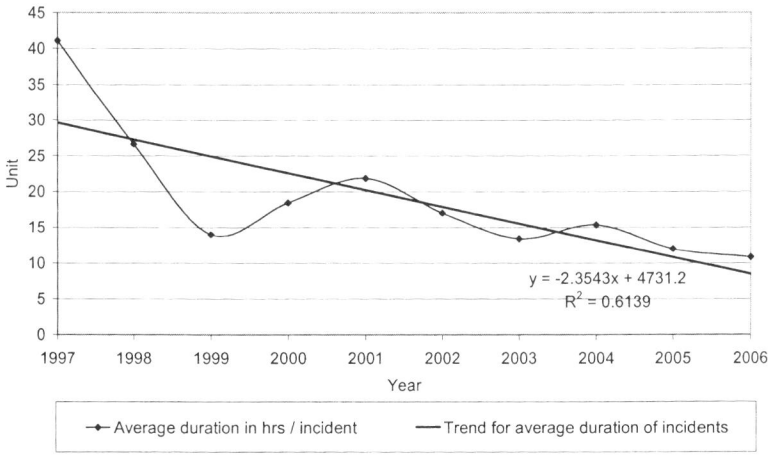

Figure 16 *Annual, average duration of incidents between 1997 and 2006*

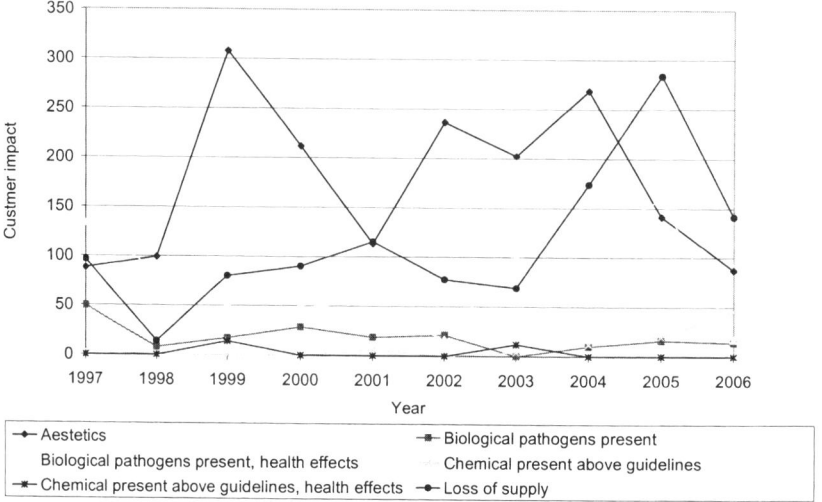

Figure 17 *Annual incident impact in specific hazard categories between 1997 and 2006*

We compared the incident impact in the different hazard categories with another and found that the highest customer impacts between 1998 and 2004 related to 'aesthetical' unpleasing drinking water quality. Since 2004, this has significantly reduced. Since 1998, a trend can be identified of increasing customer impact from incidents relating to 'loss of supply'. This trend peaked in 2005 and has, since, reduced. The third largest hazard category that affects customers during incidents relate to 'biological pathogens present with anticipated health effects'. Since 1997 a downward trend suggested an improvement in this incident category. Since 2002, the impact on customer in this category increased to a peak in 2005. In 2006, it reduced to below a 10-year average.

The frequencies of incidents for the different hazard categories are presented in Figure 18.

It can be identified that the two highest incident frequencies are associated to 'Aesthetics' and 'loss of supply'. Since 2004, the frequency of incidents associated to 'aesthetical' unpleasing drinking water quality has reduced significantly whereas the frequency of 'loss of supply' incidents has been steadily increasing since 1998.

The average, annual incident impact per incident for the respective hazard categories is presented in Figure 19. It can be identified that the annual average impact of incidents in the respective hazard categories remained largely unchanged. The only exception is identified for the average, annual incident impact related to the hazard category 'biological pathogens present with envisaged health effects'. In this category, a declining trend can be identified that suggests an improving organisational performance

- to reduce the size of the population affected during an incident, and/or

- to reduce the duration of such incident for this particular category.

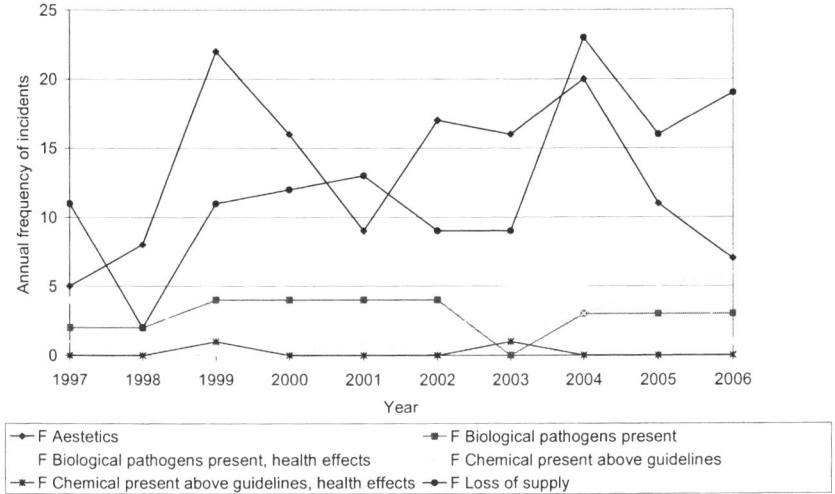

Figure 18 *Annual incident frequencies for hazard categories between 1997 and 2006*

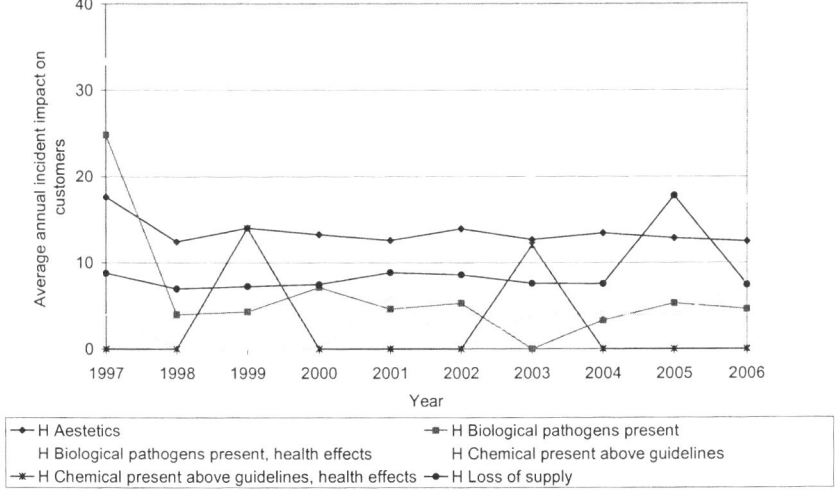

Figure 19 *Average, annual incident impact for different hazard categories between 1997 and 2006*

3.3 Concluding remarks

In the analysis of incidents that occurred in the Region between 1997 and 2006 it was identified that incidents occur rather frequently with varying impacts on customers. Every incident that occurred required organisational capacities to identify the incident, reduce its impact on customers and processes and procedures to re-instate normal operations. This is primarily a capacity within operations management that assumes the role of managing the incident.

In the following chapter, this organisational capacity is investigated in the context of the previously introduced High Reliability Organisations Framework that has been conceptualised to describe the means of creating organisational resilience.

4 GENERATING ORGANISATIONAL RESILIENCE WITH THE PRINCIPLES OF HIGH RELIABILITY ORGANISATIONS

In the study of documented incidents between 1997 and 2006 as well as observational studies in the control centre of the water utility, the capacity of the organisation to identify and manage incidents was studied.

4.1 Incident identification

Based on the analysis of 145 incidents that occurred between 2004 and 2006, the means of identifying incidents are recorded. With reference to Figure 20, it was found that the majority of incidents are notified to the water utility by customers reporting an unusual observation relating to their drinking water supply. The majority of these customer contacts refer to 'loss of supply' and 'aesthetical problems' due to discolouration. 5.52% of the incidents were identified via water quality laboratories confirming the pollution or contamination of drinking water. With a turnaround duration of 24 hrs for bacteriological test for drinking water quality parameters it has to be assumed that contaminated water has, in the meantime, passed beyond the customer tap. Therefore, the majority of incident notifications including 'customer contact' and 'water quality laboratories' indicates that customers are exposed to incidents before reactive incident mitigation can be carried out by the incident management team. Both methods of incident identification are well established business processes: in particular for customer contacts, a dedicated call centre is maintained to identify and characterise the symptoms of an incident.

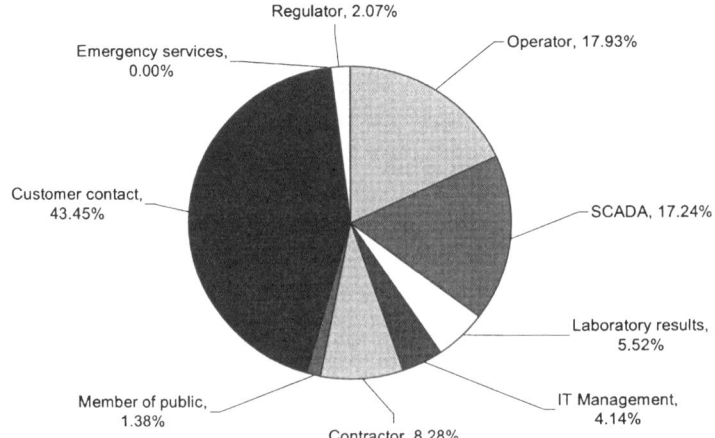

Figure 20 *Identification of incidents between 2004 and 2006*

4.2   Organisational culture of reliability

A strong organisational culture of reliability was a stipulated requirement as a bulwark against failure resulting in catastrophic consequences. In observations and staff interviews it was observed that staff in operations and incident management have a strong sense of the primary mission of the organisation. These were commonly expressed as "providing a safe and reliable drinking water for customers in line with regulatory requirements". Operations managers, engineer and operators share a common system of beliefs and perceptions when water safety is concerned. The water supply system is constantly monitored for any abnormal operating condition. Failure scenarios are usually well understood and reflected in the monitoring programme of water system assets. Past incidents are analysed and failure prediction of future incidents is facilitated in anticipatory risk assessments. In the observation of unfolding incidents, it was found that staff have a highly developed understanding of their contribution to water safety regarding their role in the technical system and in the decision making process. In particular during incidents staff act in a collaborative and collegiate manner.

Constant vigilance and concern for water safety and reliability dictates the behaviour of staff. This is particularly relevant to field operators but also control room staff who act with alertness, attentiveness and care in monitoring the healthy operation of the entire water supply system. Employees are encouraged to take responsibility, in particular where problems are identified and immediate corrective action programmes are required. On first sight of a problem with a particular aspect of the water supply system, an alarm is raised and the need for instigating the incident management procedures is assessed. With the introduction of information technology and automated monitoring and control system, the majority of asset failures are picked up by monitoring equipment and an alarm is raised. One major exception - as seen before - is the identification of water mains burst and water discolouration arising in the distribution network. Here, the organisation relies on customers to call in and report their service experience. Here too things are starting to change with the use of online, real time monitoring of flows and pressures in the distribution system in an attempt to identify potential issues before they become customer impacts.

The commitment of senior management to water safety and reliability of the organisation is communicated to all levels in the organisation and demonstrated with investments in technology, processes and personnel as long as the conflicting objectives of public health and shareholder value align. In our observations and in interviews, members of staff have communicated their strong sense for collective needs and goals. Individuals "monitor, advise, criticize and support" another, in particular during critical incidents which are immensely stressful situations and quick decisions have to be taken.

4.3   Continuous learning and intensive training

In order to facilitate continuous learning and intensive training, the operations management function review their processes and standard operating procedures (SOPs), in particular after an incident in the incident review meeting. In these meetings, the incident is scrutinised, 'lessons learnt' are identified and communicated to relevant parties in the organisation. If necessary, actions for the asset engineer to review a particular system or actions for an operations manager to review a particular procedure are formulated and their progress and completion monitored.

Staff training is extensive and focuses on the requirements for maintaining a safe system. In the organisation, operators are required to gain professional accreditation in form of college certificates as a license to operate a water supply system. This training scheme is a customised training programme for operators in that particular region. In-house training and training on the job are also important components of continuous professional development. Recently, a risk training programme was launched to provide staff with a better understanding of risk identification and assessment skills. This training programme has been recognised as industry leading and earned a number of industry awards.

The performances of tasks are embedded in formal rules, generalised guidelines and standardised frameworks. These are expressed in SOPs, risk assessments and method statements. Yet, the emphasis is not merely on adherence to SOPs but also on identifying potential sources of failure and actions to stop faults from escalating. This aspect was a particular emphasis in the risk training programme. It made participants aware of the risk perception horizon people have and develop over time. The communication of incidents to staff helps the organisation to communicate the failure proneness of the system to relevant personnel. The general interest in this training scheme demonstrates that staff maintain a commitment to continuous learning and seek the acquisition and improvement of skills. The training programme was rolled out to over 170 people in the first year and included representatives from all areas of the business including contract partners.

The water utility also learns by studying the failures, near failure and mistakes that occur within the organisation. Failures in one part of the organisation can be used as a means to study the failure susceptibility of the entire organisation or, at least, of other, similar sub-systems. Incident review meetings are designed to highlight and prompt cross organisational learnings in other parts of the business. Even minor errors and incidents provide a source for learning, which is identified using root cause analysis to discover causes and contributing factors. In this organisation, the majority of incidents affect the distribution network in form of mains bursts. Root causes for mains burst have many contributing factors such as age, material, soil condition and the operating regime. The structured collection and analysis of water mains failures enables multi-regression analysis for the derivation of risk profiles for the entire water distribution network. These models are used to prioritise maintenance and replacement programmes.

The collection of incidents on an incident database enables the organisation to develop a collective memory for failures and incidents and can help the organisation to anticipate future problems via risk assessments. With the majority of incidents being driven by technical failure, these incidents can be used to inform and enhance asset risk assessments.

4.4   Effective and varied patterns of communication

Effective communication facilitates a complex system to become more understandable, predictable and controllable. With the rapid developments of information technology, water supply systems are increasingly fitted with advanced monitoring and control instruments. They are part of an effective communication strategy to maintain safe and reliable drinking water supplies. In the organisation, the monitoring and control philosophy has been advanced to a stage where physical assets such as water treatment works are no longer operated with staff on site. Monitoring and control is performed with "Process Logic Controls" and "Supervisory Control and Data Acquisition" in remote control centres. These control centres are the hub for managing the entire water supply system. The organisation operates in information rich environments. In the first years of implementing the strategy of advanced monitoring and control, we could observe an increase of incidents due to the failure of such technologies. Where monitoring and control equipment fails, the status of a system becomes unknown. Technological developments – such as status monitoring for control and monitoring equipment – has reduced these incidents over the last few years.

Processes in water production and distribution are measured and understood, with data made transparent and available to all. An interesting observation we made is the need to manage potential overload of information during critical situation in an incident. Having the right information available at the right time in the right place is an important aspect of water utility incident management. Another important factor is the dependency on electricity to power water supply systems and operate information technologies. Some power failures have been observed as the common cause for incidents in water production and distribution systems and their monitoring and control. Over the years, the water utility has increased its dependency on the reliability of electricity from suppliers. E.g., water towers, due to their high investment and maintenance cost, are increasingly being replaced by pumping stations. Back-up power supplies are required to maintain water supply systems at all times.

During an incident, inter-personnel communication is designed as both bottom-up and top-down to ensure rapid flow of information through the hierarchy of the incident management team. Rapid dissemination of information helps the organisation respond to an incident, with corrective action aiming to prevent the escalation of the incident into an emergency.

In the design of water supply system, the organisation aims to provide multiple channels to transmit different types of information – direct and complementary – for the assessment of incident causes, effects and impact. During an incident, indirect information enhances information reliability and provides a form of information redundancy.

As mentioned before, there is a heavy reliance on customers to report their experiences to the water utility, in particular relating to incidents in the water distribution network. Efforts are underway to reduce the reliance on "end of tap" reporting for incidents. A test trial is currently planned to provide sufficient pressure and flow monitoring devices to increase the incident detection capability in an area distribution network. With this arrangement, any deviation of observed pressure and flow patterns from expected patterns will raise an alarm in the control centre so that an incident investigation team can be dispatched to investigate the source of the abnormality. This system will enable the reduction of the response time to an incident considerably.

Communicating information allows staff to shape and share the 'big picture' of the organisations vision, mission and responsibility of individuals towards reliability. This is important to integrate asset management teams into the daily operation of the water supply system. The asset engineers require the information input from operators to assess asset risks. Via risk assessments, cash may be made available for asset investment or maintenance.

In an analysis of documented incidents, the authors investigated the effectiveness of communication during the incident management response to the incident. In Figure 21, 72.41% of the incident management responses were characterised by 'effective communication'. Here, the communication between the stakeholders involved in an incident generated 'a big picture': observations, decisions and water supply systems performance were effectively communicated to all relevant staff and external bodies, which enabled comprehensive judgement on the due course of action. These actions were effectively communicated to staff and their implementation communicated back to the decision maker. In 6.90% of the incidents, the incident documentation identified aspects of excellent communication that significantly contributed to the effectiveness of the incident management response. In 13.10% of the incidents some areas of improvements were identified which meant that the incident was unnecessarily prolonged. In 6.21% of the incident, 'poor communication' had a significantly, adverse impact on the overall performance of the incident management response.

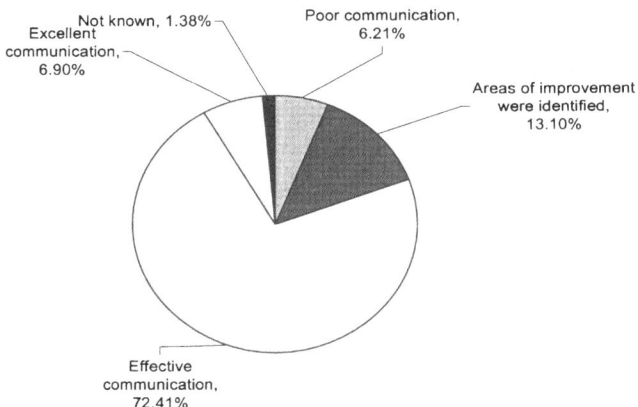

Figure 21 *Assessment of communication during incidents between 2004 and 2006*

4.5   Adaptable decision making dynamics and flexible organisational structures

As pointed out before, Perrow[15] argues that complex and tightly coupled systems can only prevent accidents with a high level of centralisation because low level decision makers have insufficient understanding of the inter-relationship between their actions and consequences on other elements of the system.[14] Perrow's definition of systems referred to technical assets such as nuclear power plants. We do not claim that technical systems in the water sector are complex technologies, yet, it is the combination of physical, information, human and intangible assets that form socio-technical systems with increased complexity. Complicated designs of technical systems, monitoring and control philosophies, significant human machine interfaces, potential for human error and difficult decision making processes in incident, operations and asset management characterise the complexity of modern water supply systems.

During an incident, the organisation assumes a centralised command and control hierarchy. This is reflected in the organisational structure in operations management in which process and performance data of the technical system are reported to a centralised control room. From this control room, the incident manager co-ordinates efforts to reduce the impact of the incident and to re-instate safe water supply. From here, the incident manager will monitor the entire system's response to the incident and the incident management efforts. The incident manager leads the incident management team within the control room but also field staff who perform the required tasks at the source of failure or within the area affected by the incident. The incident manager directs all resources at his disposal, including systems redundancy, towards reducing the incident impact and re-instating safe operations. During large scale incidents, the organisation is capable to decentralise if this is required to respond to rapidly unfolding failures. During a major storm event which had significant impact on many technical subsystems, a number of incident managers were called up to respond to particular aspects of the region-wide incidents. Although centralisation is essential in tightly coupled technical systems where interdependency is high, it was possible to de-couple the technical system so that decentralisation in the incident management response provided for action at the point of need. During large scale incidents affecting the entire region, this organisation demonstrated a centralised incident management response in order to maintain an overview of the entire system but also decentralisation where particular incident aspects could be confined to one sub-unit which is then isolated from the entire system.

The organisation requires stringent adherence to procedures and guidelines aiming for a repeatability of actions and routines. Activities based on decisions that are not defined in procedures are taken at the most senior levels, for these individuals should have the best overall knowledge of the system. Every incident that was investigated had unique and novel aspects to consider for which detailed procedures were not available. These arise out of the specific incident circumstances, e.g. the environment in which the incident occurs. Since these are unforeseen circumstances, only high level principles and guidelines are available to direct incident response efforts. Particular SOPs would not exist for such particular circumstances. In such events, the decision making process has in-built slack in order to assess and challenge decisions by a more senior member of staff. Furthermore, the incident manager has specialist staff at his disposal to guide his decisions.

With reference to Figure 22, the incident documentation was studied to identify the ability of the organisation to adapt its organisational structure to respond to the needs arising during an incident. It was found that in 88.28% of the incidents, the organisation assumed an effective organisational structure to place it in the best possible position for effectively reducing the incident impact on customers and to re-instate normal operations. In 9.66% of the incidents, the assumed organisational structure was deemed 'adequate considering the circumstances'. In this category a number of improvements could have led to better performance in reducing the impact of the incidents or the re-instating of normal operations. A number of incidents in this category reflect highly challenging or trying conditions and the incident management response demonstrated a reasonable successful outcome. In only 2.07% of the incidents, the incident management organisation was rated as 'inflexible' which suggests that the organisational structure assumed during the incident was inadequate to manage the complexity of the incident situation.

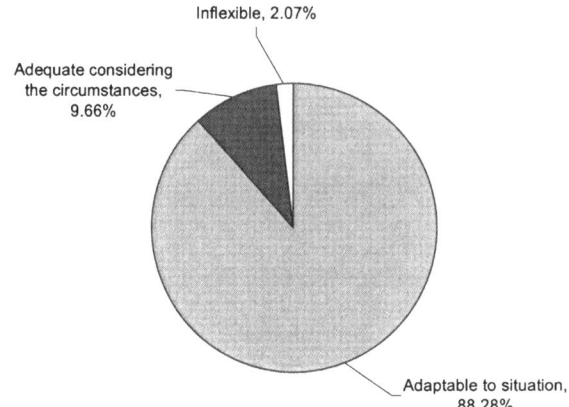

Figure 22 *Adaptability of the organisational structure to the incident situation for incidents between 2004 and 2006*

The incident assessment also focussed on decision making during the incidents. With reference to Figure 23 it was found that 64.83% of the incident management efforts could be characterised for 'good decision making'. The decision taken during the incident significantly and pro-actively contributed to reducing the impact on customers and to re-instate normal operations as soon as possible. In 24.83% of the incidents, the decision making was 'responsive to needs' meaning that the incident management efforts pursued an effective course of action by reasonably practical means. The remainder of the incidents were, in hindsight, characteristic for poor judgement, poor decision making and non-adaptive to the incident situation. These were identified as being ineffective to recover the incident situation to normal operation and provided scope to learn lessons for enhancing the incident management response. Overall, the organisation demonstrated that decision making under trying conditions effectively draws the necessary and correct conclusions from the data presented to the incident management team during an incident. This reflects on the data availability during an incident but also on the competence of the decision makers involved during an incident. In 10.35% of the incidents scope for improvements in data availability and/or competence in decision making were identified.

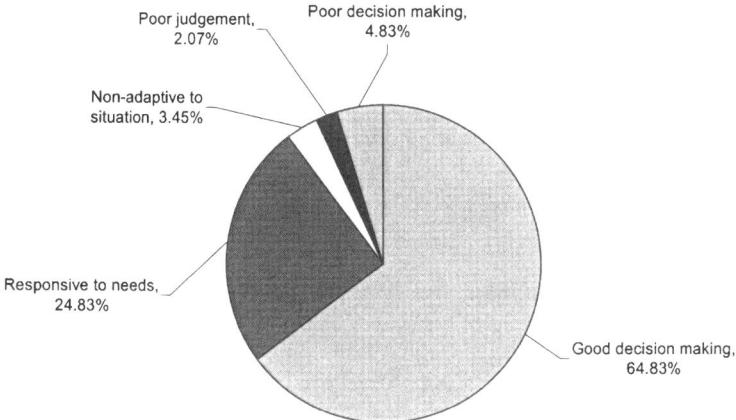

Figure 23 *Characterisation of decision making during incidents between 2004 and 2006*

4.6 System and human redundancy

The organisation maintains reserve capacity in its technical and organisational system that includes back-up functions, overlapping tasks and responsibilities. The organisation is capable to isolate the source of hazard whilst using other asset types to compensate for the loss. E.g., the distribution network has re-zoning capability to isolate a burst main and provide water supply from other sources. During major power failure incidents, the organisation can mobilise stand-alone power supply units to critical water supply assets if they have no on-site power generation.

The entire water supply system builds on duty standby systems or excess capacity to isolate a failed asset and compensate for its loss. The use of systems redundancy was investigated as part of the incident management response. In a previous study, the redundancy of assets which failed during an incident was investigated. Here, the emphasis is on alternative supplies that were used to reduce or avoid the impact of incidents on customers. We defined systems redundancy as any means of water supply capability which could be diverted to compensate for a failed asset or installation. This definition considers systems redundancy to originate from fixed installations but excludes bottled water and water tankering. The latter are commonly used to provide customers with an emergency supply of drinking water if no alternative supply can be established. With reference to Figure 24 it was found that in 55.17% of the incidents no systems redundancy was used or could be used to reduce the impact or avoid customer impact. In the majority of these incidents the water utility resorted to the supply of bottled water. In 22.07% of the incidents, the use of systems could not avoid customer impact although it had a reducing effect. In 15.86% of the incidents, the use of systems redundancy significantly reduced the impact of incidents on customers and avoided the impact for a much larger customer base. In 6.21% of the incidents, systems redundancy was available and used but had a low effect on reducing the incident impact.

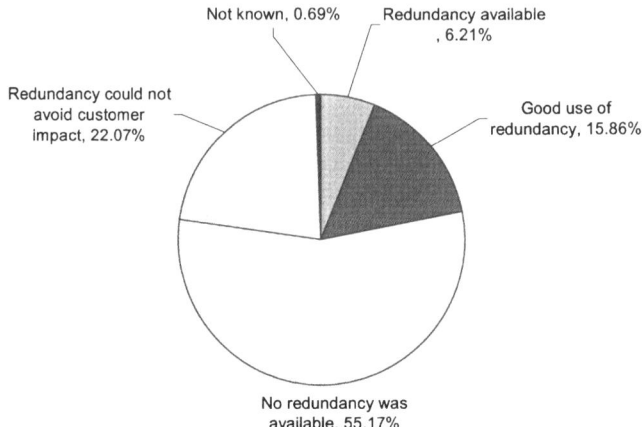

Figure 24 *Usage of redundancy during incidents between 2004 and 2006*

As mentioned before, it is important to recognise that designing redundancy for a system can be counterproductive, as back-up functions can increase technical complexity, conceal errors and lead individuals into not performing their required tasks under the assumptions that someone else takes care of his task.[20]   Although we did not find significant evidence for this, we believe that there is a potential for maintenance decisions to be deferred due to multiple technical redundancy; duty standby systems have a significantly reduced probability of failure that may lead to an assessment of low risk and, hence, low priority in maintenance spending. Such type of risk assessment considers the probability of asset failure <u>and</u> the probability of that asset failure to have an impact on customers.

4.7   Human resource management practices that support reliability

Suitability, skills and competencies are defined by the functional role that these individuals occupy in the organisation. An incident manager has to be able to cope with highly uncertain situations and demonstrate rational decision making under "trying conditions". The incident manager has to be able to communicate effectively with the staff and stakeholders involved in incidents. S/he requires the ability to demonstrate decisiveness and firm leadership to remain in control of adverse situations. S/he also requires a good understanding of the entire water supply system whilst drawing on the expert knowledge in the incident management team. On the other hand, an asset engineer requires a very different skill set. The asset engineer requires analytical skills and competencies in assessing technical systems as well as the technical means to provide and maintain safe and reliable drinking water supplies. Their job role is less reactive to incidents but rather pro-active in assessing potential sources of failure. Increasingly, the asset engineer has to consider technical systems risks and communicate them as systematic risk assessment to the custodians of the risk management process. The asset engineer requires good communication skills, in particular to communicate with operators and operations management.

Taken together, all of these dimensions have been observed to contribute to the effective management of water safety and reliability of this particular organisation. Individually, each dimension is important, but it is when acting together as a coherent configuration that incidents are effectively managed and future risks sufficiently understood and planned for.

4.8   Concluding remarks

In the study of incident management capacities it was found that the regional water supply company is well positioned to manage incidents and many HRO principles can be readily identified as management practice under "trying conditions". The observance of HRO principles significantly contributes to the resilience of the organisation to provide and maintain a safe and reliable drinking water supply.

In the previous study a large proportion of incidents were associated to asset failures which, in turn, define the technical reliability of the water supply system. In the following chapter, the management of technical reliability in the regional water supply system is investigated.

5   TECHNICAL RELIABILITY THROUGH RISK-BASED ASSET MANAGEMENT

5.1   The economic rationality of risk

In practice, water utilities are nowadays embarking on an explicit trade-off between investment cost and risk for asset investment and maintenance decision making.[27] Formalised cost risk trade-off mechanisms are becoming common practice in the water sector in affluent countries[28,27,29] that use risk registers and cost benefit analysis (CBA) in asset investment and maintenance decision making to evaluate the benefit of risk reduction in the context of the cost for asset investment and maintenance.

From an economic perspective, a water utility uses technologies to transform input factors into outputs. The three main input factors are capital, labour and natural resources. The management process considers which production factors to use, how to combine these production factors, the prices for production factors, output and risk. Similar to capital, labour and natural resources, risk can be allocated an incremental unit and a price or cost. Increasingly, risk assessments are used in the water sector to identify the units of risk in water supply systems.[24]

In that sense, risk becomes the fourth production input factor. A production function describes the quantitative correlation between production input factors and outputs.[32] In substitutional production functions the input factors can be substituted within a reasonable area of the function.[32] As with the other substitutional input factors, risk i_1 can be substituted by the production factors labour, natural resources and capital without an effect on the overall output. In Figure 25 the Cobb-Douglas function is such a production function[32] that describes the explicit trade-off between unit risk [24] and the units of assets required to reduce risks.

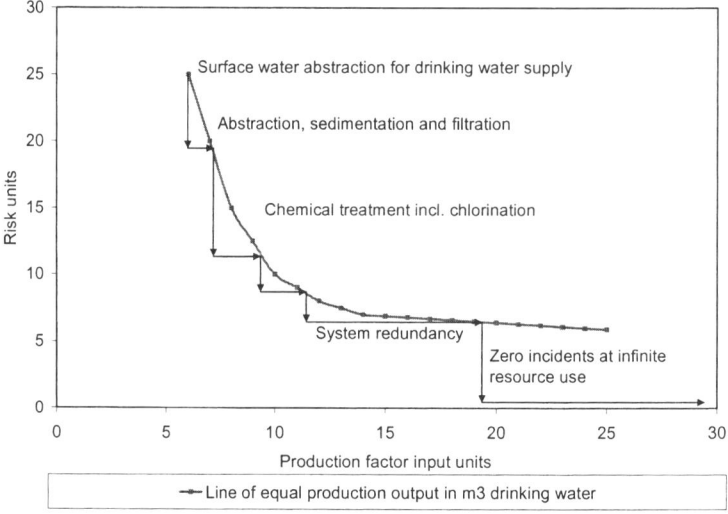

Figure 25 *The trade-off between risk and water supply system assets*

The relationship between units of risk and assets does not consider the price or cost for risk and assets. Assuming the need to maximise benefit (or profit) the optimal equilibrium between risk and assets is governed by their respective 'market' prices or costs. The determination of the risk price may consider the direct cost of incidents to the utility [28] but also the 'social' cost [33] for the affected population. Alternatively, the incremental unit of risk and their "market price" can also take the form of opportunity cost that a water utility customer is willing to pay in order for a risk event not to occur.[30, 31] In the regional water supply company, the prioritisation of investment needs are aligned with the willingness to pay of its customers. Based on this principle, Equation 2 describes the rate of technical substitution between risk (di_2) and assets (di_1) to be the negative ratio of their production input factor prices $(p_{i1}$ and $p_{i2})$.[32] It suggests that the optimal rate of substitution for production input factors is directly dependant on their factor prices[32], i.e. the price of risk. Different 'prices' for risks relating to service areas, e.g. water quality, pressure and supply reliability, are used to evaluate and prioritise investment schemes.
 This is the governing principle of economic-rationale cost risk trade-off in decision making.

$$\frac{di_2}{di_1} = -\frac{p_{i_1}}{p_{i_2}} \qquad\qquad (2)$$

5.2   Risk data quality in risk assessment programmes

Considering the vast risks to public health in the environment and the vast asset base in a water supply system that may be designed to control those risks, high consistency in risk assessments for individual assets are required to optimally allocate capital and operational expenditures across the asset base.

Despite increased use of statistical derived probability and severity data in risk assessment programmes, a significant number of assets require "manual" determination in terms of their failure modes and how that risk may impact on service provision. Probability of failure for assets may be derived from asset-specific Weibull functions; however, the asset failure does not necessarily impact on customers. The probability of incident impact on customers requires a water supply systems assessment. The acquisition of risk data and their quality involves a number of sub-processes: For the purposes of asset death related risk the regional water utility periodically conducts site surveys to collect asset data for each site and assets. It has developed an asset register to ensure consistency in data collection. An assessment record lists all elements of a facility (e.g. a Water Treatment works) down to equipment level (e.g. pumps, valves and actuators). Depending on the asset type and its function in the water supply system, the number of equipment can range from 7 for a borehole, 16 for a water pumping station, 70 for a service reservoir to 700 on a water treatment works. The assessment is undertaken utilising the expert opinion of the asset management and operation teams, who deal with these assets on a day to day basis. In terms of data quality, the organisation has introduced the use of 'technical approach' manuals as a guide to conduct risk assessments. These set out the detailed requirements of the data, formats for collection, and definitions. Staff are trained "on the job" to carry out risk assessments.

Although processes are defined and the organisation has experience in using its risk assessment methodology for assets, it relies on the competence of asset engineers to identify and assess risks. We found evidence that individuals have their own understandings and knowledge of risk that may deviate from "the common understanding". These risk perceptions are psychologically and sociologically constructed but also educationally grounded. As a consequence, human reliability is a major factor in the process that refers to the competency of asset engineers to apply the risk assessment consistently across the entire asset base and derive comparable risk assessments. In an experiment, a number of risk assessors were given the task to evaluate an asset for its probability to have adverse impact on customers. The aim was to identify the level of consistency in the assessment of risks. Across the 36 obtained risk assessments, evidence was found for inconsistencies in their quality. A varying range of risk assessments had been constructed to assess the probable consequences of failure for that specific asset and the number of assessed risks for that asset ranged from one to multiple failure scenarios. In some instances, risk was under- or overestimated. In some cases confusion was identified regarding the definitions of hazard and severity categories: One example is 'leakage' and 'loss of supply'. Both indicators had been used to assess the impact of a water mains burst. The definition for leakage, however, anticipates continuous leakage rather than being instantaneous due to a water mains burst. It was also observed that some risk assessors only derived the probability of the asset to fail but not the probability of this asset failure to have an impact on customers. This would assume that the water supply system has no standby systems and system redundancy.

Based on Equation 2, the economic effect of inconsistencies in risk assessments can be explained with the conceptualised Figure 26. At individual asset level, the over-estimation of risk attracts excess cash for risk reduction, whereas underreported or underestimated risks for assets attract deficient amounts of cash. Across the entire asset base, a standard deviation in the assessed units of risks for assets introduces a similar standard deviation in the rate of technical substitution between risk (di_2)and assets (di_1) that is governed by the negative ratio of their production input factor prices (p_{i1} and p_{i2}).[32]

Over- or underestimating as well as over- or underreporting risks distorts the economic optimal allocation of cash resources and imbalances the optimal risk-asset equilibrium. Hence, a consistent approach to risk assessments is required that builds on highly structured risk assessment processes and a common understanding and perception of risk is required that shapes the individual's psychological and sociological constructs of risks towards the organisational norm.

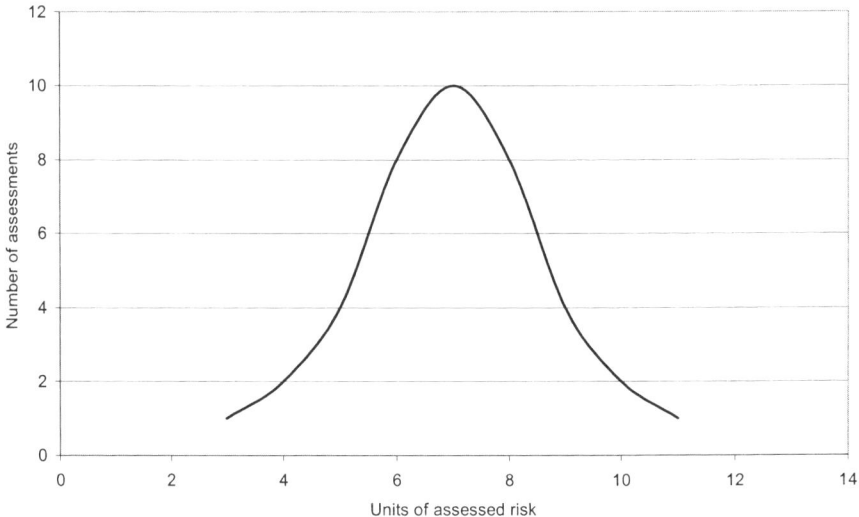

Figure 26 *Conceptual model for inconsistencies in risk assessments for identical assets*

5.3   Continuous learning and intensive training

5.3.1 Learning from failure. Failures, near failures and mistakes in the own organisation – often covert – provide opportunity to understand and assess modes of operation and decision making. Models are required to investigate incidents that facilitate the ability to enhance risk assessment. Whenever information on failures, near failures and mistakes become available, they can provide an opportunity to review assets, processes and effectiveness of operations and asset management.

Learning from failure requires skills to identify multi-causalities and interdependencies in the build up and during the incidents. Complex cause and effect relationships can be disguised by an array of contributing factors and circumstances.

A methodological problem arises in the methods of recording and analysing incident data. Based on the investigative model, the plethora of detail in the incident data is pre-filtered "to fit the model". Interpretation of incident data uses heuristic models to break down the complexity of an incident to make them more understandable. Furthermore, the investigator evaluates the incident data to derive information and knowledge based on judgement. Inadequate models and error of judgment may inhibit the identification of any new patterns arising from the incident data. Similar to incident assessment models, the structure of risk assessments use heuristic (simplified) cause and effect relationships to assess the probability of adverse effects on a defined objective. As a result, periodic reviews of the incident assessment model and the risk assessment model are required to identify new incident patterns that emerge from incident data.

   5.3.2 Risk training. The organisation has implemented staff development programmes leading to appreciation, knowledge, experience and ability to assess risks. The organisation provides information, training, instructions and supervision. Staff development is monitored, audited and reviewed. Risk training not only explores the economic-rational perspective on risk used for decision making in the organisation but also "re-frames" psychological and social construction and understandings of the organisational risk concept. The training is aimed for reducing the deviation in quality of risk assessments with a view to enhance of consistency in investment and maintenance decision making.

   The asset and operations management teams organise regular joint meetings to discuss operational issues which may have implications for asset design and maintenance. These meetings are aimed to provide an information exchange of experience between operators, operations management and asset engineers and managers aiming to disseminate the organisational understanding of risk and best practice in risk assessment and management techniques.

5.4 Concluding remarks

The organisation does not necessarily require 'state of the art' equipment but rather considers a number of factors in asset investment and maintenance decision making: Firstly, the assessed reduction of risk and its monetary evaluation are considered alongside the cash resource requirements. The organisation aims to maximise cash spending efficiency to a point where the monetary value of risk reduction balances the cash requirements for investment and maintenance.

   HROs have been described for their aim to avoid unnecessary automation.[12] To the contrary, the regional water utility has, in recent years, invested heavily in automated monitoring and control systems. The main drivers for such an investment are twofold: firstly, the availability of advanced monitoring and control systems and, secondly, the perceived efficiency of such investment over human operators. Investment in information technology has been used to substitute human assets to an extent that most water treatment works are now unmanned and directly controlled with "Supervisory Control and Data Acquisition" systems.

   HROs have been described to maintain existing technology at exceptionally high standards and there is zero tolerance of defective, substandard or malfunctioning equipment.[16] In the organisation, maintenance decision making takes a more differentiating view on capital maintenance which is in line with the "Common Maintenance Framework".[31] Maintenance decisions are based on risk assessments and a trade-off between cost of maintenance and perceived, monetary value of risk reduction. The organisation will provide cash for maintenance in circumstances where the monetary value of risk reduction exceeds cash requirements for maintenance. This process heavily relies on accurate and consistent risk assessments - accurate, for a "true" representation of "real" risks, and consistent, for company-wide, comparative assessments.

   The risk assessment procedure has been implemented in a top-down approach and requires asset managers and asset engineers to report and file risks bottom up from asset level to a strategic asset management team. An information system is available to log risks into a risk database. The database is semi-structured to guide the risk assessor in providing accurate and consistent risk assessments. The database also houses risk data derived from risk evaluation models based on quantitative risk assessments via statistical analysis designed to evaluate failure probabilities and impacts Yet, the process relies on the competency, skills and experience of asset managers and asset engineers to identify and evaluate risks accurately. Similarly, the design of quantitative risk assessment models depends on the availability of failure data and competency in designing these models. Deviations from accurate and consistent risks reported on the risk database are sought out with quality assurance procedures to ensure effective allocation of cash. Recently, a risk training programme was launched to enhance the capability of risk identification and assessment. The utility operates a challenge process on a monthly basis where risk assessments are reviewed for consistency.

The organisation does not maintain its system to highest standard per se. The asset investment and maintenance decision making process takes a more differentiated, risk-based view in resource allocation. In this point, the water utility significantly differs from the theory of HROs.

6 CONCLUSION

In conclusion, the observations and the studies conducted in the organisation suggest that many of the explored HRO characteristics significantly contribute to reduce the public health impact of incidents but also help the organisation to anticipate future risks. The utility has evolved its organisational and incident management structure to handle customer impacts effectively and as expediently as practically possible.

The regional water utility values clear objectives and ensures that they are well understood within the organisation and by its partners and has invested in it's staff to ensure that people are in the right roles and have fit for purpose skills to work effectively.

From a methodological perspective, HRO theory was specifically investigated under "trying conditions" and the effects of HRO principles as a means of generating organisational resilience during incident situations were carefully studied.

Beyond the current knowledge of HRO theory, risk-based asset management was investigated for its contribution to the technical reliability of the asset base to deliver safe and reliable drinking water.

In this study it was sought to identify high reliability functions as opposed to entire high reliability organisation. It was demonstrated that operations and incident management functions in the organisation resemble HROs. Aspects of HRO were also found in the asset management function that assesses public health risk and evaluates viable risk reduction strategies. In that sense, the asset management function supports the high reliability objective in the provision and maintenance of the asset base at a defined technical systems reliability.

Acknowledgements

The author would like to thank the American Waterworks Association Research Foundation for sponsoring this project and the regional water utility that provided the opportunity to study organisational processes and culture in their organisation.

References

1 K. E. Weick, *California Management Review*, 1987, **29**, 2.
2 T. R. LaPorte and P. M. Consolini, *Journal of Public Administration Research and*

Theory, 1991, **1**, 1.

3 T. R. LaPorte and P. M. Consolini, *International Journal of Public Administration*, 1998, **21**, 6-8.

4 K. E. Weick, K. M. Sutcliffe and D. Obstfeld, *Research in Organisational Behaviour*, 1999, **21**.

5 R. Rosness, G. Hakonsen, T. Steiro and R. K. Tinmannsvik, *The vulnerable robustness of High Reliability Organisations: A case study report from an offshore oil production platform,* 2005, available at: http://risikoforsk.no/Publikasjoner/Vulnerable%20robustness.pdf (accessed 2006).

6 K. Roberts, *Organisation Science*, 1990, **1**, 2.

7 P. Schulman, E. Roe, M. van Eeten and M. de Bruijne (2004), *Journal of Contingencies and Crisis Management*, 2004, **12**, 1.

8 M. Grabowski and K. H. Roberts, *Ieee Transactions on Systems Man and Cybernetics, Part A - Systems and Humans*, 1996, **26**, 1.

9 K. H. Roberts, D. M. Rousseau T.R. and Laporte, *The Journal of High Technology Management Research*, 1994, **5**, 1.

10 K.E. Weick and K.H. Roberts, *Administrative Science Quarterly*, 1993, **38**, 3.

11 K.H. Roberts and R. Bea, *Academy of Management Executives*, 2001, **15**, 3.

12 P. E. Bierly and J.C. Spender, *Journal of Management*, 1995, **21**, 4.

13 T.R. Laporte, *Journal of Contingencies and Crisis Management*, 1996, **4**, 2.

14 G.I. Rochlin, T.R. Laporte and K.H. Roberts, *Naval War College Review*, 1987, **40**, 4.

15 C. Perrow, *Normal accidents - Living with high-risk technologies* (2nd edition), Princeton University Press, New Jersey, 1999.

16 K. H. Roberts, *California Management Review*, 1990, 32, 4.

17 K. H. Roberts and C. Libuser, *Organizational Dynamics*, 1993, **21**, 4.

18 G.A. Bigley and K.H. Roberts, *Academy of Management Journal*, 2001, **44**, 6.

19 P. R. Schulman, *Administration and Society*, 1993, **25**, 3.

20 S. D. Sagan, (), 'Towards a Political Theory of Organisational Reliability', *Journal of Contingencies and Crisis Management*, 1994, **2**, 4.

21 B. Latane and J. Darley, *The unresponsive bystander: Why doesn't he help?*, Appleton Century Crofts, New York, 1970.

22 S. E. Hrudey and E.J. Hrudey, *Safe drinking water - Lessons from recent outbreaks in affluent nations*, IWA Publishing, London, 2004.

23 British Standard Institution (BSI), *Specification for the optimised management of physical infrastructure assets*, PAS 55-1, BSI, London, 2003.

24 L. Fewtell J. and Bartram (Editors), *Water quality: Guidelines, standards and health*, IWA Publishing, London, 2001.

25 Department for Environment, Food and Rural Affairs, *Water Industry Act 1991 - Water Undertakers (Information) Direction 2004*, HM Stationary Office, 2004.

26 The Drinking Water Inspectorate, *Guidance on the Notification of Events*, Drinking Water Inspectorate, 2004.

27 B.H. MacGillivray, P. D. Hamilton, J. E. Strutt and S.J.T. Pollard, *Critical Reviews in Environmental Science and Technology*, 2006, **36**.

28 G. Lifton and P. Smeaton, 'Asset risk management in Scottish Water', in ICE/CIWEM *Risk and reward in asset management delivery - who is best prepared for the challenges ahead?* at London; ICE/CIWEM, London, 2003.

29 S. J. T. Pollard, J. E. Strutt, B. H. MacGillivray, P. D. Hamilton and S. E. Hrudey, *Trans. IChemE, Process Safety Environmental Protection* , 2004, **82**,B6 (Special Issue: Risk Management).

30 P. Abell in *Proceedings of the AWWARF International Workshop "Risk analysis*

strategies for better and more credible decision making"ed. Pollard, S. J. T. et al., Cranfield University, AWWARF, 2006.

31 UK Water Industry Research Limited, *Capital maintenance planning - A common framework*, Report Ref. No. 02/RG/05/3, UKWIR, London, 2002

32 T. Bonart and U. Peters, *Mikrooekonomie kompakt; Einfuehrung, Aufgaben, Loesungen*, Betriebswissenschaftlicher Verlag Dr. Th. Gabler GmbH, Wiesbaden, 1997.

33 P. Hughes and E. Ferrett, *Introduction to Health and Safety at Work*, Butterworth-Heinemann, Oxford, 2003.

A SIMULATION TOOL FOR CONTAMINANT WARNING SYSTEM DESIGN AND EVALUATION

W. Einfeld[1], S. A. McKenna[1] and M. P. Wilson[2]

[1]Sandia National Laboratories, P.O. Box 5800, Albuquerque, NM 87185-0734
[2]GRAM and Associates, 8500 Menaul Blvd. NE, Suite B-335, Albuquerque, NM 87112

1 INTRODUCTION

Water utilities are faced with the ongoing need to provide safe, contaminant-free water to their customers. Contamination sources can include compromised source water that penetrates treatment barriers or malevolent attacks from disaffected groups or terrorists. Residual disinfectant levels within a water distribution system can in some circumstances address the effects from such events; but in others, the residual disinfectant levels can be totally depleted without complete inactivation of the pathogen or toxic chemical that is in the system, thereby posing a public health risk. In nearly all contamination event scenarios, early detection and response can help reduce public health risk and impact. To that end, contaminant warning systems (CWS) offer advantages over more traditional and slower means of contaminant detection.

A CWS design evaluation tool has been developed that enables an assessment of overall system performance under various contamination scenarios using simulation techniques. Using such design tools, a variety of contamination scenarios can be simulated using hydraulic flow models and utility-specific pipe layouts. The location and types of sensors can then be explored to assess cost, operations and maintenance burden as well as risk reduction benefits likely to be derived from installation and use of the overall CWS system.

The concept of a CWS incorporates a number of key elements as shown in Figure 1. A key aspect of a system is a network of sensors that directly sample treated water within the distribution system. In light of the wide spectrum of contaminants that might be encountered in a contamination event, the approach most likely to be successful is the use of indicator species or parameters that can provide an alert to a significant change in the "baseline" water quality state that may warrant further investigation. Ideally, such an alert would be followed with a second-tier analysis that would provide additional detail with respect to the contaminant species; however, the current state of on-site, second-tier technology is generally insufficient to   provide   the   desired   level   of   detail   in   a   short   time

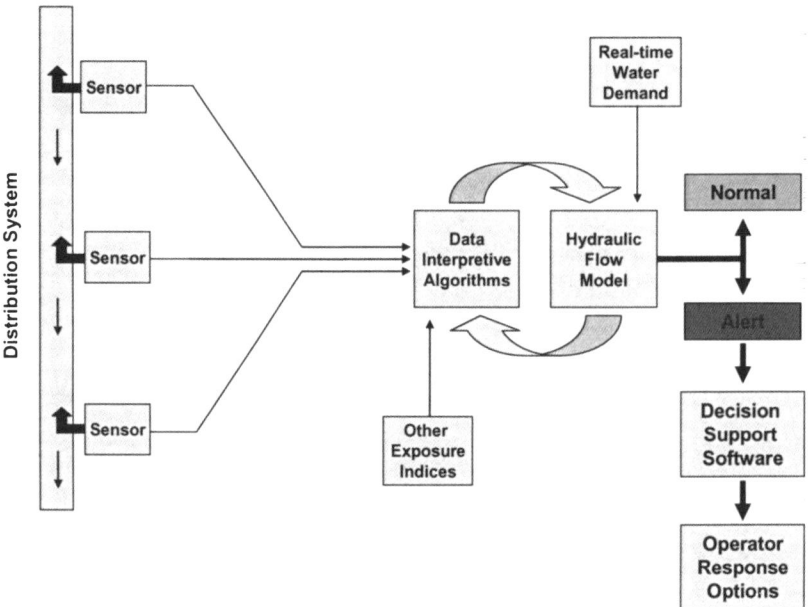

Figure 1 *Conceptual diagram showing the key elements in a Contaminant Warning System.*

frame. A number of reviews have been published that categorize commercially available sensors for application in a CWS (EPA 2005, Foran and Brosnan 2000, Conio et al. 2002, States et al. 2004). A network of sensors can be utilized to provide additional confidence in the declaration of an "alert." Significant baseline sensor data deviations encountered at more than one sensor location provide additional confidence and can help to minimize the occurrence of false positives.

Surrogate sensors suitable for use in a CWS at this point in time are limited to those normally used for routine water quality assessment. Water quality parameters that could be linked to a contamination event include pH, turbidity, oxidative-reductive potential, residual chlorine, UV absorption, and total organic carbon (TOC). Furthermore, software routines can be included in a CWS that enable baseline "state change" detection with increased reliability over simple sensor-signal alarm threshold methods. These algorithms can be used to apply statistical methods involving one or more sensors at multiple locations to increase the degree of confidence in the designation of a contamination event (Hart et. al. 2007).

Ideally, a CWS will also include a hydraulic model of the distribution system being measured that runs in parallel with the sensors' data detection algorithms. The hydraulic model can be used to predict contaminant transport throughout the distribution system as a function of time and can be used to anticipate and confirm sensor detects that would be expected to occur at a later point in time downstream from an initial detection point. The best hydraulic models should also include a dynamic means of estimating water demand at any particular time of the day since water demand directly influences water flow volume and velocity throughout the pipe system. In a functioning CWS, the hydraulic model can be further used in a predictive mode to indicate the likely spread of contaminant, both spatially and temporally. Networked sensor data when combined with system flow data can be used in data inversion routines to gain information on source of contaminant introduction into the system. The degree to which the hydraulic model is calibrated to the current state of the system will impact how much information it will be able to provide during a contamination event.

The diagram shown in Figure 2 illustrates how the sensor simulator functions in concert with a change detection module as the two principal components of the tool. The input data stream is a contaminant pulse, usually with an added water quality baseline. The baseline is the background signal from water quality measurements made in the absence of contamination and includes the normally observed day-to-day changes in water quality. The sensor simulator is configured to mimic sensor performance characteristics (total organic carbon and residual chlorine sensors were examined in this study) with respect to key operational parameters such as noise, drift, and response time. The raw data signal is processed through the sensor simulator and the extent to which the raw signal is influenced by the sensor operational parameters is observed on the output side of the sensor simulator. An example of how the input signal is influenced by sensor operating characteristics is shown in Figure 3.

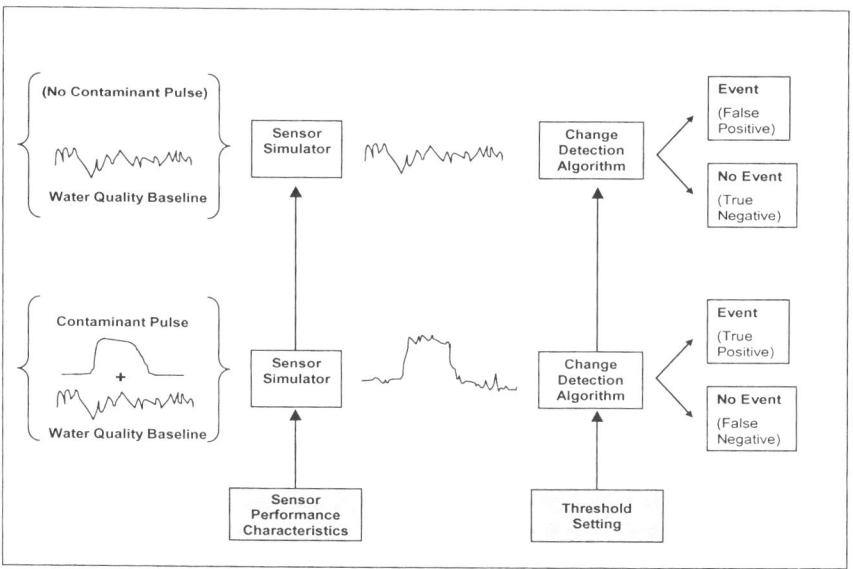

Figure 2 *A schematic diagram showing flow of raw data through a sensor simulator and change detection algorithm.*

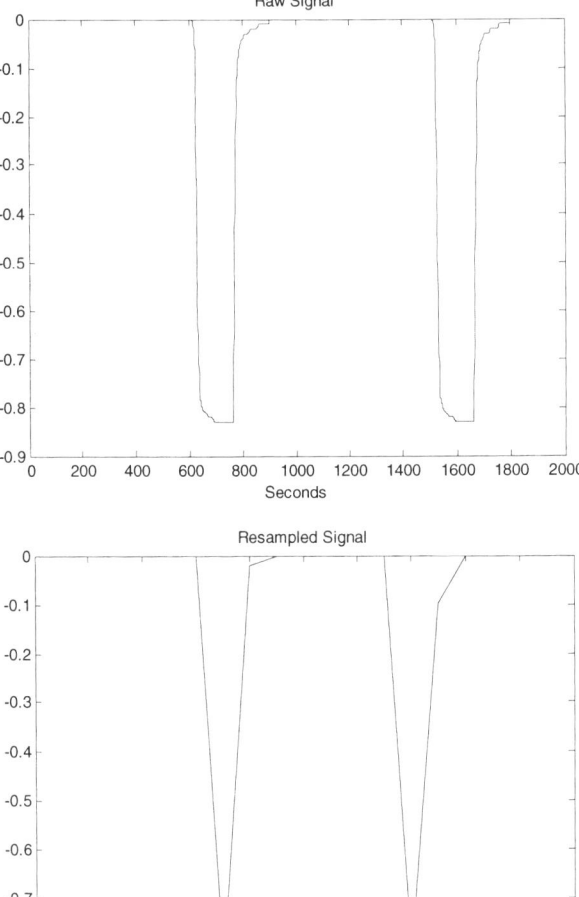

Figure 3 *Output from the sensor simulator showing how the sensor's sampling interval can change the sensor output data. The top figure shows the continuous input signal and the bottom figure illustrates the signal changes that occur when a sensor has a sampling interval of 1 sample per 2 minutes.*

A second module features signal change detection algorithms that process the sensor data in order to detect contamination events. The results at the output side of the change detection algorithm can be categorized as true positive, false positive, false negative or true negative. Assessing the frequency of each of these four possible outcomes under specific contamination scenarios and sensor configurations enables comparison of the overall performance of various sensor options within this simulation environment.

2 METHODS

Study objectives were to develop a simulation tool to assess overall CWS performance under various contamination event scenarios and to incorporate specific sensor performance characteristics in the analysis in order to more realistically assess how sensor operational parameters influence overall CWS performance.

2.1 Sensor Types and Parameters Investigated

This study concentrated on two sensor types: residual chlorine and total organic carbon. Testing by the EPA Water Test and Evaluation Center has shown that these two water quality parameters are desirable surrogate indicators of a contamination event (Hall et al. 2007). Tables 1 and 2 include summary descriptions of the sensors evaluated in this study for residual chlorine and total organic carbon measurements, respectively. The tables include a description of the operating principle of each sensor along with performance specifications published by the sensor manufacturer. Ideally, it is desirable to verify these performance parameters through independent testing; however, such data are expensive to obtain and as a result are limited. For this effort we used manufacturer-provided specifications as provided with no further verification testing.

2.2 Sensor Simulator Algorithms

The sensor simulator was built using a variety of algorithms and routines available from a commercially available math routine software package (Mathworks, Natick, MA). High-level language and interactive software tools within these packages allows the user to perform computationally intensive tasks faster than with traditional programming languages. The sensor simulator also includes a graphical user interface that facilitates simplified entry of sensor performance characteristics and test conditions by the user. A variety of algorithms were incorporated into the simulator to accept and simulate the effects of key sensor attributes such as: sampling rate, response time, drift, noise, resolution, and others.

Table 1   *Residual Chlorine Sensors Selected for Study*

Sensor Manufacturer	Sensor Model No.	Operating Principle	Performance Specifications
Endress Hauser	CCS140/141	Electrochemical with polymeric membrane	• Measurement Range: Model 140 (0-20 ppm or mg/L), Model 141 (0-5 ppm) • Sampling Rate/Cycle Time: 2 samples/second • Response Time: 30 seconds • Response %: 90 • Noise Value: 0.5% of measured value • Drift Amount: <0.5% of sensor range • Drift Period/Unit Time: 1 month • Resolution: 0.01 mg/l
HF Scientific, Inc.	CLX	Colorimetric	• Range: 0-10 mg/l • Sampling Rate/Cycle Time: 1 sample per 90 seconds • Response Time: N/A • Response %: N/A • Noise Value: +/- 5% • Drift Amount: 0 • Drift Period/Unit Time: 0 • Resolution: 0.01 mg/l

2.3 Description of the Simulation Process

A summary of the overall simulation process used in this study is described as follows:

- Conduct EPANET runs for short- and long-pulse contamination events with aldicarb (a carbamate pesticide) using pipe layouts from a real water distribution system.
- Obtain the contaminant pulse shape at selected locations downstream from the injection point using EPANET data outputs.
- Re-scale the downstream pulse shape to a threshold, no-health-effect dose level for the target contaminant (in this case aldicarb).
- Re-scale the observed contaminant concentration levels to both TOC and residual chlorine sensor equivalent responses using published sensor response functions, e.g. the expected sensor response level as a function of aldicarb concentration (Byer and Carlson 2005).
- Superimpose these re-scaled contaminant pulses onto a baseline data trace (from either a residual chlorine or TOC sensor) obtained from measurements at selected US water utilities. The contaminant pulse data and baseline data are added together to produce a composite signal. (See Figure 2).
- Process combined signal through sensor simulator that is configured with performance characteristics for the sensor of interest (e.g., sampling rate, noise level, drift).

Table 2 *Total Organic Carbon Sensors Selected for Study*

Sensor Manufacturer	Sensor Model No.	Operating Principle	Performance Specifications
Shimadzu	TOCN-4110	Combustion/infrared analysis	• Range: 0-5 ppm, 0-1,000 ppm • Sampling Rate/Cycle Time: 1 sample per 4 minutes • Response Time: N/A • Response %: N/A • Noise Value: +/- 2% Full Scale • Noise Sigma: 1 • Drift Amount: +/- 2% Full Scale • Drift Period/Unit Time: 1 week • Resolution: .01 ppm
Endress Hauser	CSS70	UV Absorption	• Range: 0.4-60 ppm, 20-900 ppm • Sampling Rate/Cycle Time: 1 sample per 40 seconds • Response Time: 60 Seconds • Response %: 90 • Noise Value: 2.5% of upper range • Noise Sigma: 1 • Drift Amount: -2% Full Scale • Drift Period/Unit Time: 1 month • Resolution: .01 ppm
HACH	Astro TOC	UV-assisted oxidation/Infrared analysis	• Range: 0 – 5 up to 20,000 mg/l • Sampling Rate/Cycle Time: 1 sample per 8 minutes • Response Time: 8 minutes • Response %: 90 • Noise Value: +/- 4% of full scale • Noise Sigma: 1 • Drift Amount: 2% full scale • Drift Period/Unit Time: 1 month • Resolution: .001 mg/L
S:Scan	Spectro:Lyser TOC	UV Absorption	• Range: 0 – 20 mg/L • Sampling Rate/Cycle Time: 1 sample per 30 seconds • Response Time: N/A • Response %: N/A • Noise Value: +/- 0.3 mg/L • Noise Sigma: 1 • Drift Amount: 0.2 mg/L • Drift Period/Unit Time: 1 month • Resolution: 0.001 mg/L

- Iteratively process the sensor-modified output signal from the sensor simulator through the event change detection module. Vary the threshold value (i.e. the event change detection software "sensitivity" setting) with each iteration and compile the event detection statistics (e.g., true positive, false positive, true negative, false negative) for each threshold setting.
- Compile the data from the iterative runs into a format suitable for analysis using a Receiver Operating Characteristics (ROC) approach. Further analysis of ROC curve data serves as an indicator of overall sensor/change detection algorithm performance under the chosen contamination scenario.
- Repeat the simulations for other concentrations, sensors, contaminant pulse lengths, and baseline conditions.

2.4 Analysis Using the Receiver Operating Characteristic Method

The receiver operating characteristic (ROC) approach was used as an analysis tool to assess and compare overall sensor performance in this study. The ROC concept was developed during World War II to improve the performance of radar detection algorithms. Since that time the approach has been further developed and is used extensively in the medical diagnostics field to optimize the interpretation of diagnostic tests (Swets 1988). An example of ROC data in graphical format is shown in Figure 4. The ROC curve is a graphical representation of the frequency of true detects versus the frequency of false positives over the variable threshold range of the particular sensor/event detection module being evaluated. For each simulation run and at a specific threshold setting, a contaminant peak is either detected (true positive) or missed (false negative).

Additionally, the sensor/event detection module may detect a peak when in fact there is not one present (false positive) or correctly report a no-detect (true negative). At each event detection module threshold setting, the occurrences of any of these four test results are compiled using a sensor data stream that has contaminant peaks interspersed with sections of no peaks. Water quality baselines may be added to the sensor signal or not, depending on the desired characteristics of the simulation run. To generate all the data needed for a ROC curve, the threshold setting of the event detection module is incremented and the run is repeated until detection statistics are compiled over the entire threshold range. These data are then plotted in a ROC curve format where the x-axis is defined as the probability of a false positive (FP) and the y-axis is the probability of a true positive (TP).

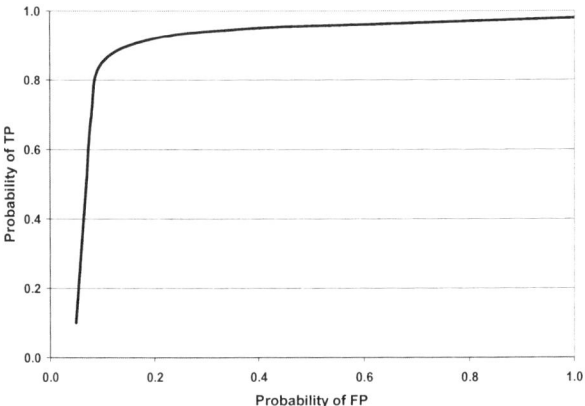

Figure 4   *Example of a Receiver Operating Characteristics (ROC) curve.   The curve represents the balance of true positive detection rates versus false positive rates over the sensitivity range of the combined sensor and event detection module.*

An additional statistic that can be derived from the ROC curve analysis is the area under the curve or AUC. This value serves as an overall indicator of the ability of a particular sensor/event detection module to accurately detect contamination events and minimize false positive events. An AUC value of 1 signifies a perfect system that has a true positive detection rate of 1 and a false positive detection rate of zero at its optimum threshold setting. An AUC value of 0.5 corresponds to a diagonal line from the lower left to the upper right corners in Figure 4, and represents a useless sensor/event detection combination since at any threshold setting the true positive rate and false positive rate are the same. Values of the AUC between 0.5 and 1 give a measure of the overall effectiveness of a sensor/event detection combination. Values close to 0.5 indicate poorer performance whereas values close to 1 reveal better sensor/event detection module performance.

3 RESULTS

Summaries of residual chlorine and total organic carbon sensor performance in combination with two event detection modules are shown in Tables 3 and 4. The forward linear predictor (FLP) event detection algorithm involves the calculation and analysis of a moving average of the sensor data whereas the threshold method simple looks for the excursion of the sensor signal above of a preset alarm point (Bras and Rodriguez-Iturbe 1993). The values tabulated are areas under the ROC curve, an indication of overall system performance. Study results are summarized for a contamination scenario with aldicarb, a carbamate pesticide of moderate toxicity. Additional results are published elsewhere (Einfeld et. al. 2008). Simulation results shown here are taken from an analysis that includes a 30-minute contaminant injection at three dose levels with added background sensor baseline data taken from water utility measurements. The highest dose evaluated is just below the no-health-effect threshold for aldicarb (IRIS 2007). The two lower dose levels serve as surrogate cases for more toxic classes of contaminants that might be encountered in a contamination event.

Sensor/detection module performance degraded with decreasing contaminant concentration levels. At the highest dose level, performance of all combinations was relatively good with ROC curve areas very close to 1. At the lowest dose level examined, all sensor/change detection combinations provided essentially no value (ROC curve areas at or below 0.5) as reliable detection devices in a CWS. In this case the probability of a false positive is essentially equal to the probability of a true positive. Differences in sensor system performance were observed and the most influential operational parameter of the sensor was the sampling interval. Those sensors with a higher frequency sampling rate performed better than those with lower sampling rates.

The event detection algorithms performed equally well for the dose levels investigated. While the more advanced statistical change detection algorithms performed on a par with the simple threshold method in this simple case, they are expected to outperform the threshold approach when used with a CWS design that includes multiple sensor types and locations (McKenna et. al. 2006). The limited scope of this study did not include the evaluation of more complex CWS designs involving multiple sensor types and locations within a water distribution system.

This study illustrates the potential advantages of simulation tools that can be used during the design of a contaminant warning system for a water utility. Through the use of such tools, inter-comparisons of candidate sensor performance can be made under a variety of realistic contamination scenarios that will enable system designers to fully optimize the warning system design prior to component acquisition, installation and deployment.

Table 3   *Performance Measures of Two Residual Chlorine Sensors in Response to Simulated Contamination Events*

Residual Chlorine Sensor	In-Pipe Aldicarb Concentration (mg/L)	Change Detection Method	
		FLP	Threshold
HF Scientific	High – 31	0.96	0.99
Model: CLX	Mid – 1.2	0.67	0.70
	Low – 0.05	0.52	0.56
Endress Hauser	High – 31	0.98	0.99
Model: CCS141	Mid – 1.2	0.83	0.67
	Low – 0.05	0.56	0.53

Table 4   *Performance Measures of Four Total Organic Carbon Sensors In Response to Simulated Contamination Events*

TOC Sensor	In-pipe Aldicarb Concentration (mg/L)	Change Detection Method	
		FLP	Threshold
Hach	High – 31	0.88	1.00
Model: Astro TOC	Mid – 1.2	0.68	0.80
	Low – 0.05	0.55	<0.50
S:Scan	High – 31	0.97	0.98
Model: Spectro:Lyser	Mid – 1.2	0.56	0.63
	Low – 0.05	0.52	0.50
Shimadzu	High – 31	0.91	1.00
Model: TOCN-4110	Mid – 1.2	0.70	0.69
	Low – 0.05	0.52	<0.50
Endress Hauser	High – 31	0.97	0.99
Model: CSS70	Mid – 1.2	0.74	0.74
	Low – 0.05	0.52	0.51

Acknowledgements

The authors thank David Hart for his assistance with development of the change detection algorithms used in this investigation.   Thanks also to Victoria Cruz for running the many simulations through the change detection module, collating the results, and generally assisting with troubleshooting and simulator operations.

This work was co-funded by the American Water Works Research Foundation as a Partnership Project (No. 2978) and the U.S. Environmental Protection Agency through an Interagency Agreement (DW8992192801-B).

Sandia is a multiprogram laboratory operated by Sandia Corporation, a Lockheed Martin Company, for the United States Department of Energy under contract DE-AC04-94AL85000.

References

1 EPA, Technologies and Techniques for Early Warning Systems to Monitoring and Evaluate Drinking Water Quality: A State-of-the-art Review, 2005,EPA/600/R-05/156. Available: http://www.epa.gov/NHSRC/pubs/reportEWS120105.pdf

2 J.A. Foran and T. M. Brosnan,Early Warning Systems for Hazardous Biological Agents in Potable Water. *Environmental Health Perspective,* 2000, 108(10): 993-995.

3 O.M. Conio, E. Chioetto and E. Hargeshriner. Organic Monitors in *Online Monitoring for Drinking Water Utilities.* Edited by E. Hargeshiriner, Conio and J. Popcovia, Denver, CO: AwwaRF and CRS Proaqua, 2000.

4 S.J. States, J. Newberry, J. Wichteman, J. Kuchta, M. Scheuring and L. Casson, Rapid Analytical Techniques for Drinking Water Security Investigations, 2004, *Journal AWWA,* 96:152-164.

5 D.B. Hart, S.A. McKenna, K.A. Klise, V. Cruz and M. Wilson, CANARY: A Water Quality Event Detection Algorithm Development Tool, 2007, submitted to: ASCE World Environmental and Water Resources Congress, Tampa, FL, May 15-19th.

6 J. Hall, R.B. Marx, P.C. Cafauber, E. R. Krishnan, R.C. Haught, J.G. Herrmann, Online Water Quality Parameters as Indicators of Distribution System Contamination, 2007, *Journal AWWA* 99:1:66.

7 D. Byer, and K. H. Carlson, Real-time detection of intentional chemical contamination in the distribution system, 2005, *Journal AWWA,* 97(7):130–133.

8 J.A. Swets, Measuring the accuracy of diagnostic systems, 1988, *Science*, vol.240, 1285-1293.

9 R.L. Bras and I. Rodriguez-Iturbe, *Random Functions and Hydrology*, Dover Publications, Inc., Mineola, New York, 559 pp, 1993.

10 W.S.A. Einfeld, S. A. McKenna and M. P. Wilson, *A Simulation Tool for the Assessment of Contaminant Warning System Sensor Performance Characteristics.* Denver, CO: American Water Works Association Research Foundation, 2008 (in press).

11 IRIS (Integrated Risk Information System) 2007. Aldicarb. Available: http://www.epa.gov/IRIS/subst/0003.htm

12 S.A. McKenna, K.A. Klise and M.P. Wilson, Testing Water Quality Change Detection Algorithms, in Proceedings of the 8th Annual Water Distribution System Analysis Symposium, Cincinnati, OH, August 27-30, 2006.

CBRN MODELLING: APPLICATION TO WATER CONTAMINATION

I. H. Griffiths

RiskAware Ltd, Colston Tower, Colston Street, Bristol BS1 4XE, UK

1 BACKGROUND

The modelling of Chemical, Biological, Radiological and Nuclear (CBRN) hazards has been an active research area for more than seventy years. This has involved experimental and theoretical investigation throughout the world and it has resulted in sophisticated and well-validated modelling capabilities. The advent of modern computers has transformed the performance possibilities of these capabilities and allowed approaches to be developed that would have been computationally prohibitive previously. The research effort continues with many areas being investigated, from source terms to the fundamentals of meteorology and atmospheric dispersion, the inclusion of complex environmental factors such as buildings, through to effects modelling. A substantial amount of effort is being targeted at applying the capabilities within decision aids.

The research and development has been particularly driven by military defence needs, with the aim of providing capability to protect military personnel (and others) in the event of a CBRN event and to allow the military to continue to conduct its missions despite the threat of CBRN use. Research has also been carried out to support capability development for homeland security and civil environmental regulatory applications. There has been a high degree of cross-collaboration between all the research efforts.

Important drivers in recent years for the research into CBRN modelling within the defence community has been the development of approved operational systems for use on military infrastructures, which have required well validated models and a high degree of software testing and verification. In the US, there are three official Department of Defense (DOD) procurement activities, or programmes of record, namely: the Joint Effects Model (JEM – a CBRN hazard modelling capability),[1] the Joint Warning and Reporting Network (JWARN – an information management system that links detector output to hazard warning, making use of the JEM capability),[2] and the Joint Operational Effects Federation (JOEF – a capability that allows the evaluation of CBRN releases on military personnel, equipment and operations).[3] In the UK, the Ministry of Defence (MOD) is currently procuring the Integrated Sensor Management System (ISMS – a system for networking detectors and fusing their output) and was procuring the CBRN Battlespace Information System Application (NBC BISA – a CBRN information management system with hazard modelling capability).[4] Other nations are developing or considering similar capabilities.

The CBRN defence modelling requirements are focused on a number of applications:

- Hazard prediction – this involves the modelling of a CBRN release, its subsequent transport and diffusion in the atmosphere and the estimation of the footprint of the plume in contact with the ground (usually integrated over time to provide the dosage, which relates to the amount of material a person would breath in for example).
- Information management – this involves gathering and processing information (including the collation of sensor output), identifying CBRN release events, developing the best possible picture of the CBRN situation now and into the future (using the hazard prediction capability) and providing support to the decision maker in making optimal decisions.
- Operational analysis and studies – this involves identifying optimal solutions to operational questions. It is used for informing procurement decisions, carrying out capability analysis and conducting planning studics Example questions it may answer include determining how many sensors are required to protect a military base, identifying the performance requirements for a piece of CBRN equipment, and so on.
- Modelling and simulation – this involves carrying out complex realistic simulations running in real-time. Numerous models may be linked together for the simulation as well as human players. It is typically used for training and conducting wargaming experimentation.

2 RELEVANCE TO WATER CONTAMINATION

Water contamination is a matter of importance at both a local level and national level for major incidents. There are clear parallels between the water contamination monitoring and CBRN defence modelling applications, which will be discussed in this paper. The paper is intended to give a brief overview of some of the results of the CBRN modelling research and the capabilities that have evolved. It is far from exhaustive but it does aim to describe several of those that have potential relevance to the water contamination application.

The extensive research and development invested in CBRN modelling may be relevant to water contamination problems in two main ways:

- As an input of a CBRN hazard to water systems.
- As a resource for techniques and capabilities that may be adapted and reused.

The CBRN modelling capabilities are able to provide a high fidelity prediction of the CBRN hazard that includes modelling capabilities for:

- The source term, including secondary sources (e.g. evaporation from pools of liquid agent resulting from the deposition of liquid droplets);
- Meteorology, particularly at smaller scales using data input from large scale national weather prediction models;
- Atmospheric dispersion;
- Casualty estimation;
- Operational effects.

Increasing sophistication is being added, for example the effects of buildings on CBRN plumes dispersing in urban environments. When combined these provide a sophisticated hazard modelling capability (see examples shown in figure 1) that forms the core of the various CBRN defence applications. The output of this modelling capability can be used as an input to water contamination modelling system. This would perhaps be most relevant for supporting the response in the event of a CBRN or toxic release that affects the water supply system, or for studying what the effects of such an event and evaluating mitigation and response strategies.

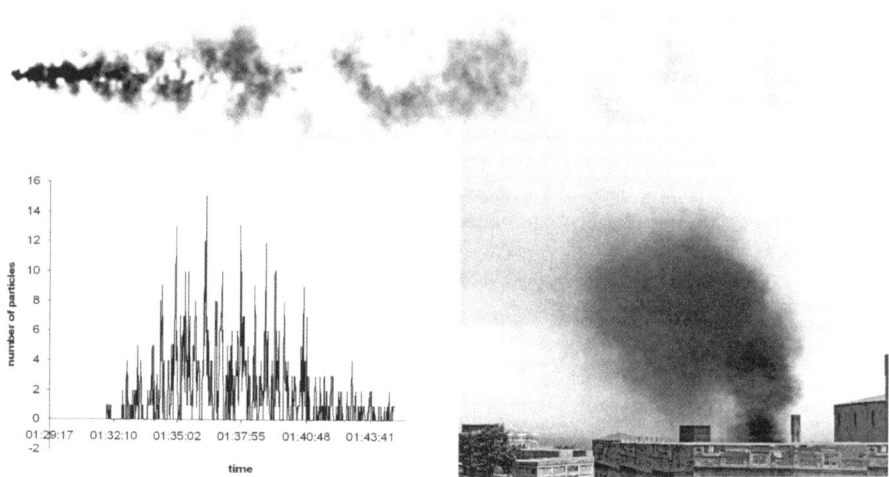

Figure 1   *Samples of the output that can be produced from advanced CBRN hazard models. Top: a dispersing plume synthetically generated; bottom-left: a high-resolution concentration time series simulated at a sensor location; bottom-right: a modelled smoke plume. All of these can be modelled in real time.*

This paper will focus on the second topic: presenting some of the techniques and capabilities that may have the potential to be adapted and applied to water contamination monitoring. Many areas of research are being pursued that may be applying and adapting techniques, which may be of relevance to water contamination. The paper explores two that are perhaps the most relevant: optimal sensor placement and sensor data fusion. Often the CBRN modelling research has taken established techniques from other domains (finance, risk management, industrial process management, general mathematics, etc) but the lessons learned in applying them to an application with some similarities may be of interest and highly informative. It is likely and hoped that there will also be research that has been carried out in the water contamination and quality monitoring field that will be of relevance to CBRN defence, so one of the intentions of this paper is to stimulate cross-fertilisation of ideas between the two areas.

3   SENSOR PLACEMENT AND OPTIMISATION TECHNIQUES

The optimal placement of sensors is of importance in CBRN defence. CBRN sensors are a limited resource and their placement needs to be optimised to provide the most useful information. A similar need exists with water monitoring equipment for identifying water contamination and water quality. The approaches that have been developed for CBRN sensor placement are relevant for the following common reasons:

- This highest priority information is generally likelihood of detecting a release, but another important factor is warning time (the duration from determining an event has occurred to the agent hitting an area of interest such as a military base). For the water contamination monitoring application a finite number of monitors may be available and the aim will be to best place them to ensure that contamination events (from either short or longer duration releases of contaminant) are identified and the minimum number of people are affected. It can be seen that the underlying aims are essentially the same for the two applications.

- CB detectors are typically point sensors, which measure the concentration of agent at the location of the instrument, and they have been the focus of CBRN sensor placement optimisation. Water contamination monitoring equipment measures contamination levels at the location of the sensor so they are also point detectors. The increases the potential exploitation of approaches to solve one problem by the other.

- Solutions developed for the CBRN need to be able to handle a heterogeneous mix of detector types, which will also be the case for water contamination monitoring where a number of different sensor equipment models and even types may be available. This results in flexible approaches that are independent of specific sensors and sensor technologies and so more cross-applicable.

In the past CBRN sensor placement involved applying rules of thumb and other heuristic approaches that have been developed for CB sensor placement. However, by treating it as an optimisation problem recent research has successfully developed computer-based capabilities that will automatically identify the best sensor locations against user-defined goals and within user-specified constraints. The approach is generally made up of two stages. First, a library of possible event simulations are computed – this may be many thousands of different runs with varying conditions such as release parameters (material, mass and location), wind speed and direction, etc. Second, an optimiser is used to calculate the best locations for the sensors by comparing the results for placing different combinations of sensor location within each of the simulated scenarios. A number of optimisation algorithms could be used. Originally genetic algorithms were generally applied but alternative approaches have been found to perform better. Indeed the best way is yet to be fully established although it may be that a range of approaches with different strengths and weaknesses will be necessary.

The main challenge to the sensor placement development is computation time. To be robust, the optimisation requires many simulations (in fact tens of thousands are typically required) because of the range of possible event scenarios and conditions and the need to ensure sufficient coverage of these using Monte Carlo sampling. Much of the development is focused on methods for handling the large numbers of simulations efficiently when optimising. Approaches have also been successfully developed that minimise the numbers of simulations needed through careful experimental design and use efficient memory handling;[5] however overall run times are still on the order of days. The SPARTA tool (shown in figure 2) has been developed that applies broadly similar efficiencies in implementation but also very significantly reduces the number of distinct modelling simulations that need to be computed to bring execution times down to several minutes.[6]

The resulting capabilities are typically calculation engines linked to a user interface. The optimisation core is generally a distinct component from the CBRN simulation model and the sensors are separate models with parameters defined in the user interface. This means that the optimisation core is generic; it is highly likely that with only limited modification it could be applied to optimising water contamination monitoring equipment.

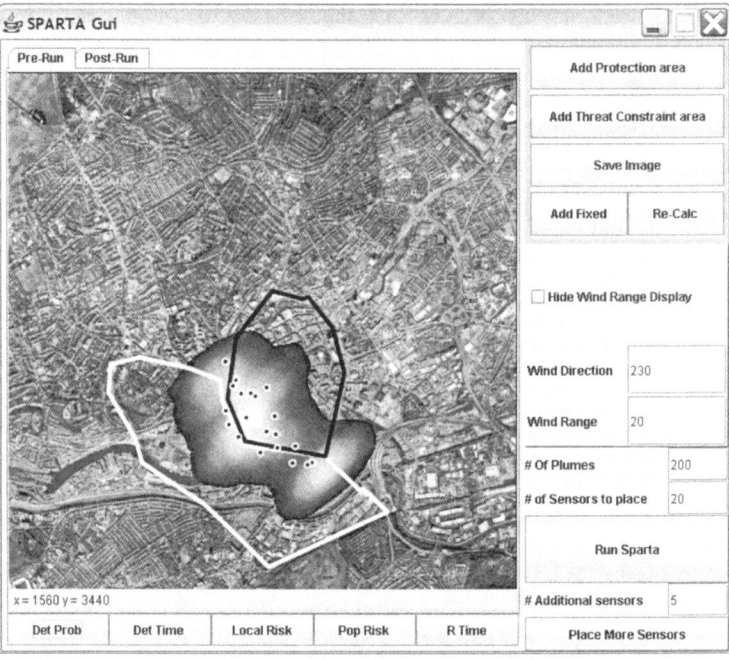

Figure 2   *User display from a sensor placement optimisation capability. The black area shows the area to be protected and the white area is where releases may occur; the shaded grey region is an aggregation of all hazard clouds that hit the protection area; the black dots show the optimal sensor positions for 20 sensors. (Output taken from RiskAware's SPARTA tool.)*

It would require a simulation capability for the water system and models for the water quality sensors. Moreover the simulation multiple run controllers are typically generalised capabilities that could sample across the range of possible scenario parameters and run any model, including a water system model. This means that the CBRN sensor placement systems that have been developed contain components and capabilities that could be reused in a water monitoring sensor placement system, or they could even be modified to provide a full sensor placement capability for water contamination monitoring.

4 INFORMATION FUSION AND SOURCE TERM ESTIMATION

The fusion of CBRN information has been the subject of intensive international research. The aim is to increase situational awareness by fusing together available information to aid decision makers. There are several ways the problem can be broken down and here we consider network fusion and source term estimation, which are two of the main areas.

4.1 Network Fusion

Biological detectors currently have a high false alarm rate. This is because of the difficulty in distinguishing small amounts of biological material against a complex background of non-biological material (e.g. dust, salt, vehicle exhaust fumes etc) and then identifying biological pathogens against other non-dangerous biological species (e.g. pollen, bacteria, etc) present within the background. There are numerous technologies being pursued to identify these differences and a suite of approaches may be used in a single biological detector; however, false alarm rates are still higher than desired and further research and development is being conducted around the world.

The premise of network fusion is that by fusing together output of a network of geographically dispersed sensors, it is possible to better distinguish actual biological agent against the natural background. In this way the networked sensors can perform with a lower false alarm rate than the individual sensors. In the UK the Ministry of Defence is procuring the ISMS to deliver such a networked capability.

Initial work involved development of simple rules for determining whether the network should enter alarm state, e.g. the network will alarm when two detectors alarm. These may be extended to include geospatial relationship, for example by considering how detectors alarm in relation to wind speed and direction. More advanced approaches have involved using signal processing techniques, often taken from financial data modelling, such as generalized autoregressive conditional heteroskedasticity (GARCH).[7] The estimated background noise is removed from each data series (using for example GARCH), the individual signals are then shifted based on spatial and temporal information (the wind blowing over the array of sensors) and then overlaid to identify common peaks, after which thresholding can be applied to the overlaid signals. However, all of these approaches may have a limitation in that the biological background is being transported in the air along with any biological agent and so making distinguishing the two more difficult. It is likely that an approach is required that incorporates some of the physical phenomena of dispersing particles as well as signal processing techniques.

The problem for water contamination modelling may be broadly similar with a number of factors (e.g. turbidity, pH, chlorine, total organic carbon, etc) collectively indicating a water quality issue or contamination event in the same way as a group of biological sensing factors are used for bio-detectors. One individual water monitoring sensor may false alarm but by fusing the response of a network of sensors and identifying correlations, the false alarm rate could be reduced, as with biological aerosol detectors. As the use of financial data analysis techniques has shown some success in biological sensor network fusion, its application to water contamination modelling may prove fruitful. Indeed some of the efforts to tailor these techniques to a network of battlefield biological sensors may be exploitable in the water contamination application. One important point that has emerged from CB sensor network fusion that will be relevant to the water contamination application is the desirability for the fusion algorithms to operate on the sensor measured data and not simply on alarm status. If only alarm status data is available to the fusion algorithms their performance will be seriously compromised and much of their power will be lost. This will require increased data transmission across the network but the network fusion algorithms can potentially be distributed with some pre-processing at the individual sensor to reduce the data volume sent to network hubs for the full fusion algorithms to use.

4.2   Source Term Estimation

Network fusion is only part of the problem: it helps determine whether an event has happened, which can be useful to initiate a response, but it does not determine the source of the incident. In CBRN modelling the source term is needed to predict the future trajectory of the dispersing hazard so that mitigating action (such as wearing personal protective equipment or evacuating a location) may be taken. However, CBRN releases may be clandestine and most plumes of agent are invisible, so the only indication of a release may be the output of CBRN sensors or even the onset of symptoms in people exposed, which for biological releases may be some time after the release event. There is therefore a requirement to fuse the sensor readings to determine the most likely source release that resulted in the observations in a source term estimation tool (see figure 3).

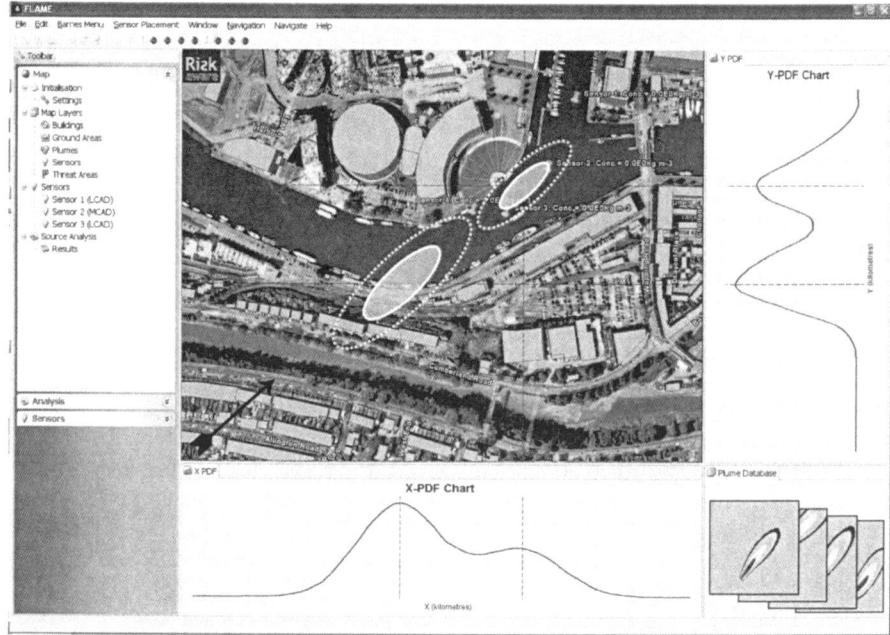

Figure 3   *Example of estimated source terms (light coloured ellipses showing regions of highest probability) overlaid on a map display. (Image taken from prototype user interface for a conceptual source term estimation capability.)*

The situation for water contamination is likely to be broadly similar with the first indication of a problem being water quality monitors triggering or complaints from water users. The priority will be to warn people so they can stop using the contaminated water but also highly important will be to identify the source of the contamination so that it can be dealt with. The close analogy between the two applications means that any methods developed for one is likely to be of great interest to the other.

Source term estimation is an area that has been actively researched by the CBRN modelling community. Three main approaches have been pursued:

4.2.1 Rules Based Methods. These use geometric rules to triangulate the position of potential sources from the detectors alarms by assuming a constant rate of cloud expansion. They also use simple rules that attempt to identify alarms that are consistent with identified sources or else requiring a new triangulation calculation to be carried out to determine an additional source. The method is very fast but is limited both in its robustness (its ability to handle multiple observations and identify consistent events) and realism (it ignores important physical phenomena such as actual dispersion rates and terrain effects, etc). Rules based methods for source term estimation were mainly seen as an interim solution while the more sophisticated approaches were developed (see below) but currently they are closer to being operational capabilities if their limitations are felt to be acceptable. Although this approach has proved fairly successful for CBRN airborne releases, a completely different set of rules would be required for water contamination and it may not be possible to develop such a set that will provide an acceptable solution.

4.2.2 Adjoint methods. The adjoint approach provides a fast method for source term estimation. The idea of adjoints is widespread in mathematics. For atmospheric dispersion models the adjoint method provides an inverse relationship between the model input and the model output.[8] It means an adjoint formulation of the model provides an inverse relationship between the release term and the sensor measurement. The advection-diffusion equation of the adjoint has reversed time, reversed velocity but positive diffusion, so it is similar to running the model in reverse. The method involves mapping a sensor response back to the source by solving the adjoint system, with the calculated adjoint concentration giving the relationship between the source and the sensor output as illustrated in figure 4. For each sensor output reading an adjoint dispersion calculation is carried out and this provides a complete field (in space and negative time) of release possibilities. Additional sensor output readings allow the algorithm to calculate release possibilities that are consistent with the readings and so identify the source location.

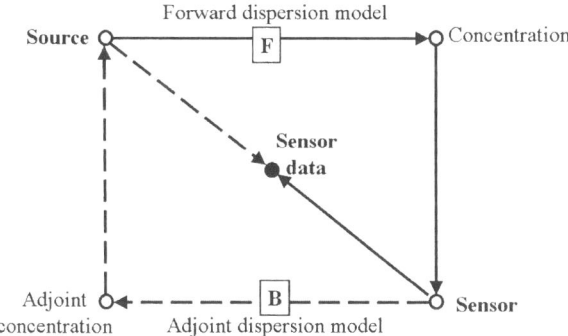

Figure 4 *Illustration of the relationship between the source and sensor, and the forward dispersion model and adjoint dispersion model. (Due to Sykes[8].)*

The adjoint approach can effectively determine the regions where the source (or sources) was most likely released from. It also performs well when the number of sensor readings is not large; however, a new adjoint dispersion calculation is required for each sensor reading which would become very computational demanding with high frequency sensor output. Whether it is suitable for applying to water contamination will depend on if an adjoint model for water and contaminants in flow systems can be produced.

4.2.3   Multiple Simulations with Bayesian Inference.   The third approach involves using Monte Carlo simulation to run large numbers of forward calculations. Initially a set of random hypotheses for the source term (size, mass, location and release material) are generated. Forward calculations are made for each of these. As sensor information is received, it is compared against the model simulations and those that match the observations are selected as the likely source. Bayesian inference is used as this is ideally suited to updating the belief in the likelihood of the various hypotheses based on the observations and these updated hypotheses are used when further information is obtained from the sensors. Care is required to avoid degeneracy so for example at each update the lowest probability hypotheses are removed and new random hypotheses are added as well as resampling of the current hypotheses. With this approach the hypotheses home in on the most likely releases, creating a probability density distribution from which the peaks are usually used as the source terms.

The Bayesian approach does require very large numbers of forward calculations to be carried out as the algorithm progresses. A great deal of effort has gone into improved sampling techniques to reduce the number of simulations and reuse calculated values; however, current approaches still have significant computational requirements. The approach is very versatile and powerful and can incorporate other uncertainties (e.g. background noise, sensor reading error, the weather, etc) and can give true probabilities for events have occurred which can be important for effective risk management.

The overall approach has general applicability and has been applied in numerous domains. The work in CBRN hazard modelling is perhaps most relevant to water contamination modelling owing to the similarities in the two applications. Moreover the inference engines are separate from the hazard modelling and could be modified to use a water system forward model and models for water quality monitors.

4.2.4   Issues and Future Development.   It can be seen that a range of capabilities are being developed although this is still a very active area of research in CBRN source term estimation and further progress is required before an operational capability is obtained. All the current approaches have significant advantages and disadvantages. There are efforts to bring the methods together either as a suite of methods to be applied dependent on the conditions or as a hybridised approach.[9]

An important issue for the Bayesian approach and also the adjoint method is that they greatly benefit from having the detailed sensor output potentially for multiple channels of parameter readings rather than a single alarm status. The techniques can do very little with a single alarm on-off reading as it inevitably provides little information for the algorithms to use. It is therefore important that water monitoring sensors have the facility to output detailed readings and not simply an alarm status.

5 SUMMARY

This paper has provided a very brief overview of two areas of CBRN modelling that have been very actively pursued in recent years: optimal sensor placement and sensor data fusion (including network fusion and source term estimation). Significant progress has been made in both of these areas in CBRN defence, many lessons have been learned and adaptations and extensions to theory and technique have been developed. The work and techniques may be of relevance to water contamination monitoring.

All of the approaches discussed rely on a reliable and validated forward model. For CBRN modelling this is used to simulate very large numbers of possible releases and must execute extremely rapidly and, for practical application, offer automated operation. For water contamination and quality monitoring an equivalent forward model of flow within the water supply system with similar capability would be necessary. Even with the types of sampling reduction techniques developed for CBRN modelling, it would still need to be capable of running tens of thousands of simulations in an hour and include optimisation in memory and data handling. However, this many simulations are likely to be required to obtain reliable and statistically robust results, as it has been for CBRN modelling.

The CBRN modelling applications have also shown the importance of sensors outputting to the data fusion algorithms their actual measured values for parameters and not simply an alarm status (as may be obtained from thresholding the output time series or some similar signal processing technique). Without this more complete information, the performance of the algorithms is significantly degraded. It would be expected that this will also be the case for data fusion within water contamination monitoring to determine whether an event has happened and where the source is. Inevitably this does place an increased burden on the network but techniques are being developed to alleviate this, such as distributing the fusion algorithms.

The results of the CBRN modelling research have shown a great deal of progress and efforts are still continuing. There is also research into other areas such as methods for automatically processing and manipulating input data such as buildings for optimal use by models, handling uncertainty and uncertainty reduction in model predictions, more advanced decision aids and consequence management tools, the presentation of complex risk information, and information management on fragile and limited bandwidth networks. Some of this work may result in techniques that could have relevance to water contamination.

The underlying commonalities between some of the applications of CBRN modelling and water contamination monitoring, which this paper has attempted to emphasise, suggest that collaboration and leverage of effort between the two areas could be highly beneficial to both research communities and result in improved capabilities for end users.

References

1 "Joint Effects Model" fact sheet, DOD JPEO. Available at http://www.jpeocbd.osd.mil/documents/IS-Joint_Effects_Model_(JEM)_(L).pdf.

2 "Joint Warning and Reporting Network" fact sheet, DOD JPEO. Available at http://www.jpeocbd.osd.mil/documents/IS-Joint_Warning_and_Reportin_Network(JWARN)_(L).pdf.

3 "Joint Operational Effects Federation" fact sheet, DOD JPEO. Available at http://www.jpeocbd.osd.mil/documents/IS-Joint_Operational_Effects_Federation_(JOEF)_(L).pdf.

4 Defence Industrial Strategy, UK Ministry of Defence, 2005, p 115. Available at http://www.official-documents.gov.uk/document/cm66/6697/6697.pdf.

5 V. Rapley and P. Robins, "Source Term Estimation and Sensor Placement", presented at *2007 Chemical Biological Information Systems (CBIS) Conference and Exhibition*, 2007.

6 M.D. Bull and R. Gordon, "Sensor Placement Algorithm for Rapid Theatre Assessment (SPARTA)", presented at *2007 Chemical Biological Information Systems (CBIS) Conference and Exhibition*, 2007.

7   T. Bollerslev, "Generalized Autoregressive Conditional Heteroskedasticity", *Journal of Econometrics*, 1986, **31**, 307-327.

8   R. Fry, R. I. Sykes and R. Kolbe, "Chemical/Biological Source Characterisation", presented at *Science and Technology for Chem-Bio Information Systems*, 2005.

9   A. Keats, F.-S. Lien and E. Yee, 2006: "Source determination in built-up environments through Bayesian inference with validation using the MUST array and Joint Urban 2003 tracer experiments", *Proceedings of the 14th Annual Conference of the Computational Fluid Dynamics Society of Canada*, 2006, **14**, 8.

PLANNING, PREPAREDNESS AND SECURITY OF THE ALTERNATIVE WATER SUPPLY

K. Silcock

Managing Director, WATER DIRECT, Earls Colne Business Park, Colchester, CO6 2NS

1 INTRODUCTION

Water Direct has since 1996 been specialising in the provision of quality assured drinking water in various forms, for when the normal piped supply is either unavailable or unusable. During that time we have developed bench marks for due diligence in terms of best practice and quality assurance of the alternative supply, whether it be in a bottle, a bowser, or a bulk road tanker.

Alternative Supply is all about moving water from where it is available to where it is needed, in whatever form is appropriate and without affecting its quality. We do nothing else; and we have undertaken this on behalf of our clients quietly and effectively for over a decade.

But then the floods came! And now Alternative Water Supply is a hot subject. The flags are flying high and the spotlight has been switched on, with a Cabinet Office Review, a Water UK Review, reports from DEFRA, DWI, OFWAT and numerous conferences on flood prevention, dealing with floods and how to ensure people have safe clean drinking water should the worst happen again. Not to mention the plethora of opportunists that now think they can make their fortunes selling water to Water Companies!

My first message therefore is BEWARE; of over reaction.

Water supply losses and/or contaminations happen somewhere almost every day; and it is always a damned inconvenience. Unless you supply it through a pipe, water is heavy and awkward to transport, especially quickly and whilst maintaining its inherent quality. But it is drinking water and, in the interests of safeguarding public health, you cannot take short cuts.

We were well positioned to immediately provide Severn Trent Water with more than 3 million litres of bottled water, 120 static tanks deployed onto the streets and our dedicated drinking water tankers to fill and sustain them for 9 days. We do not have a crystal ball, but we were ready and we were prepared – as we always are.

My objective now is not to preach on how best to provide an alternative supply, although I could, but to raise a few questions so that you will think about what is best for you in ensuring due diligence.

My subjects for you therefore are:
1. What should you plan for?
2. How prepared must you be?
3. Why rehearse?
4. How should it be done?
5. Security – what does it mean?

Which all forms part of;
6. Due diligence

Deploying the alternative supply needs to be done properly. Planning, preparedness and practice precede a perfect performance!

Or should!

2 WHAT SHOULD YOU PLAN FOR?

Flooding? Recommendation No.1 to Water Utilities from the Water UK Review Group on Flooding was to *"...put in hand a thorough review of their emergency response and contingency plans on the assumption that the sheer scale of future floods may make current plans inadequate"*

But 2 years ago it was **drought**! In April 2006 we spent considerable time looking at the feasibility of transporting huge volumes of clean water into London – and then in May it rained so hard that all the planning for drought just evaporated.

So what about **contamination**? The threat from terrorism is ever present these days; but accidental contamination can happen at any time.

And that brings in the question of simultaneous events. There is currently a lot of pressure to plan properly for another major event; but, what about normal **system failure**? Which believe it or not, can and does happen somewhere every day.

How many people should you plan for? 350,000? Surely that is when Mutual Aid kicks in. But can even that cope properly with an event of this size or with more than one 'major' event simultaneously?

The Pitt Review calls for a – *"...central stockpile of equipment"*. Recommendation No 13 of the Water UK Review Group on Flooding was *"...to review the state of preparedness of the industry for future events; in particular the mutual aid scheme"*, including the *"...compatibility and readiness of assets, staffing and catering for simultaneous events"*.

But this is nothing new. Water Direct initiated an Alternative Water Container Bank of prepared and quality assured assets 6 years ago, which already serves as a 'central stockpile of equipment'. And it can already cope with simultaneous events.

At the time of Mythe, Severn Trent was not one of Water Direct's contracted clients. We had just finished deploying bottled water to doorstep for 2,000 properties for Sutton & East Surrey Water when they called us.

But on a best endeavours basis we sustained for 9 days the equivalent of Severn Trent Water's Local Response Plan; as well as serving 2 hospitals, 3 care homes and a number of commercial needs in the area, including supermarkets so that they could maintain provisions to the people affected by flooding.

Simultaneously we were put on standby for other potential flood related events with 3 more Utilities. We sustained this support without compromising any of our contracted client's response capabilities.

This alternative water bank is not only for Water Utilities. Our members also include the MOD, Local Authorities, Highways Agencies, Emergency Services and many commercial organisations that have recognised the risks to their business of an unplanned water loss.

Our own regional based systems have already been replicated to provide a Nationwide Bank of compatible and quality assured alternative supply assets.

The larger the Bank, the more resilient and cost effective it becomes and the more 'local' branches we can open.

So much for the big stuff! But what about your obligations under legislation, such as your Local Response Plan, for a population of up to 50,000? How about DEFRA's current expectation under the Security & Emergency Measures Direction of 10 litres per person per day?

This 10 litres per person per day is already a **minimum** volume in the SEMD. But what if the minimum is raised to 20 litres per person per day? How will you cope? You are already thinking about that, I'm sure!

But what about frequent day to day events? Systems fail or become contaminated of their own accord all the time, affecting relatively few people. **And they are all just as important!**

So what is manageable and realistic? You are more likely to face an event of 500 or 5,000 than 350,000, or even the 50,000 that you have planned for. The key is to have a system that works. Use the sector approach. Perfect your systems small scale, by sector and then train your teams and practice. This way you can then replicate the system, if and when things get bigger.

The question then is what should you use to provide the alternative supply? The industry trend, post Mythe, is leaning toward reliance on using only bottled water in the first day. But why? Water UK clause 3.2.6 – *"In the short term, until bowsers and tankers can be deployed and kept well supplied, bottled water is needed..."* Considering the present level of preparedness of many Utility's, this could be to provide time to get bowsers and tankers ready for the second day.

But is that the right decision, or just an easy or inexpensive option? Whilst this solution is eminently sensible in terms of speed of response, and could reduce standby costs, it should not be used as an excuse for holding unprepared assets! And it relies very heavily on adequate volumes of quality assured bottled water being **immediately available**. But where will you obtain sufficient quality assured bottled water that quickly?

Once you have decided what you will deploy, where will you deploy it to? How many vulnerable customers do you have and where are they? Is the list current? Does it include all vulnerable sectors currently required? How will you get the water to them? Will you deliver to doorsteps or to collection points?

Who else will you deliver alternative water to? Schools, Hospitals, Care Homes, Prisons?

Do you have the resources in place to deliver to them, or will everyone be expected to collect it from a drop off point? Is doorstep delivery **practical**? It would certainly ensure that each customer gets their 10 litres per day. Is collection by each customer from a central point **acceptable** and will everyone get their share? How much will be lost to pilfering or wasted through vandalism?

Your decision may depend on what form of alternative supply you are delivering and whether your customers are able to collect it for themselves.

So if to a collection point, where exactly will you place the pallet, bowser or static tank? There are IT programmes available to draw 200 metre circles on a map, and the blue dot in the middle is where you place your pallet of bottled water, or your tank, to provide 10 litres per person per day with no further than 100 metres to walk to collect it. But do they go far enough? Do you have sufficient data for planning?

How many people will one pallet of bottled water serve? The size of bowser or tank you use will affect the deployment density. Have you considered vertical population such as tower blocks?

What if the blue dot is on a roundabout, or in someone's garden, or a cemetery? (That tank will stay full at least!) Is there a convenient roadside verge, or lay-by? Do you know the topography of the area and whether it is even possible to manoeuvre your delivery vehicle and tanker down the roads, especially when cars are parked there?

Have you pre-agreed with land owners that you can use their premises or car parks as a rendezvous or collection point? Or do you work on the basis of calling upon them if and when you need to because it will probably never happen?

Most importantly, who will take responsibility for the final location? The water utility, the sub-contractor delivering for you, or the driver who makes the decision to drop it at the roadside?

What if there is an accident, or at worst, heaven forbid, someone gets killed because of where the asset was actually dropped? The new corporate manslaughter law came into force on 6th April!

Have you been diligent in your planning process, or was it restricted because **you had no budget?** It is reported that the Mythe Incident cost Severn Trent Water and its Insurers upwards of £30 million.

3   HOW PREPARED MUST YOU BE?

The Water UK Review Group on Flooding, Clause 3.2.3, stated: *"In some cases emergency equipment provided through mutual aid could not be confirmed as clean and ready to deploy directly".*

Recommendation No. 6 of that review to Water Utilities was: *"Water Companies and any other organisations with a responsibility to provide equipment under Mutual Aid should ensure that this equipment is kept in a roadworthy and clean condition at all times to ensure that response times to emergency events are kept to a minimum".*

Recommendation No. 9 was: *"Water Companies should review the efficacy of their emergency supply assets...(eg, bottled water supply chain, numbers and locations of bowsers and tankers).*

This part is spot on. But then it went on to say: *"Dialogue with supermarkets and other bulk providers of bottled water should also ensue to determine how best to ensure adequate supplies during emergency events"*

To rely on Supermarkets to fulfil your obligations is, quite frankly, ludicrous. This is not a criticism of the Water UK Review Group. Their objective is to make utilities review the resilience of their supply chains. That is absolutely critical. But how can you show due diligence if you are not in control of your own supply chain? I will talk more on this issue later.

So, how fast do you need to respond? Your level of preparedness must ensure that you meet the requirements of the Security and Emergency Measures Direction serving the populations stipulated in your Local Response Plan within 24 hours and 72 hours. DEFRA will have audited your plans against this; but have they ever been put into practice and tested?

In day to day deployments Water Direct is often contracted to have alternative supplies deployed to the customer within 4 hours; and in some cases less than 2 hours. That takes detailed planning, substantial preparation and practise. But what do you need to prepare and have available?

3.1 Is bottled water available?

How much do you need? How much have you bought and put into storage so that it's ready to deploy? What if you don't use any before it goes out of date? What is the shelf life and has that been validated? Or do you rely on a 'just in time' call off from a stock that **may not be there when you need it**?

If your bottled water is stored, is it quality tested regularly? If so, what is it tested for and how often? What happens if a sample fails, are there quarantine procedures in place? Will the testing regime show due diligence for conformity with the Water Supply (Water Quality) Regulations 2000, as amended, at the time and place of deployment to your customer? It has to conform with those regulations and it is the undertaker's responsibility to prove conformity! Will a Supermarket provide that evidence for you? Who will be responsible if things go wrong?

How is the water labelled and packed? Is it suitable for all users, ie, baby's formulae, and what about immunocompromised customers? Does the labelling carry appropriate information?

Is the bottled water you intend to use available and accessible 24/365? Is it within your territory, or will it be coming from the mountains of Scotland, or Wales, or Italy, because it was cheaper? What is the lead time? 4hrs? 12hrs? 24hrs? How quick will you need it?

Do you have appropriate resources and logistics ready to load, deploy and, most importantly, unload and distribute at the destination?

3.2 Are your Bowsers and Tanks ready?

A bowser or tank that is not clean, disinfected and sampled is not ready to deploy. The Water UK Review Group clause 3.2.3 says: *"It takes 24-48 hours to fully clean, drain and sample bowsers before use.* [Assuming lids and taps are intact and don't leak]. *Having to do this at the outset of an event will increase response times in deploying the equipment and hence in getting alternative water supplies to affected communities."*

And if they are stored on soft or rough ground it will take longer and you will also need an all terrain telehandler to access them. Do you have one? Are *your* bowsers and tanks clean and disinfected, ready to deploy immediately?

We were able to respond to Severn Trent last July, as we do to all our customers, **immediately** because all our vessels and tankers are dedicated for drinking water and are kept in a ready to deploy condition.

Are the 'bags in box' vessels a good alternative? Have you tested them in anger? Can they be safely deployed full? Do you have enough bags for refilling? How will they be deployed for the second and third fills? How 'resource efficient' are they?

How will you deliver your vessels? Having clean prepared tanks, or even filled bags in boxes, is no good unless you have adequate and appropriate resources available to deploy them – such as self offload vehicles and certificated driver/operators. In a large event it is unlikely that you will have enough of your own resources and you will therefore have to ask for help.

So do you have robust outsourced contracts with assured 24/365 availability? Have you tested the outsourced supplier's response, without prior warning? Do they actually have the stock, assets, trucks, drivers and rapid response capability that they **say** they have? And will it be available to *you* when *you* need it?

3.3 How will you then fill and replenish your vessels?

Section 3.4.2 of the Water UK Review Group on flooding stated *"It also quickly became clear that there is a shortage of appropriate tankers, particularly mid-sized tankers, available to the industry"*.

What is 'appropriate'? A tanker that is fit for purpose and immediately ready to carry wholesome water, one would presume! 'Mid-sized' refers to tankers that are small enough to go down residential streets to fill deployed vessels.

Water Direct's response to the Review is quoted in this same clause: *"Dedicated drinking water tankers must be available for immediate despatch, again, clean, disinfected and sampled for assuring quality at the time of deployment and use"*

Why must tankers be dedicated for drinking water? Quite simply in the interests of safeguarding public health; and because everything else you do, and everything else you use, **is**! Why should the tankers be any different? You wouldn't deliver milk to your customers through your water pipes in the mornings and then water later that day, would you? The risk of contamination from a tanker (or its hoses) that is not dedicated for drinking water can be high. The lead time to prepare, clean, disinfect and then await sample results of 'food grade' tankers is completely inappropriate for emergency response.

How many tankers do you have, or can you access, that are appropriate and ready? Are they the right size for filling tanks in residential streets? Are they dedicated for drinking water or were they hauling milk, or cooking oil this morning?

Are their hoses compatible with your equipment? Do those hoses comply with Regulation 31 and are they clean, disinfected and sealed before use? If on a 'food grade' tanker I suspect they won't be!

Wherever you obtain your tankers, who will operate them for you? Normal HGV drivers are not sufficiently trained for this. **This is drinking water**! You will need trained technicians certified under the EUSR National Water Hygiene Scheme for working on restricted operations. Do general logistics companies or "Driver Hire" have any of those? All of our Field and Office staff is multi-skill trained and certificated – even me! Now this leads me to ask;

3.4 What about quality assurance?

Clause 3.5 of the Water UK Review Group on flooding mentioned that *"...various issues of conformity with drinking water regulations arise"* and *"There should be a clear statement of the role of DWI and the water companies in emergencies."*

Well the Drinking Water Inspectorate's position has always been very clear and this was reiterated at this conference yesterday. The quality of water provided by a water utility to its customers as an **alternative to the piped supply**, whether in a bottle, container or tanker, must be wholesome as defined within the Water Supply (Water Quality) Regulations 2000, as amended 2007.

To quote the speaker, I believe correctly, **"You can't give them just anything"**!

Under the amendment to the Water Act (2003), DWI has the powers to take forward a prosecution of the water undertaker, or any relevant person, if they did not take all reasonable steps and exercise all due diligence in securing the supply of wholesome water. So, potentially, everyone in that supply chain is liable and could be culpable.

You are not exercising due diligence unless you can validate wholesomeness and therefore conformity with Water Supply (Water Quality) Regulations and if you take shortcuts or make assumptions you are taking an avoidable risk.

To consider that, actions taken in good faith during an extraordinary event, such as Mythe, can now be assumed to be acceptable practice, is a very unwise and possibly dangerous assumption. It is complacency bordering on recklessness.

For the safeguarding of Public Health the quality standards set within the industry, for good purpose, cannot and must not be compromised.

So you have perfect plans that have been audited by DEFRA. Fantastic – but how far does that audit currently go and do your plans work in practice? Recommendation No. 15 of Water UK's Review states: *"Water Companies should rehearse emergency plans on a regular basis. This should include physically moving equipment..."*

We have been involved in [only] 4 physical exercises with contracted clients within the last 6 months; two for street deployments and two for Hospitals. Have you rehearsed your plans?

4   WHY REHEARSE?

Because you have a responsibility to meet the requirements of the Security and Emergency Measures Direction, whatever the scale of the event. In any plan the theory is always great, but does it work in practice? The issues that arise in rehearsal will help you iron out otherwise unforeseen problems that may just prevent you from meeting your responsibilities in the real event. For instance, how much do you rely on outsourcing? Have you tested the contractor's response?

Can your suppliers and sub-contractors **actually do** what they said they could do when you awarded the contract? Is providing an alternative drinking water supply service their core business, or are they just haulage companies? Either way, try calling them at 4.30 on a Friday afternoon, or 3 o'clock on a Sunday morning, because that's when most events occur.

Have they preloaded their vehicles with washing machines for Monday's scheduled run? Has the milk or cooking oil been properly cleaned out of their tankers? How long will it take to disinfect and sample those tankers before they can be used? Are those tankers of an appropriate size for your needs? Do they have any uncommitted vehicles and drivers [whether trained or not] that are awake, sober and with available driving hours?

If you don't rehearse there is a very good chance that it won't be 'alright on the night'!

5   SO HOW SHOULD IT BE DONE?

Water Direct does nothing other than provide, manage and deliver alternative water supplies that are wholesome and conform with the Water Supply (Water Quality Regulations) 2000, as amended. As this service has evolved many aspects of it have become best practice and all processes and procedures are fully auditable for due diligence. The following précis is intended to help you with assessing your own processes.

5.1 Bottled Water

What an odd commodity. You can now buy a bottle of a mountain stream from Fiji, or a melting Icelandic glacier. The current annual UK consumption of all bottled waters is 2.2 BILLION litres.

Mythe took 1% of the annual UK consumption of bottled water in one week and 2% overall. Which was fortunate for the Bottled Water Industry as the lousy summer weather had affected their normal sales; which in turn was ironic as in a normal summer there would have been little, if any, for Severn Trent to buy! It's only by good fortune that 350,000 people were not without water during hot weather or because of a drought!

So would that method of procurement be reliable in planning terms, and what quality assurance was provided?

You could assume that because the bottled water you procured for your emergency is commercially available then it is OK to give to your customers. But the commercial market gives customers the choice of whether to buy or not, and a choice of which brand or type to buy, or not. When you provide that water to your customers they have no choice.

Furthermore, and more importantly, you take on responsibility for the quality of that water when you provide it to your customers as an alternative to the piped supply. So have you taken all reasonable steps and exercised due diligence in ensuring that water is wholesome, as defined by the Water Supply (Water Quality) Regulations 2000, as amended 2007? Do you know what information is required to validate conformity?

Is bottle size important? Of course it is. A vulnerable customer will not thank you for delivering a 5, 10 or 19 litre [cooler] bottle to their doorstep. They won't be able to lift it. Small bottles are convenient and light, but cost per volume is high. In consultation with our Industry clients we settled on 2 litre bottles in 6 packs 12 years ago; and they work very well.

Should you procure Natural Mineral Water, Spring Water, Table Water or "Purified" Water? The topic of whether any of these classifications is better than the others is too protracted for this forum. In case you are wondering, tap water in a plastic bottle has a comparatively short shelf life. We have tested it.

We use a number of mineral and spring water sources that have been audited and approved. The more **natural** the water, with minimal processing from source to bottle, the better the shelf life and long term aesthetic quality. That is what is important for **an emergency buffer stock.**

But where do you get bottled water from?

5.1.1 Bottled water producers? To procure bottled water direct from a producer in an emergency is a risk not worth taking. A production release criterion is normally 3 days to ensure all quality tests taken at time of production are clear and reported. To uplift in advance of that release period could result in a recall at best if samples fail, depending of course on what the failure is.

If that unproven or even contaminated water has already been given to customers the fall out does not bear thinking about – but this precise situation happened last year when Anglian Water took bottled water straight from Hadham Water's production in a hurry. The source had became contaminated with E.coli and this had not been detected – until after the bottles had been distributed to the customers whose piped supply was not usable. A fine example of how diligent you must be with the alternative supply!

That situation could and should have been prevented with proper procedures and processes, using adequate and available quality monitored buffer stocks. Ironically, that very management service that we currently provide to other Utilities was developed in conjunction with Anglian Water 10 years ago. Their scientific people were stringent on the processes and procedures that must be adopted for ensuring due diligence in the provision of bottled water to their customers in an emergency. And all credit to them. Hadham Water was actually the first source we used for them and the only one initially, until they had a source contamination that caused 6 lorry loads of stock to be contaminated with coliforms. This contamination did not show up in testing at the time of production but through our regular monitoring and management of stock this sporadic problem was detected well before any of that water was deployed.

The subsequent production plant audit undertaken by us, with Anglian Water scientists, at that time identified that there was an intermittent risk of contamination to the source from 'farm yard' surface run off during periods of heavy rain!

We provided our managed service to Anglian Water for 6 years, using multi-source provision to ensure the Hadham scenario did not recur. But then the 'procurement professionals' decided that they wouldn't pay for such a service, because it was cheaper to just buy bottles of water direct from a single producer when they needed it. The water quality people involved at that time appeared to ignore their obligations under the Water Supply Regulations, stating that testing at time of production would be all that was required [under the bottled water regulations]. But even that seems to have been abandoned on this occasion! Where was the due diligence then and did anyone take all reasonable steps to ensure the water they supplied to their customers was wholesome? Who took responsibility for that procurement decision and who will be culpable for the outcome?

This also emphasises that to use a single source of production is an unnecessary and serious risk. To our knowledge there is not a single production source in the UK that, despite what they may tell you, has never had a quality or production problem that will affect the critical resilience of your supply chain. Even if they haven't, yet, they could at any time. We use **multiple approved sources** for our buffer stock in a rotation.

So, where else could *you* get bottled water from?

5.1.2   Supermarkets? In a water loss event the bottled water on supermarket shelves will disappear faster than you can look up their phone number in your Emergency Manual. To quote Severn Trent from clause 3.2.6 of the Water UK Review Group report…
"…the loss of mains supplies is likely to lead to panic buying of bottled water and hoarding…" And… *"Bottled water flew off the shelves…"*
Supermarkets procure on a 'just in time' basis from producers who earn their living from the supermarkets, not from infrequent and entirely unpredictable water company emergencies. Neither will be in a position, or necessarily willing, to commit any serious quantity of product to a water utility that may only need it on rare occasions. Their own core customers will and do take priority. Why should Supermarkets help you out? Water is *your* business, not theirs. Whilst Mythe was an exception, it may be only a once in 100 or 1000 year event.

5.1.3   Stockists? Are there any bottled water 'stockists'? Who *would* stock bottled water? Supermarkets don't. Producers stock only what is required to supply their regular customers' needs and that depends upon the weather. They will produce stock in advance of expected warm weather, the sale of which will follow the daily temperature variation almost exactly. That is why Severn Trent was so fortunate last year. But it is not a reliable supply chain.

Water Direct holds considerable buffer stocks of quality assured bottled water for our utility and non-utility clients, which forms our Nationwide Bottled Water Bank. The protocols of this Bank arrangement allow each depositor to draw on overdrafts above their own account volume for major events. There is sufficient stock immediately available for simultaneous events.

The stock is especially packaged for its purpose and is labelled as "Courtesy Water for drinking and cooking". It is therefore less prone to pilfering and can be shared across borders with no 'corporate' conflict.

This facility is already available nationwide as a resilient source of appropriately packaged wholesome water in bottles that meets the requirements for due diligence in every respect.

Is the bottled water you procure wholesome **at point of delivery;** ie, when the cap is removed by the customer? Even if it was available from Supermarkets or Producers, what quality assurance could they provide? The answer is none. Producers test only at time of production and this is not sufficient for ensuring wholesomeness at the point and time of delivery to customers, as we have demonstrated many times. There is no requirement under the bottled water regulations for any further testing. So Supermarkets do not test it. It's not within their stewardship for long enough and they do not have to comply with any regulations in this regard. The Natural Mineral Water Association has stated that its members cannot guarantee quality after a period in storage.

Our bottled water is tested regularly in store to monitor quality parameters and ensure that wholesomeness is assured at the point and time of delivery to customers. That's what the Water Supply Regulations require. We have a decade of data to support our validation.

We have been inspected by DEFRA auditors under the SEMD. Our facilities and processes are open to inspection by DWI.

5.2 *Courtesy* **Water for drinking and cooking**

Our Courtesy labelled bottle has been developed in conjunction with Water Industry colleagues and clients. It is generically labelled to enable 'cross border' stock sharing between many utilities in the spirit of the 'central stock pile' and carries an agreed customer information statement. Packaging is robust for multiple handling and weatherproofing and the extended best before date is validated from extensive sampling.

Bottled water is best delivered to doorstep. From a decade of experience this is the only way to ensure that the right customers receive it with minimal losses or wastage. 2 packs per property equates to 10 litres per person, based on the current national average occupation of 2.4 people per property.

To leave it on a street corner for collection will lead to theft, hoarding and failure to actually deliver the required volume to each customer.

Doorstep delivery is achievable with planning and preparedness; and for less than 500 properties it is far cheaper than using bowsers or tanks. We serve this size event frequently.

We have also serviced larger events with bottled water only, delivered to doorstep. For example:

In 2002 – approximately 6,000 properties, 2 deliveries per day for 4 days with ample reserves to sustain for longer

In 2007 – approximately 2,000 properties, with daily deliveries over the same weekend that the Mythe event occurred

5.2.1   Buffer stocks. We hold stocks in numerous warehouses and various geographic locations across the country. It is real, it exists and it is immediately available. It is quality tested by batch every month to ensure conformity with Water Supply

Regulations at point and time of delivery to customers. The logistics capability to deliver and offload it when and where you need it is also available.

The whole supply chain is managed by us and batches are used in date order to alleviate wastage; although they have a verifiable long shelf life. We are in the process of validating an increase from 24 to **36** months, based on extensive microbiological, chemical and aesthetic sampling of archived product from all sources.

However, it is not just an emergency stockpile. Regular use for day to day events helps refresh the stock and **reduces costs** by eliminating waste through ageing.

An added advantage of membership is the often 'free' renewal of stocks. One member of our bottled water consortium did not use a single bottle of their own stock for 4 years. We renewed it for them twice over by managing the overall stock effectively to ensure none of the batches in the Bank went out of date.

This saved that particular member more than £20,000 and emphasises the value of our service. Last July they had an incident which took their entire stock in one weekend.

Excess stock can be returned to the bank from a deployment, subject of course to quality verification to ensure it has not deteriorated whilst out of the stores. Of the 50 million litres of bottled water procured by Severn Trent for Mythe, a vast quantity was left over after the piped supply was fully restored. Disposal of that excess proved rather challenging as the producers and supermarkets who sold it to them would not take it back!

5.3 Bowsers or Static Tanks

Bowsers take up far more storage space than an equivalent volume of static tanks and require appropriate vehicles to tow them full. They also need to be roadworthy and fitted with wheel clamps when deployed. Not necessarily for theft prevention but for Health & Safety! Vandals will let the handbrakes off to push them down the hill! Do you have wheel clamps for every bowser? Are they stored with the bowsers?

Whether you have bowsers or static tanks how many do you need? This is a simple function of the volume you need to deploy divided by the vessel volume; taking account of how many times you plan to refill them each day; although in practice that may be determined by how quickly the customers empty them.

Have you got enough? Are they prepared and ready to use? Are they accessible? Being ready for deployment or not is the same whether a bowser or static tank, although bowsers must be roadworthy too. If the condition of the vessel is such that **you** wouldn't drink water from it then you should not deploy it.

How quickly do you need to deploy? If a large deployment is needed in a hurry then bowsers are impractical. Loading onto flatbeds takes longer than static tanks and volume for volume, you would get at least 5 times more with static tanks than bowsers on the same vehicle.

If you need to deploy lots of vessels quickly then bowsers should be left in the yard. They are OK for specific uses, such as farms, but for large scale fast deployment to street, forget them. They will most likely be stolen anyway

Where will the vessels be placed? I mentioned deployment planning software earlier, but unless you know precisely where they are going to be placed then you risk missing your deployment commitments.

Once mobilised, the logistics will need to run smoothly. Do you have detailed and accurate information available for the deployment teams? Local knowledge is worth a great deal. Your contractors may not know the quickest ways around the flooded roads, or even speak English! Do you have personnel with local knowledge available to assist them?

Once a vessel is placed, log its actual location. [This can be done automatically with the IT systems we are developing with IMASS]. Apart from refilling and uplift, you will need that for updating your plans.

5.3.1   Deployment of vessels. As an example, on a small 7.5 tonne flatbed truck it is possible to carry only two 1,100 litre road bowsers. The same vehicle can safely carry eight 1,800 litre 'CD' tanks or sixteen 1,100 litre Western cube Aquastax. If most of your assets are bowsers, will you now plan to replace them with static tanks?

Our articulated wag 'n' drag vehicle can deploy between 34,000 and 500,000 litres of vessel volume in one run – depending on the type of vessel of course! This is based on either CD, Aquastax, vertical cylindrical tanks or pillow tanks. We use it for bulk transportation into a rendezvous point for cross loading, but without it's trailer it can still access most residential areas for direct deployment.

5.3.2   Prepared vessels in storage. Vessels held in our [alternative water container] Bank are uniquely identified, clean, prepared, sampled, sealed, accessible and available – with a full audit trail for each individual vessel. This audit trail includes a record of every deployment it has been sent on, the tanker used to fill it, the hydrant fill point used to fill it from, all maintenance and repair activities, chlorination dates and sample results.

There are a number of different tanks currently used within the industry. So which tanks are best?

5.3.3   The Aquastax 'bigfoot'. These have a capacity of 1100 litres, which is a very small volume for the footprint it takes up. Its design enables stacking well in the yard, reducing storage costs, but it is difficult to load onto transport with a fork lift. Fork slots are from one direction and this way round the width prevents side by side formation on a truck. There is no stability to lift them from the end without fork slots so they need to be lifted by the crane, slowing the deployment time down.

5.3.4   Titan tanks. There are lots of these within the industry in 1100 or 2500 litre capacity, but they are difficult to clean because of the awkward shaped moulding. They are also difficult and time consuming to handle safely as they have no fork slots or lifting eyes and have to be strapped for lifting with a crane. They have separate feet which can go missing or get damaged and need to be strapped together with the tanks or carried separately.

We were once called out by a major Utility to provide our drinking water tankers to fill their Titan tanks, which were being deployed by their contracted logistics provider. We were able to witness the deployment method as our tankers arrived at the incident an hour and a half before the tanks did! They arrived on flatbeds [without cranes] and in large vans and were literally 'kicked' off/out of the back as they drove along the road. Needless to say, when we then pumped our water into the tanks they leaked!

Presumably 'Procurement' made the decision to award the logistics contract to a company who had experience, capability and adequate resources, and displayed a duty of care? Or were they just the cheapest? There is always a reason why things are cheap.

What did that one event **cost** the Utility in terms of inadequate service to their customers; failure to provide required volumes and the replacement of the split and damaged tanks, not to mention the cost of our tankers for an 'aborted' delivery?

5.3.5   The bag in box. These are now available in 1,000 litre capacity and purport to negate the requirement for pre-disinfection – assuming the bags comply with regulation 31 and have a certificate of conformity to prove they do not need disinfecting when you buy them. What is the shelf life of the bags if not used? Can that life be validated?

The bag-inbox has a purpose, but in a large incident is resource intensive to erect and deploy. Have you tested their efficacy? Like the Titan tank it does not have lifting eyes so must be lifted with straps, which inhibits the quantity per truck when transporting full. It is far better suited to small events. The Water Direct Aqube version has a compliant refillable bag.

5.3.6   The Aquastax cube tank. This is becoming the most popular new tank currently available. We collaborated in the design of it. It is easy to clean, store, handle and deploy safely, with four way fork slots and lifting eyes. They can be efficiently stacked for economic storage and are resource efficient in deployment, although may be unstable if carried full. Have you tested it?

5.3.7   Civil Defence (CD) tanks. Like the 'Titan' there are still many of the original CD style tanks held within the industry. But there is also a redesign of that tank that we commissioned 2 years ago. It has improved features and security. It is easy to clean, store, handle and deploy safely with an option of one or two piece stands. It is resource efficient with a high volume to footprint ratio.

5.3.8   Larger vessels. Water Direct also stocks larger vessels in its alternative water Bank. These are not generally used by utilities for residential customers but are ideal for larger users such as hospitals and commercial organisations. We have 5,000 and 10,000 litre vertical static tanks, which are resource efficient in terms of volume per footprint and Pillow tanks of up to 50,000 litres, which fold into a pallet box for storage and delivery.

All of these are ideal for creating a buffer between the delivery tanker and the building system where a surrogate pressurised water supply can be connected to the facility. We used this approach for hospitals and care homes, as well as commercial users, in Gloucestershire. The ideal scenario is to pre-plan the connections by undertaking a risk assessment under our Contingency Planning service. This enables fast connection from our alternative supply to internal systems, thus achieving a virtually seamless changeover.

5.4 Water tankers

Have you got any? Do you own your own water tankers? Are they available under a contract or just a best endeavours response? Are they an appropriate size and dedicated for drinking water only? Will they be available for you when you call?

If the tankers you use are 'food grade', are the hoses Regulation 31 approved, suitable for drinking water, clean and disinfected, and the right size for your equipment?

Are the tankers clean, disinfected, sampled and ready to deploy immediately? That may depend on how you procure your tankers and where from. Can you access them 24/365 or do you rely on a just in time best endeavours response from a logistics company? Have you tested their response? Has the milk or cooking oil been properly cleaned out of them? How long would that take? More importantly, would **you** drink the water from them? Is it **only** water in them?

Are the tankers an appropriate size and how is the water discharged from them?
All our tankers have on-board pumping facilities as well as multi-outlet sparge pipes and gravity outlets, all compatible with conventional water utility equipment.

In terms of size, we have:

5.4.1   2,000 litre road bowsers. With 4x4's that will go anywhere. They are ideal for small deliveries, congested areas, farms and difficult terrain.

5.4.2   10,000 litre purpose built dedicated drinking water tankers. They are small enough to go into any residential area.

5.4.3   15,000 litre purpose built dedicated drinking water tankers. They are still small enough to go into any residential area but with more discharges per load

5.4.4   Articulated dedicated drinking water tankers up to 26,000 litres. These are used for a multitude of purposes transporting bulk volumes, such as reservoir transfers and as mother ships for replenishing smaller tankers on an incident.

If there is no clean water work for any of these tankers **they do not go out** of the yard.

6   SECURITY – WHAT DOES IT MEAN?

Part of your planning is assessing risk. What are the risks in providing an effective and adequate alternative water supply in an emergency that remains wholesome and conforming with the Water Supply (Water Quality) Regulations 2000, as amended, throughout the whole process?

Does security mean protection against loss, theft, or vandalism? You will, of course, take all necessary precautions. Does it mean securing against terrorist attack? This is a real threat to all of us which other speakers have discussed.

It was once put to me at a conference like this that a terrorist bent on causing *actual* harm would *say* that a reservoir had been contaminated, but they would have actually contaminated the alternative supply! How secure are the alternative supply assets?

But what about the security of your supply chain and the threats to your resilience? There are two main categories of risk here.

6.1   Operational risks.

This re-emphasises rehearsal of your plans.
- Are your assets available, adequate, prepared, suitable and with the capability to meet all your requirements?
- What happens if your bottled water production source stops working, or a sample of your stock, if you have any, fails a test?
- What do you do when you cannot access the warehouse where your stocks are held; or if the producer/supplier saw an opportunity to sell off your stock without asking or even telling you, just when **you** need it?
   - This happened to two Utilities that we are aware of, who both used the same single production source supplier during the Mythe incident
- Are all your vessels properly maintained, clean, disinfected, sampled and ready to deploy? And are the vessels you might borrow under Mutual Aid ready to use?
- Do you have arrangements in place for appropriate logistics to deploy bottled water and those vessels in line with your Local Response Plan responsibilities?

- What is your 'Plan B'; a contingency plan for if your alternative supply or logistics provider cannot perform on the day? Maybe it is to call Water Direct?
- Do you have tankers that are appropriate and ready to deliver **drinking water** quickly?
- Do you have drivers for them?
- Are they trained and certificated for working on restricted operations?
- Can you uplift quickly once the event is over; and clear up the mess?
- What will you do with any surplus bottled water, which if deployed without proper managed plans could be considerable?
- How quickly can you replenish bottled water stocks and re-disinfect the vessels ready for the next event? This may be tomorrow, or even **today**!
- Have you budgeted for this preparedness on an ongoing basis?

This brings me to the biggest potential risk to not only the integrity and functionality of your supply chain, but also the whole issue of due diligence.

6.2 Financial risks.

This can endanger your entire response. All of what I have discussed here and raised questions about, and advised on what may be the best way of doing things, costs money; sometimes a lot of money – although in relation to Mythe, it is insignificant.

But this cost is your insurance premium, the response to which you will not be able to benefit from unless you pay it!

You could take the view that you have enough money to throw at an event should it occur; but the service you need will most likely be unavailable at that time unless you have properly planned for and adequately resourced it in advance; so you still may not achieve your objectives. Nothing is free.

As I said before, water is heavy and awkward to transport, especially quickly and without affecting its wholesomeness.

So, one serious threat to the security of your supply chain, and your due diligence, is internal budget constraints. With a fixed revenue, one way for a utility to increase profits is to cut expenditure; so how can you get money for proper planning, for cleaning your vessels, for buying and storing bottled water stocks, for investing in a water tanker or two, or for outsourcing that service to a competent provider, when the likelihood of an event happening to you is 'so remote'?

OR IS IT?

You must strongly express to the 'bean counters' the importance of **due diligence**. Responsibility for this is everyone's but the buck stops with the decision makers at the top, who ultimately control the budgets. Have you given those decision makers all the facts that they need to be aware of to make the right decisions?

Probably a more serious threat to your resilience is the procurement process. With all due respect, a 'procurement professional' will go out to the market for the cheapest possible service, because that is their job and often reflects how they are remunerated. Will they therefore look at best value; or just lowest price? Do they understand the service they are buying and the importance of all the elements and consequences of due diligence? Have you explained it to them beyond any doubt? Will they take account of what you say?

Or will they procure products or services that are not actually fit for the purpose they were purchased to perform, because lowest price was the deciding factor? What does that cost you in the long run?

Due diligence must be exercised in the tender construction and assessment processes to ensure that the product or service provider actually has the appropriate experience, capability and adequate resources to comply fully with all your requirements, to be diligent in safeguarding public health and in achieving your responsibilities under legislation.

By the time they fail to perform it will be too late!

Everyone wants good value for the money they spend. But why risk everything by not spending enough, or spending your money unwisely on something that may not do what you need it to do when the 'proverbial' hits the fan.

7 DUE DILIGENCE

It is imperative that you take all reasonable steps and exercise due diligence at every stage of the procurement process, to safeguard public health and ensure that you can fulfil your obligations under legislation.

I thank John Ruskin for his ode to "**Value**" which summarises what in my view is the essence of **due diligence**:

"It is unwise to pay too much, but it is also unwise to pay too little. When you pay too much you loose a little money; that is all.

When you pay too little you sometimes loose everything, because the thing you bought was incapable of doing what you bought it to do.

The common law of business balance prohibits paying a little and getting a lot. It cannot be done.

If you deal with the lowest bidder, it is well to add something for the risk you run. And if you do that you will have enough to pay for something better".

PROCEDURES FOR THE DECONTAMINATION OF BUILDING PLUMBING SYSTEMS

S.J. Treado[1], M.A. Kedzierski[1], and V.J. Gallardo[2]

[1]Building and Fire Research Laboratory, National Institute of Standards and Technology, Gaithersburg, MD, 20837, USA
[2]Office of Research and Development, National Risk Management Laboratory, U.S. Environmental Protection Agency, Cincinnati, OH, 45268, USA

1 INTRODUCTION

Adequate supplies of clean, safe drinking water are a prerequisite for buildings and their occupants, including both commercial and residential structures and facilities. In the past, there have been occasional concerns regarding insufficient treatment of water supplies or undesirable migration of contaminants from plumbing system materials into water distribution systems, usually due to some change in environmental or operating conditions. More recently, another issue has arisen regarding the potential for accidental or intentional introduction of contaminants into water distribution systems, and the need for effective methods for dealing with such events. This paper describes recommended procedures for responding to building plumbing system contamination incidents and restoring the water system to safe operation. The recommendations are based on analysis of the results of a measurement and modelling research project investigated contamination and decontamination issues related to building plumbing systems.

The general classes of contaminants that are considered include chemicals such as fuels, solvents, pesticides and poisons, and biologicals, such as bacteria, spores and toxins.[1] Considering the wide range of plumbing system materials, including copper, PVC and iron pipe, solder, rubber gaskets and sealants, there are many potential combinations of contaminants and substrates that are of interest. The presence of chemical deposits and biofilms typically found on the interior of plumbing components present another consideration.[2] In addition, plumbing system designs can vary widely, and flow obstructions, water tanks and other water-using appliances can significantly complicate the analysis.[3]

2 BACKGROUND

In order to reach the objective outlined above, a combination of detailed static and dynamic measurements was conducted, along with a computer simulation effort, aimed at identifying the tendency of various contaminants to accumulate in building plumbing systems, and enabling the determination of effective methods for eliminating or rendering innocuous any accumulated contaminants, thereby ensuring restoration of a safe water supply. The basic measurement methodology involved exposing a particular plumbing material substrate, component or system to a water/contaminant mixture followed by a flushing or other decontamination activity, while periodically monitoring or collecting samples of the substrate and/or water to evaluate for the presence of the contaminant or any residual materials. The details of these measurements are beyond the scope of this paper, but are available elsewhere.[4,5,6]

In addition to the measurements of contaminant behaviour in building plumbing systems, the impact of plumbing system design and operation on decontamination strategies was investigated. Since building plumbing systems are typically composed of a complicated network of water supply piping, fittings, valves and fixtures, along with a sanitary drain system and water-using appliances, the potential for contaminant accumulation, and the associated strategies for removal, require careful consideration of real-world factors.

3 POTENTIAL CONTAMINATION SCENARIOS FOR BUILDING PLUMBING SYSTEMS

In general, contaminants may enter a building water supply anywhere upstream, as represented by the four locations shown in Figure 1.

1. Contaminants that are introduced far upstream from the building, either before or near the water treatment facility, and travel a significant distance through the water distribution system to reach the building service line
2. Contaminants that are introduced into a water main supplying multiple branch lines, including one that supplies the building
3. Contaminants that are introduced near to, but outside of, the building, via the building service line
4. Contaminants that are introduced to the building water supply from within the building

The reason that these scenarios are differentiated has to do with the fact that the methods and the amount of contaminant introduced would likely be different in the various cases. As a result, the duration and level of contaminant concentrations in the water supply system would vary, and the portions of the water distribution system, including the building plumbing system, that are affected would be different. Thus, the response to the contamination events, including methods to control the spread of the contaminant, and procedures used to remove the contaminant would differ depending on where the contaminate was introduced.

In scenario 1 above, the effects of dilution and water treatment will likely reduce, but not necessarily eliminate, the impact of most contaminants on downstream building plumbing systems. The closer the point of contaminant introduction is to the building, the higher the concentration and the greater the potential for contaminant accumulation. However, longer lengths of affected water supply lines will require longer flushing times.

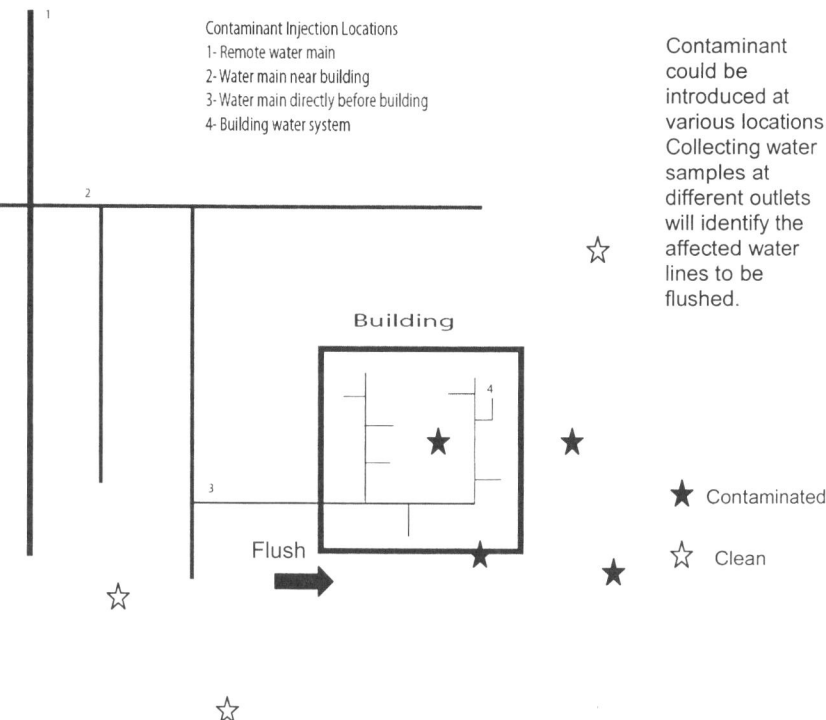

Figure 1 *Analyzing water samples from various locations will enable affected portions of the plumbing system to be identified*

Most contamination events will be recognized based on a consumer complaint (odor, color, taste), illness or reaction, a sensor reading (unlikely), or a verbal or written threat. The first step would be to determine if there is an actual contaminant present in the water supply, what the contaminant is, and the extent of the contamination. The point of contaminant introduction can be deduced by collecting water samples from a range of locations, and mapping the water lines that are found to be contaminated, working upstream until unaffected water lines are found, as shown in figure 1, where point 3 has been determined to be the point of contaminant injection. Once the affected water lines are identified, there remain a number of concerns, as follows:

•It will be difficult to precisely determine how long the contaminant has been in the water supply system, or how much of the contaminant was introduced

•Water samples can be collected/analyzed to identify the contaminant, but the sample contaminant concentration will likely be less than that at the point of injection

•Contaminant may have accumulated in the water supply system, so simply flushing out the contaminated water may not remove all of the contaminant

•Additional flushing or special procedures may be needed to restore a safe water supply

•

Based on the above concerns, it is advisable to assume a worst case exposure condition, which would be a highly concentrated contaminant that has been introduced to the water supply system and had the chance to interact with substrates in contact with the water. Those contaminants that accumulate on plumbing system surfaces exposed to contaminated water will need to be eliminated before the system can be restored to safe operation and use. The accumulation may be due to the combined effects of a number of mechanisms, including different types of adsorption, chemical reactions and sedimentation on/with pipe materials and pipe scale. It may be possible to remove some small sections of piping or other components for laboratory analysis, but that is not always practical.

The magnitude and the location of contaminant accumulation will be a function of the characteristics of the contaminant/substrate interactions and the exposure conditions. For example, some substrates are more conducive to contaminant accumulation, and longer exposure times and higher concentrations are more likely to lead to greater accumulations. Contaminants may or may not be soluble in water, and may be more or less dense than water. Soluble contaminants will dissolve up to a point, and mix with the water so as to come into contact with most of the plumbing system surfaces. Insoluble contaminants will either float (specific gravity, SG<1) or sink (SG>1), thereby preferentially coming into contact with the upper or lower inside surfaces of pipes and tanks (see Figure 2). Contaminants with sedimentary characteristics, such as bacteria and spores, will sink due to gravity when water is not flowing, and may collect at the bottom of plumbing system components.

4 DECONTAMINATION PROCEDURES

The two basic steps in restoring a building plumbing system following a contamination scenario are to first safely purge the system of the contaminated water, and second, to flush or treat the system to eliminate any accumulated contaminant. The most effective methods for removing the accumulated contaminants will be a function of the contaminant, plumbing system materials and design. The preferred method, in terms of simplicity and cost is, of course, conventional flushing using water directly from the water distribution system. Other methods may be faster or more effective at cleaning, however. In general, the three choices that need to be made are:

1. *Which flushing fluid should be used?*
2. *How should the flushing fluid be introduced?*
3. *How should the system be flushed?*

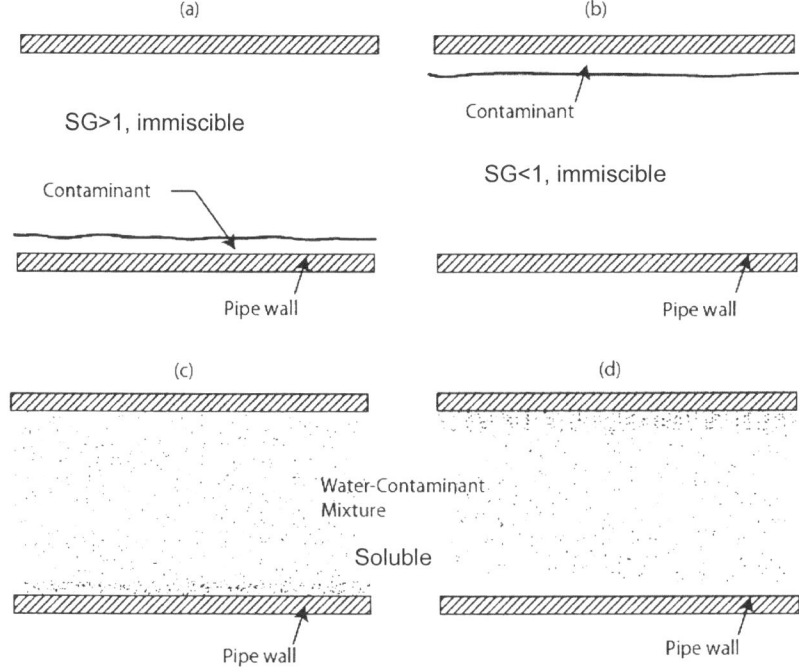

Contaminant/substrate exposure can vary with solubility and density

Figure 2 *The contact between the contaminant and plumbing system components will depend in part on the contaminant properties.*

The fluid could be:
- water as is available from the distribution system,
- water that has been treated with additional chlorine or other disinfectant,
- water that has been treated with surfactants or other chemicals to neutralize or react with the contaminant, or
- hot water or steam
- germinant solution to promote spore germination

The fluid source could be:
- from the distribution system,
- from a reservoir supplied by the distribution system, or
- from tanker truck.

The flushing method could be:

- conventional flushing, effuse down drain,
- conventional flushing, effuse collected for disposal,
- flood system and let stand, then drain effuse,
- flood system and let stand, then collect effuse for disposal,
- high velocity pumping,
- steam injection, or
- pulsating flow.

Generally, the most effective method for removing *chemical* contaminants is by continuous flushing, since accumulated contaminants tend to become entrained in the water due to turbulence, advection and diffusion; thus removal is primarily a function of the amount of clean water passing through the system. This approach works best for pipe sections, but is less efficacious for water tanks or reservoirs, especially those that have their outlet at the top. Flush water velocities are very low in tanks with large diameters relative to pipe diameters, which is usually the case, so the interaction of flush water with any contaminant that has accumulated on surfaces exposed to the water is slight. However, given enough time/water, water soluble contaminants will tend to be removed from the plumbing system. The amount of time/water required to reduce contaminant residuals to a safe level depends upon many factors, including the type and severity of the contamination, the design of the plumbing system and the residual levels considered to be safe, and is beyond the scope of this paper. Previous contamination events and measurements with immiscible organic substances suggest that flushing times on the order of days may be required in some cases.[7,8]

For contaminants that are immiscible, flushing with hot water will help dissolve the contaminant and thereby shorten required flushing times. Immiscible contaminants that float are best removed from water tanks by flushing out through the top, rather than draining from the bottom, since the latter procedure allows a high concentration of the contaminant to come into direct contact with the sediments that tend to accumulate at the bottom of the tank, making removal more difficult. In contrast, immiscible contaminants that sink should be drained from the bottom of the tank, and the sediments flushed out if possible.

In the case of bacterial contaminants, they are best attacked by flooding the plumbing system with disinfectant such as chlorine, and letting it stand to kill or otherwise disable the bacteria, followed by a short flushing to restore clean water.[9] For spores, the plumbing system should be flooded with germinant solution, such as an inosine or L-alanine solution, which will encourage the spores to germinate making them susceptible to killing by disinfection in the same manner as bacteria. In both cases, use of hot water enhances the decontamination effectiveness. Also, since the disinfection and growth media solutions only need to flood the system, as opposed to a continuous flushing, only a modest amount of solution is required.

Flushing with water from the water distribution system is the simplest method, but it is difficult to control water velocities, since they are dependent on water pressure, which will vary with location and as a function of the water demand. Sequential flushing of individual water lines will provide the highest water velocities, and therefore the greatest scrubbing. Flow control devices, such as faucet aerators and showerheads, should be removed before flushing and cleaned individually. It is important to ensure that the number of water lines being flushed simultaneously does not overwhelm the capacity of the building drainage system, since drainage lines are sized based on the expectation of some probabilistic usage pattern. It is possible that additional chlorine or other disinfectant or cleaner could be introduced into the flushing water, either directly from the distribution system, or from an auxiliary source closer to the contaminated water system. Care should be taken to ensure that the concentrations of disinfectant or cleaner will not damage the plumbing system materials.

Another concern is the possibility that a contaminant might become an inhalation hazard due to volatilization into the air after exiting a faucet or other fitting above a sink or tub. Where that is a consideration, avoiding this problem might require special flushing precautions, such as a direct connection between water supply outlets and drain lines. Following flushing, all sinks, tubs and other surfaces that have been exposed to the contaminated water should be thoroughly cleaned.

Water-using appliances present another set of challenges regarding decontamination, and the greatest concerns lie with those that involve water that may be consumed or come into contact with building occupants. Chief among these are water tanks, such as hot water heaters, that tend to accumulate sediments and deposits, and are difficult to flush due to their large volumes and corresponding low flow rates. Cleaning water tanks may require direct draining and filling with special cleaning solutions using one of the techniques described below. In some cases, it may not be possible to eliminate all of the accumulated contaminant, and water lines, fittings, fixtures or appliances may need to be replaced. Appliances that have no drain provision, such as residential ice makers, will not be able to be flushed so will need to be removed and cleaned or replaced. Appliances such as dishwashers and clothes washers cannot be flushed per se, but can be cleaned through operation, or disconnected to allow their supply lines to be flushed.

Figure 3 shows a schematic depiction of a possible configuration for flushing a building water supply system without using water directly from the distribution system, by connecting to an external water spigot or other similar connection point. It may be necessary to bypass or disable any back-flow prevention device to allow water to flow in the reverse direction. The flushing water can be pre-treated as is appropriate with disinfectant or cleaner before injection. If the valve at the water meter is closed, the flushing water will be forced under pressure through the water heater and both the cold and hot water supply lines to any fitting or fixture (e.g. faucet, shower, etc.) that is open. In this configuration, the flush water would be directed into the sanitary drains. It could also be collected and transported for disposal. The water supply line leading from the service connection would need to be flushed separately in the normal flow direction.

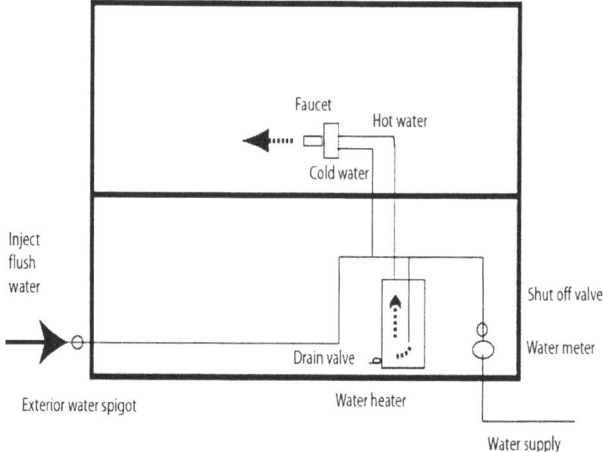

Figure 3   *Injecting flush water through an exterior water spigot or similar point will allow flushing of both hot and cold water lines, and water heaters*

Figure 4 shows a similar configuration for decontaminating a water heater tank, or for using the water heater an injection point for flushing the building water supply system. This operation would be similar to that described previously, although in this configuration, cleaning fluid could be both injected and extracted directly from the water heater via the drain valve. This could be repeated as many times as necessary.

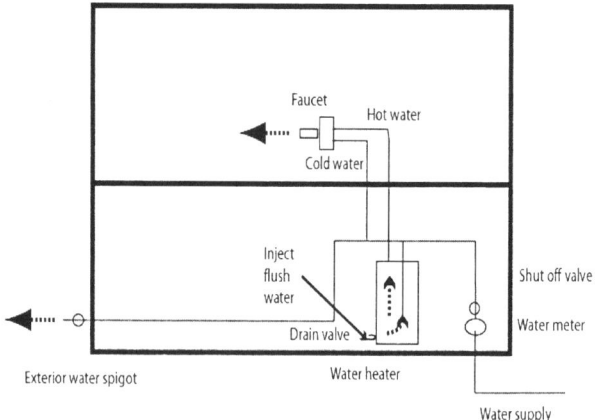

Figure 4   *Injecting flush water through the drain valve of a hot water heater can directly flush the tank and water lines.*

5 RECOMMENDED DECONTAMINATION PROCEDURES FOR BUILDING PLUMBING SYSTEMS

In general, lacking more specific information about a particular contaminant, the recommended decontamination procedures can be determined according to the type of contaminant, based on the following categories listed in table 1.

Table 1 *General Decontamination Procedures Based on Contaminant Type*

Contaminant Category	Example	Key Methods
Soluble chemicals	Strychnine, Cyanide	For pipes and tanks- Continuous flushing with water, water buffered with chlorine, or water mixed with cleaner
Immiscible chemicals with specific gravity less than one	Diesel fuel, Gasoline	For pipes- Continuous flushing with water, water buffered with chlorine, or water mixed with cleaner. For tanks- Flush through drain valve at bottom of tank or water spigot
Immiscible chemicals with specific gravity greater than one	Phorate	Continuous flushing with water, water buffered with chlorine, or water mixed with cleaner. For tanks- Drain through drain valve at bottom of tank, and fill with cleaning solution. Repeat as needed
Sediments or particles	-	For pipes- Continuous flushing with water, drain from cleanouts where available. For tanks- Drain and flush from bottom
Bacteria	E. coli 0157:H7	For pipes and tanks- Flood system with water and disinfectant and let stand, followed by short flush. Repeat as needed
Spores	Bacillus Anthracis	For pipes and tanks- Flood system with germinant solution and let stand to allow spores to germinate, followed by short flush
Toxins	Ricin	For pipes and tanks- Continuous flushing with water, water buffered with chlorine, or water mixed with cleaner

The following is a summary of the recommended list of steps for dealing with a potential contamination scenario involving a building plumbing system following a complaint or other indication of a problem.

1. *Collect and analyze water samples to determine if the complaint is associated with the presence of a contaminant, and identify and measure the contaminant concentration*
2. *Determine the extent of the contamination*
3. *Isolate the contaminated water piping to prevent propagation to uncontaminated piping*
4. *Try to locate the source or point of introduction of the contaminant*
5. *Determine if the contaminated water can be flushed into the waste water system*
6. *Assess volatilization potential of contaminant if exposed to atmospheric pressure within a building*
7. *Determine maximum drainage water flow rate per building to prevent overloading the drainage system*
8. *Flush with water at the appropriate rate, if considered safe*
9. *Estimate contaminant accumulation within the plumbing system*
10. *Select appropriate decontamination procedure*
    a. *If water- continue flushing*
    b. *If cleaning agent or shock chlorine- select injection point, flush with solution*
11. *If waste water cannot be discharged into the drainage system*
    a. *Collect waste water*
    b. *Back flush where possible*
12. *Verify effectiveness of decontamination effort*
    a. *Analyze water samples*
    b. *Analyze pipe samples*
13. *Determine if remedial measures are needed to restore plumbing system components*
    a. *Clean/replace faucets, valves, aerators, tanks, hoses*
    b. *Possible surface restoration*
14. *Replace and dispose of any components that could not be decontaminated*

6 SUMMARY

This paper presents a discussion of methods for decontaminating building plumbing systems and restoring them to safe operation, based on generic contaminant characteristics. More specific recommendations are being developed that link decontamination procedures to particular contaminants or groups of contaminants with similar characteristics. It is hoped that these recommendation will prove useful as a starting point for a set of comprehensive guidelines that support general response plans for effective recover from water supply system contamination events.

References

1   EPA, National Primary Drinking Water Regulations, http://www.epa.gov/safewater/mcl.html#mcls, U.S. Environmental Protection Agency, 2005.
2   L. Mays, Water Distribution Systems Handbook, 2000, McGraw-Hill, New York.
3   J. Wingender, and H.C. Flemming, Contamination potential of drinking water distribution network biofilms. 1: Water Sci Technol. 2004, **49**, (11-12), 277-86.
4   S.J. Treado, Technical Issues Related to the Measurement and Modeling of Contaminant Transport and Accumulation in Building Plumbing Systems, NISTIR 7253, National Institute of Standards and Technology, Gaithersburg, MD, 20899, 8/2005.
5   S. J. Treado, M. Kedzierski, S. Watson, D. Bentz, N. Martys, and K. Cole, Report of Phase One Measurement and Analysis of Building Water System Contamination and Decontamination, NISTIR 7351, National Institute of Standards and Technology, Gaithersburg, MD, 20899, 9/2006.
6   S.J. Treado, The Decontamination of Building Plumbing Systems- Analysis and Procedures, NISTIR 7448, National Institute of Standards and Technology, Gaithersburg, MD 20899, 9/07.
7   M.A. Kedzierski, "Development of a Fluorescence Based Measurement Technique to Quantify Water Contaminants at Pipe Surfaces During Flow," 2006, NISTIR 7355, U.S. Department of Commerce, Washington, D.C.
8   Morbidity and Mortality Weekly Reports , Chlordane contamination of a public water supply- Pittsburgh, Pennsylvania, 1981, Morbid Mortal Wkly Rep 33, 687-693.
9   V. Reipa J. Almeida, and K. Cole, Long-term monitoring of biofilm growth and disinfection using a quartz crystal microbalance and reflectance measurements, National Institute of Standards and Technology, Gaithersburg, MD, 20899, 10/2005.

LESSONS LEARNED FROM SUMMER FLOODS 2007. PHASE 1 REPORT –
EMERGENCY RESPONSE PREPARED BY WATER UK'S REVIEW GROUP ON
FLOODING

P. Mills

Water UK, 1 Queen Anne's Gate, London, SW1H 9BT, UK

CONTENTS
1   Summary
1.1   Recommendations of the Phase 1 report – Emergency Response
1.2   Recommendations for the water companies
1.3   Recommendations for Water UK
1.4   Phase 2 report – Longer Term Issues
2.1   Introduction
2.2   Background to the summer floods
2.2.1   Meteorological background
2.2.2   Expected impacts of climate change
2.3   Lessons Learned and Recommendations
2.3.1   Assumptions
2.3.2   Preparedness
2.3.2.1 Emergency command structures and local resilience forums
2.3.2.2 Availability and sharing of critical data
2.3.2.3 Cleanliness and readiness of bowsers and tankers
2.3.2.4 Location of emergency centres and supplies
2.3.2.5 Loss of key assets
2.3.2.6 Emergency water supplies
2.3.3   Communications
2.3.4   Mutual Aid Scheme
2.3.4.1 Compatability of bowsers and other equipment
2.3.4.2 Provision of appropriately sized tankers
2.3.4.3 Provision of personnel
2.3.5   Public health and recovery
2.3.6   Emergency plans
2.3.7   Drainage responsibility and sustainable options
2.3.8   Resilience
2.4   Timetable and actions
2.5   Phase 2 – Longer Term Issues
Key References
Appendix 1 – Membership of Review Group on Flooding
Appendix 2 – Organisations providing substantial response

NB Because of editing constraints, the paragraph numbers given in this copy of the report do not
match the paragraph numbers in the original report.

1 SUMMARY

1.1 Recommendations of the Phase 1 report – Emergency Response

Whilst the water industry is used to dealing with floods and other emergencies and individual companies have contingency plans in place, there are clear lessons to be drawn from the extreme summer 2007 flooding events for the management of emergencies and their immediate aftermath. Many of these require the attention of water companies and the water industry as a whole, including working closely with regulators, consumer bodies, health agencies and emergency response organisations. Probably the most important conclusion to be drawn is that the sheer scale of the summer floods overwhelmed the existing emergency and flood management plans that the relevant water companies had in place.

Our first and principal recommendation is therefore that water companies put in hand a thorough review of their emergency response and contingency plans on the assumption that the sheer scale and severity of future floods may make current plans inadequate. *(Recommendation 1)*

The following sections outline the remaining recommendations made by the Review Group. These have been grouped to indicate those actions for the individual water companies and those for Water UK on behalf of the water industry.

1.2 Recommendations for the water companies

The following recommendations are made specifically for the attention of individual water companies.

1.2.1 Preparedness. Water companies should ensure that they are appropriately involved with all key agencies in planning, training and rehearsing for critical incidents. These would include public health bodies, social services, statutory consumer organisations and animal welfare agencies in addition to the emergency response organisations, government departments and command structures. In particular we recommend that the relevant water companies be included in the Gold and Silver Emergency Command Structures in order that all parties can familiarise themselves with the working methods of such structures, ensuring that understanding of the roles and responsibility is constantly refreshed and takes account of staff turnover. (Recommendation 4)

Water companies, highways authorities, private asset owners and other organisations with responsibility for provision and maintenance of data should review the data and information that are available within the sector and that could be securely shared amongst key stakeholders to better aid the planning and response process. Areas where data may not be available should be identified and solutions proposed to redress these gaps. (Recommendation 5)

Water companies and any other organisations with a responsibility to provide equipment under Mutual Aid Scheme should ensure that this equipment is kept in a roadworthy and clean condition at all times to ensure that response times to emergency events are kept to a minimum. (Recommendation 6)

Water companies should take a fresh look at the potential vulnerability of their key assets, including the risks from other utility service failures, and then harden those sites as best they can against the higher levels of risk now emerging. In the short-term, this may involve the deployment of temporary measures. In the longer term substantial investment may be required. (Recommendation 7)

Water companies should review the efficacy of their emergency supply assets to cope with such minimum levels (e.g. bottled water supply chain, numbers and location of bowsers and tankers). Dialogue with supermarkets and other bulk providers of bottled water should also ensue to determine how best to ensure adequate supplies during emergency events. (Recommendation 9)

Water companies should ensure that they maintain a full and up to date register of key stakeholders and contact lists for organisations responsible for vulnerable consumers, and of any special communication requirements they may have. These registers should be accessible by overflow call centres and emergency response teams. (Recommendation 10)

The registers should highlight in advance the appropriate actions required for each group of vulnerable consumers, including farm animals. Companies should also consider any limitations of bottled water supplies for consumers with specific medical needs or infants. (Recommendation 10)

1.2.2   Communications. Water companies should undertake to review their communication strategies for addressing customers, the wider public, other agencies and the media during emergencies to ensure that they are suitable for widespread service failure taking account of:

- The possibility of widespread disruption to channels of communication;
- The need for clear, simple and up to date information and advice;
- The likely weight of calls on web-sites and call centres;
- The demands on senior management time;
- The roles of the emergency services and other agencies;
- The particular needs of vulnerable consumers.   (Recommendation 11)

Water companies should consider developing proforma standardised text and vocabulary to ensure that messages to consumers are consistent across and between water companies and that this consistency is maintained in the event of the use of emergency or overflow call centres. (Recommendation 12)

1.2.3 Mutual Aid Scheme. Water companies should rehearse emergency plans on a regular basis. This should include physically moving equipment within individual company areas and ensuring with other companies that provisions under the conditions of Mutual Aid Scheme are available. These rehearsals should include the emergency response organisations. Such rehearsals should include scenarios allowing for disruptions to access to sites and locations due to flooded roads and facilities. (Recommendation 15)

1.2.4 Public health and recovery. Water companies should give post-event clean up operations further consideration as an opportunity to recover service to customers including in situations where responsibility is not directly attributable. (Recommendation 17)

1.3 Recommendations for Water UK

The following recommendations are made for the attention of the industry as a whole and should be actioned through Water UK as the industry's representative body.

1.3.1 Assumptions. Water companies and Water UK should work closely with the Met Office and the environmental agencies further to develop specific industry requirements for weather information and advanced severe weather warnings and to obtain a better understanding of the potential severity of rain storms that might give rise to large scale flooding events. (Recommendation 2)

Water companies and Water UK should continue to work with the environmental agencies to build on existing generic and company specific flood forecasting tools to include, for example, depth as well as extent of flooding, and in particular to identify key infrastructure, such as treatment works, pumping capability, and the siting of emergency centres and supplies, that may be at greater risk than currently understood. (Recommendation 3)

1.3.2 Preparedness. Through Water UK water companies should review with drinking water regulators and public health organisations the likely scale of consumers' requirements for water during emergency events and how this requirement may change throughout an event. We recommend that plans for the provision of emergency drinking water supplies should take as their starting point that each person should be supplied with a minimum of 20 litres a day (i.e. twice the current assumption). (Recommendation 8)

1.3.3 Mutual Aid Scheme. Water UK should use its existing emergency planning and security network to review the state of preparedness of the industry for future events; in particular the industry's Mutual Aid Scheme should be reviewed with a view to ensuring:

- The technical compatibility of assets;
- The number and readiness of such assets;
- The means of deploying and managing staff made available under the Scheme;
- The resilience of the scheme to cater for simultaneous events. (Recommendation 13)

Through Water UK the water industry should address the standardisation of emergency supply equipment to ensure that in the event of an incident equipment from other companies or organisations is compatible. (Recommendation 14)

1.3.4 Public health and recovery. Through Water UK the water industry should establish a standard approach to the temporary use of non-potable water to restore sanitation supplies, clearly outlining the conditions and situations in which it should be considered as an option. To deliver this standard the industry will need to work at a national level with government, statutory consumer organisations, public health bodies and drinking water regulators. (Recommendation 16)

1.3.5 Timetable and actions. The Review Group's final recommendation is that an appropriate group is established to oversee the actions on these recommendations, and those of other reviews, and to identify an appropriate method of reporting progress. This group should be governed by Water UK and should take its membership from water companies and other organisations with the appropriate knowledge and influence to complete the actions required. (Recommendation 18)

1.4 Phase 2 report – Longer Term Issues

The Review Group will carry out a phase 2 of its review into flooding considering the longer term issues facing the sector. The scope of the report will remain flexible but will cover the questions around long-term policy and investment issues highlighted so far. This will include but not be limited to:

- Climate change impacts and implications for investment;
- Resilience of water infrastructure and assets;
- The allocation of responsibilities for flood prevention and remediation;
- Sewers – automatic right of connection, sewer flooding, ownership / private sewers, suitability of design standards in a changing climate;
- Impact of EU Flooding Directive;
- Public expectations in the event of future flooding.

2 THE REPORT

2.1 Introduction

Through Water UK, the UK water industry is undertaking a review of the unprecedented flooding events that affected parts of the UK during 2007. A Review Group on Flooding (Review Group) was established under the independent chair of Sir John Baker. A full list of members of the Review Group is given in Appendix 1.

The report is intended to cover the whole of the UK water industry and as such recognises that different structures exist within the different administrations. The report uses generic language and terminology that should be read as applicable to each model unless specifically noted otherwise.

The water industry is used to handling floods; the events of summer 2007, however, were different. Each company has in place plans for dealing with floods or other emergencies (as part of the requirement of the Security and Emergency Measures Direction (1998)) but these were overwhelmed by the intensity and severity of the rainfall and the consequences.

"The scale and speed of the floods that affected people in summer 2007 came as a shock. In many cases, this reflects people's limited awareness of risk, especially of surface water flooding, and limited engagement in preparedness planning" (Pitt Review, 0)

"Existing, well rehearsed, emergency plans were used and were generally considered fit-for-purpose but the scale and complexity of the summer incidents challenged the plans and highlighted where there were gaps or weaknesses. (Health Protection Agency response to Review Group)

The review examines the experiences of those who were involved in the exceptional events, and considers from the point of view of the water industry, its customers and all water consumers what went well, what went less well, and what lessons can be learned for the inevitable next time when drinking water supplies and waste water services come under threat from extreme weather events.

The review is not concerned with awarding praise or blame in relation to any of the people or organisations caught up in the floods, but focuses on the issues of how to ensure that the impacts of floods on water customers can be further mitigated next time. The Review Group is also considering wider policy issues such as the vulnerability of services and infrastructure to climate change, rising consumer expectations about the reliability of water supplies, the reduction of risk, and the possible further upgrading of the water and sewerage infrastructure.

The Review Group has invited contributions to its thinking from all key stakeholders, including all water companies, regulators, relevant government bodies, NGOs and business and academia contacts. Responses have been received from the organisations listed in Appendix 2 and the Review Group is grateful for their input.

The outcome of the review is to be published in two phases:

This Report (the Phase 1 report) focuses on the industry's immediate responses to the flood emergencies and the recovery of normal services, and makes recommendations to both the industry and its stakeholders designed to ensure that the lessons that can be learned from the 2007 floods can be applied so as to ensure that, next time round, the water industry's response is even better prepared and carried out fully and effectively.

A second report will be produced in spring 2008 that will address the wider strategic, infrastructure and policy issues.

The focus of these reports is not to review again the specific individual events of 2007 but to draw out from the reactions and responses to these events the implications of such events for the industry as a whole. Local reports have already covered the Yorkshire and Humberside events (0, 0) and the Gloucestershire event (0, 0) and the Review Group has benefited from direct discussion with Yorkshire Water (YW) and Severn Trent Water (STW) fully to understand the specific events in those companies' areas.

The Review Group also recognises that concurrent reviews are being carried out by other bodies - in particular Environment, Food and Rural Affairs (EFRA) Committee, the Cabinet Office, the Water Services Regulation Authority (Ofwat), and the Environment Agency (EA). Each of these organisations has been contacted during Phase 1 for their initial thoughts and the Review Group will seek further discussions in Phase 2.

2.2 Background to the summer floods

Floods are natural and inevitable events whose effects can be devastating in terms of property, livelihood, and human and animal welfare. Non-coastal flooding is defined as occasions when land that is usually dry is covered with water as a result of a river overflowing or breaking its banks (fluvial flooding) or heavy rain (flash, pluvial or surface water flooding). Society can try to take precautions to reduce the circumstances in which heavy rain leads to flooding and the consequences of flooding when it occurs but we cannot prevent floods. Specifically, town planners and agencies can try to manage the manner in which water is moved around towns, and drainage authorities and water companies can try to manage the removal of run off and waste water.

The water industry is spending £16.8 billion on capital programmes to deliver infrastructure improvements in the current planning period, 2005-10. Of this it is spending £1bn on reducing the risk of sewer flooding. However, because of the diverse and numerous causes of flooding, it is not possible for the water industry or any other single body or agency to ensure that flooding never occurs.

It is simply not credible that the nation could afford the cost of ensuring there would be no more floods - a judgement has to be reached about how much prevention is worthwhile. That judgement needs to be revisited from time to time to see if the balance is right: rapid climate change, which will change the severity, frequency, intensity and impact of floods, suggests that that judgement needs revisiting now.

The meteorological events of the summer of 2007 were extraordinary and unprecedented. They resulted in flooding in many parts of the UK in June and July that was seen as unprecedented in both scale and location. Homes, businesses and agriculture were all adversely impacted on a scale that was outside the assumptions on which emergency plans had been prepared.

The most severe flooding occurred in:

* Northern Ireland on 12 June;
* East Yorkshire and the Midlands on 15 June;
* Yorkshire, the Midlands, Gloucestershire and Worcestershire on 24 - 25 June;
* Gloucestershire, Worcestershire, Oxfordshire, Berkshire and South Wales on 19 - 20 July.

The Environment Agency report into the flooding (0) notes that over 55,000 properties were flooded during these events and over 140,000 properties in Gloucester faced disrupted drinking water supplies. The Association of British Insurers (0) further notes that, in total, 165,000 insurance claims were made with an estimated cost in the region of £3.0 billion.

The immediate cost to the water industry in responding to the events has yet to be fully assessed by the companies. Not only will costs include the direct costs of dealing with the incidents but also additional staff costs and costs to other companies for providing support and logistics through the Mutual Aid Scheme.

The potential future cost to the water-bill payer of upgrading the water infrastructure to reduce the risk of flooding in the future and to make water supplies and sewerage more robust when floods occur will be significant, so debate will be needed between all the parties - water companies, government, other agencies and local authorities, and consumer representatives - to help determine where the balance between spending and protection is to be drawn.

2.2.1 - Meteorological background. The Met Office reported to the Review Group that the cumulative rainfall total in the UK for May, June and July 2007 was unprecedented. Records show that 414mm of rain fell across England and Wales, making it the wettest May to July since records began in 1766. This is in the order of 2.5 times the 1971-2000 average for England and Wales of 186mm.

Locally the effect was even more marked with some weather stations recording over four times the average rainfall for the time of year. Figure 1 shows the volumes of rain experienced across the UK between May and July 2007 and the variance from the average.

The prolonged nature of the wet weather meant that rain was falling onto ground that was already saturated, causing increased surface water run off. In urban areas paved surfaces prevent rainwater infiltrating the soil, increasing surface water run off and adding to the flooding.

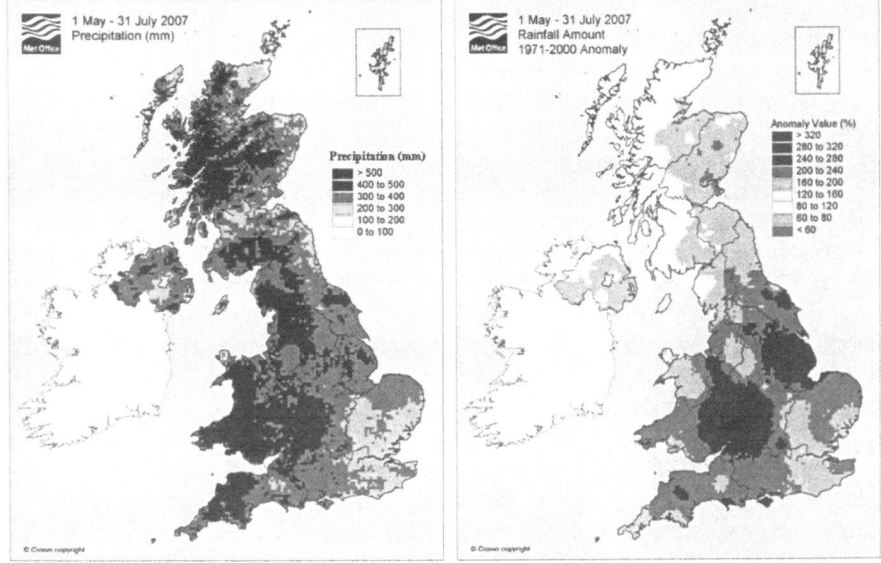

Figure 1 *Precipitation (mm) 1 May - 31 July 2007 and anomaly from the 1971 – 2000 average (from Met Office response to Water UK Review Group on Flooding, 16 November 2007)*

On 24 -25 June, a deep and slow-moving area of low pressure brought a prolonged period of heavy rain, causing widespread surface water flooding in parts of Yorkshire and the Humber, Derbyshire, Lincolnshire and Worcestershire.

During the period of heavy rainfall pumping stations in the area continued to operate as designed. However on 26 June, flood water overwhelmed a pumping station in Hull called Bransholme surface water pumping station (SWPS). The station operated throughout 25 June but was itself overcome by flood water on the morning of 26 June. Temporary pumping was introduced the same day.

Over the course of the two days Hull County Council reported that 7,800 properties were flooded and over 1,300 businesses had been affected by the flooding in the city.

Heavy rain moved northwards across the UK from late on 19 July throughout 20 July. Widespread surface water flooding occurred on the morning of 20 July across Southeast England, and later in the day across the Midlands as the system moved north-westwards, causing widespread disruption to the motorway and rail networks.

This heavy rainfall resulted in flooding which inundated the Mythe Water Treatment Works (WTW) which sits at the confluence of the Rivers Severn and Avon. This inundation required Severn Trent Water (STW) to commence a controlled shut down of the works commencing on Sunday 22 July. The phased and planned Mythe WTW shut down resulted in the loss of water supplies to some 160,000 properties, though some 20,000 of these were able to be switched to an alternative pressurised supply from an adjacent WTW (0). As a result some 350,000 water consumers had no piped drinking water for periods as long as 15 days whilst the recovery effort took place.

Measures to provide an alternative water supply to these consumers were taken through the provision of tankered and bottled water for the duration of the incident.

The response to this event is considered the biggest mobilisation of equipment of this type faced in peacetime Britain.

2.2.2 - Expected impacts of climate change. In order to set these floods in context, the Review Group met with UK Climate Impacts Programme (UKCIP) on the predicted impacts of climate change on flooding at the start of its review.

The scientific expectation is for an increased frequency of extreme weather events, including localised heavy rainfall, especially in the winter months. But whilst the trends may be clear, a large degree of uncertainty will remain about when severe events will occur, exactly where they will occur, and whether they will tend to bunch or occur on some more evenly spread pattern. These uncertainties are highlighted by the fact that the 2007 floods occurred after heavy summer rain when climate change forecasts predict for the UK generally drier summers and wetter winters but with more extreme events within this overall picture at any time of the year.

No specific event can be attributed to one particular aspect of climate change. Nevertheless, it is reasonable to assess the incidence and severity of future risks in terms of the return rate of similar events in the light of general understanding of the implications of climate change for weather in the UK.

It is also likely that rainfall events will intensify into shorter time periods but there is great uncertainty around where or when they will happen. This highlights how natural variability in weather can confuse or mask predicted climate change impacts.

UK climate models, especially with the improvements expected with the latest climate models being developed by researchers (e.g. the UKCIP08 models), are likely to provide increasingly useful tools for water resource management planning where the long term impacts and trends can be taken into account. However, for planning for flooding they are more likely to be useful as general statements that various intensities of events could be expected to become more frequent and thus drive generic policy or investment decisions, rather than giving the ability to predict just where or when a severe storm might strike.

The water industry is already carrying out much work on the impacts of climate change. The outputs of this research and more specific detail on climate change impacts will be presented in the Phase 2 report. At the same time, water companies are already factoring the implications of climate change into their strategic planning and need to ensure that they take account of the most up to date modelling and science available. The industry should build on this good work to ensure that it has a common understanding of the impacts of climate change for operational and emergency planning.

2.3 - Lessons Learned and Recommendations

First and foremost the water industry has learned that whilst it has plans in place for dealing with emergency events these can be overwhelmed by the sheer scale of the weather conditions such as those that affected the UK in June and July 2007. The events seen and the responses required involved the mobilisation of support and relief not previously seen in peacetime Britain.

The brunt of major floods is inevitably borne by the people caught up in them who may have to leave their homes, will see property damaged or destroyed, businesses where they work forced to shut down, livestock displaced or drowned and crops damaged or destroyed. For those directly affected, a flood is distressing and disruptive.

Two pieces of consumer research were carried out after the summer 2007 flooding; by GfK NOP Social Research for the Pitt Review (0) and by Accent for the Consumer Council for Water (CC Water)1 (0). The outcomes of this research are revealing, and give some strong messages about the expectations of consumers in the future.

Customers do by and large understand that what happened was exceptional. In the STW region the public felt that STW were operating "under difficult and extremely unusual conditions" (0). They do accept that a lot was done by the authorities, including water companies, to deal with the emergency. However, they consider that the failure of assets under extreme weather events should not have happened and, when such weather events happen again, they expect the industry to have learned lessons to deal with the situation.

Loss of essential services causes distress and can lead to panic. In the GfK NOP (0) findings consumers rated water and power as top priority services that need to be restored or maintained in the event of an emergency of this nature. Consumers generally accepted the provision of alternative supplies on occasions where supplies were lost. Indeed:

" People appreciated the work carried out by those who provided them with water, but there were also reports of the scarcity of water causing arguments and tension in local communities" (Pitt Review, 0)

Criticism was made of organisations, including water companies, that there was little evidence of existing contingency plans either being in place or acted upon when the emergency developed. Under the Security and Emergency Measures Direction (SEMD) 1998 water companies do have plans in place to react to and deal with emergencies. These plans are approved by government and audited regularly. On this occasion however they were overtaken by the scale and unprecedented nature of the rains and floods.

Recommendation 1:
Water companies put in hand a thorough review of their emergency response and contingency plans on the assumption that the sheer scale and severity of future floods may make current plans inadequate.

To assist the water industry to achieve this level of expectation this Phase 1 report highlights key areas where lessons can be learned and existing water company practices improved: assumptions need to be revisited, plans revised, emergency operations upgraded and rehearsed, communications overhauled, and relationships with other emergency authorities put on a firmer footing.

2.3.1 – Assumptions. A recurring theme in many contributions made to the Review Group is the scale of the meteorological and flooding events of the summer of 2007, in particular in the intensity of the rain and the significance of preceding wet conditions on infiltration of rain into the ground. This rainfall was on a scale well beyond the existing assumptions or expectations of most of the relevant agencies, including the water companies. The emergency plans that the water companies had in place worked to the extent and in the way they had been planned for but were soon swamped by the sheer scale of the events. This was highlighted in Gloucester but was also evident in the Yorkshire and Humberside events.

---

[1] The CC Water research only focuses on the events in Gloucestershire associated with the loss of supply from the Mythe WTW.

In their submission to the Review Group the Met Office noted that they successfully issued early warnings of severe weather. However, they noted that there was evidence that the underlying message was confused primarily over the distinction between "flood warnings" and "severe weather warnings". The difference is that severe weather warnings and flood warnings do not necessarily occur simultaneously.

Recommendation 2:
Water companies and Water UK should work closely with the Met Office and the environmental agencies further to develop specific industry requirements for weather information and advanced severe weather warnings and to obtain a better understanding of the potential severity of rain storms that might give rise to large scale flooding events.

The water industry accordingly needs to stay close to the Met Office and UKCIP to understand future climate trends in greater detail and to review its current assumptions as to the scale, frequency and consequences of future potential floods. It should share its thinking and assumptions in detail with the other relevant bodies (local authorities, environment bodies, drinking water quality regulators, health authorities, etc.) to ensure that future preparations are relevant, coherent and consistent, and based on a common understanding.

Recommendation 3:
Water companies and Water UK should continue to work with the environmental agencies to build on existing generic and company specific flood forecasting tools to include, for example, depth as well as extent of flooding, and in particular to identify key infrastructure, such as treatment works, pumping capability, and the siting of emergency centres and supplies, that may be at greater risk than currently understood.

2.3.2 – Preparedness. There has been some criticism from consumers and other agencies caught up in the floods that the level of preparedness at outset of the 2007 events was not as full as it could have been, resulting in a slow or inadequate reaction to the unfolding situations. All water companies have existing plans for dealing with emergencies and these plans include being able to call on the support of other water companies through the industry's Mutual Aid Scheme. These plans are reviewed on a regular basis. However, the sheer scale of the summer 2007 floods meant that these plans left the water companies concerned unable to respond as effectively as they had expected to and planned for. Full preparedness is essential. In particular, attention needs to be given to:

- Emergency command structures and local resilience forums;
- Availability and sharing of critical data;
- Cleanliness and readiness of bowsers and tankers;
- Location of emergency centres and supplies;
- Loss of key assets;
- Emergency water supplies.

2.3.2.1 Emergency command structures and local resilience forums
The water industry's role within the formalised Gold and Silver command structures was raised as an issue by a number of respondents. A Gold - Silver -

Bronze command structure is used by emergency services to establish a hierarchical framework for the command and control of major incidents and disasters. Within this structure water companies are classified as Category 2 responders and should be consulted as appropriate.

The experiences during summer 2007 showed a patchy and inconsistent picture in the level and timing of involvement of water companies in the developing emergencies. The degree of participation of water companies in emergency command structures ranged from none to full. The points at which water companies were invited to attend the command structure meetings also varied (and not just dependent on the type of event). Yet experience in summer 2007 showed that once a water company was directly incorporated into the emergency command structure and reported to the command leader then both communications, understanding of needs, and decision-making rapidly improved.

It is evident that participation in and training with local resilience forums2 will allow the development of working relationships with other emergency responders and local authorities during normal working conditions that will have benefits in the event of an emergency. Emergency response plans need to be exercised regularly to be fully effective. Water companies clearly ought to participate in such rehearsals and rehearsals need to be sufficiently frequent to ensure full familiarisation by relevant personnel, allowing for staff turnover.

The process of co-ordinating advice to the public was carried out by the establishment of Scientific and Technical Advisory Cells both locally and nationally (STAC). These were convened by the Health Protection Agency (HPA) as a Category 1 responder to develop agreed advice to the command structure that had the support of all organisations involved.

The creation of this group facilitated the production of consistent public information messages and minimised the risk of contradictory or confused advice being issued. The STAC included representation from the relevant water company, the HPA, the Food Standards Agency (FSA) and the drinking water quality regulator.

In general, this approach was seen to successfully co-ordinate public health advice and this arrangement should be a model for future events. It would be helpful if appropriate regulatory bodies present on a STAC ensure that the advice given assists the delivery of their statutory duties. In their submission to the Review the HPA noted that:

"Effective partnership working contributed to the response and the engagement of the water company on the STACs was a positive development." (HPA response to Review Group)

The HPA also however noted that whilst this was generally a success story there were causes for concern over a lack of clarity on statutory roles and responsibilities.

---

[2] The principal mechanism for multi-agency co-operation at the local level is the Local Resilience Forum. Local Resilience Forums are generally based on local police areas (with the exception of London), and bring together all the organisations that have a duty to co-operate under the Civil Contingencies Act, along with others who would be involved in the response to an emergency.

Local Resilience Forums ensure effective delivery of those duties under the Civil Contingencies Act that need to be developed in a multi-agency environment. In other words they ensure that preparing for emergencies is done in a co-ordinated, effective way by all local responders working together. http://www.preparingforemergencies.gov.uk/government/local.shtm

This was evident during the efforts to restore supplies following the disruptions at the Mythe WTW. Section 3.5 provides more consideration on this issue.

Recommendation 4:
Water companies should ensure that they are appropriately involved with all key agencies in planning, training and rehearsing for critical incidents. These would include public health bodies, social services, statutory consumer organisations and animal welfare agencies in addition to the emergency response organisations, government departments and command structures. In particular we recommend that the relevant water companies be included in the Gold and Silver Emergency Command Structures in order that all parties can familiarise themselves with the working methods of such structures, ensuring that understanding of the roles and responsibility is constantly refreshed and takes account of staff turnover.

2.3.2.2 Availability and sharing of critical data

The water industry maintains comprehensive asset and location data for their above ground treatment assets and their underground clean water distribution infrastructure. The same cannot be said for the sewer and drainage networks due to the extent of private sewers and other asset owners. The data that would be required for a truly integrated planning process will require good quality data from water companies but also from other organisations such as environmental bodies, highways agencies and local authorities as well as private land owners.

Another area where better information was needed was in relation to records of the availability, location and readiness of emergency equipment, precisely what was deployed where, and what was still available in reserve.

During the course of the emergency response to the Mythe WTW disruption Water UK attempted to keep a record of the additional equipment provided by water companies under the Mutual Aid Scheme to help the STW effort. This information was then provided to the Department for Environment, Food and Rural Affairs (Defra) to assist in briefing ministers. It was difficult to ascertain how many bowsers were being held in reserve by companies and how many would still have been available had the magnitude of the event increased further. Water companies should maintain their own information and make it available through a centrally-maintained database during times of emergency.

Recommendation 5:
Water companies, highways authorities, private asset owners and other organisations with responsibility for provision and maintenance of data should review the data and information that are available within the sector and that could be securely shared amongst key stakeholders to better aid the planning and response process. Areas where data may not be available should be identified and solutions proposed to redress these gaps.

2.3.2.3 Cleanliness and readiness of bowsers and tankers

In some cases emergency equipment provided through the Mutual Aid Scheme could not be confirmed as clean and ready to deploy directly. In their submission to the Review Group Water Direct3 noted that they believed most utility stocks are not

---

3 Water Direct was a best endeavours responder to Severn Trent Water during the flooding incident in Gloucestershire in July 2007.

in a condition that enables them to be immediately deployed. In particular Water Direct stated that:

"Dedicated drinking water tankers must be available for immediate despatch, again, clean, disinfected and sampled for assuring quality at the time of deployment and use." (Water Direct response to Review Group)

It takes 24 - 48 hours to fully clean, drain and sample bowsers and tankers before use. Having to do this at the outset of an event will increase the response time in deploying the equipment and hence in getting alternative water supplies to affected communities.

Recommendation 6:
Water companies and any other organisations with a responsibility to provide equipment under Mutual Aid Scheme should ensure that this equipment is kept in a roadworthy and clean condition at all times to ensure that response times to emergency events are kept to a minimum.

2.3.2.4 Location of emergency centres and supplies

The scale of the summer 2007 events meant that some centres from which an emergency event was intended to be managed, and some of the equipment intended to be deployed, became the victim of flooding and could not be deployed. At the same time, especially in the events around Gloucester and Tewksbury, the widespread nature of the floods severely impacted transport and traffic, making it difficult for staff to carry out their assigned emergency roles. Again, the assumptions made in emergency plans as to the viability in all circumstances of locations, equipment and personnel need to be reviewed to take account of the possibility that future floods may be more extensive, the water deeper, and the duration longer than may be being currently assumed.

2.3.2.5 Loss of key assets

STW lost a key water treatment works to the floods and as a result 350,000 people were without mains drinking water for 15 days. Castlemeads electricity sub-station was shut down due to flooding, leaving approximately 42,000 people without power which was not restored for 24 hours. More significantly the Walham electricity sub-station was also at significant risk of flooding. If this had failed the whole of Gloucestershire and part of Wales and Herefordshire would have been without power.

Gloucestershire County Council Overview and Scrutiny Management Committee noted that:

"Only a concerted effort involving the Fire and Rescue Service, the Military, the Environment Agency, and National Grid prevented the flooding of this sub-station." (Gloucester County Council, 0)

In none of these cases was such a loss (actual or potential) anticipated because none of the assets had come under serious threat in previous floods. The loss of such assets to floods was not assumed in emergency plans.

Recommendation 7:
Water companies should take a fresh look at the potential vulnerability of their key assets, including the risks from other utility service failures, and then harden those sites as best they can against the higher levels of risk now emerging. In the short-term, this may involve the deployment of temporary measures. In the longer term substantial investment may be required.

2.3.2.6 Emergency water supplies

The water industry's current planning assumption is that in the event of a breakdown in mains drinking water supplies consumers should be provided with ten litres of drinking water per person per day by alternative means. This is stipulated in the Security and Emergency Measures Direction (SEMD 1998). For bulk water, bowsers and tankers are a key means of provision. In the short-term, until bowsers and tankers can be deployed and kept well-supplied, bottled water is needed, and in some situations, bottled water may also be required longer term, for example to those consumers who may have physical difficulty accessing bowsers or where specific health requirements dictate.

Experience of the summer 2007 loss of drinking water supplies on a large scale suggests that the ten litres per head assumption does not in practice meet consumers' expectations. In their analysis of the event STW calculated that the demand for water during the event was closer to twenty litres per person per day. It also needs to be recognised that loss of mains supplies is likely to lead to "panic buying" of bottled water, and hoarding of water. As STW told the Review Group, bottled water "flew off the shelves" in the initial stages of their emergency, stabilising only when the public acquired confidence that when they went to the bowsers there would be water there.

Water companies need to revisit their assumptions as to what consumers water requirements are in an emergency and their supply chain arrangements for bottled water and for deploying and for refilling bowsers. Learning specifically from the experience in Gloucestershire we suggest the requirement to provide ten litres/head needs to be at least doubled.

Water companies currently maintain registers of vulnerable consumers within their supply areas. The use of these registers allows prioritisation of provision of emergency water supplies in the event of an incident. However, questions were raised as to the extent and suitability of these arrangements. Companies need to consider the type and extent of data that is held on these registers and consider working with local health authorities, consumer organisations and other community groups to ensure that they are maintained in a secure, comprehensive and up to date manner.

Two issues particularly arose:

- Not all bottled waters are suitable for the reconstitution of infant formulae. The HPA noted that the communication of this issue to consumers was not effective. It was suggested that health bodies and drinking water regulators work to develop an agreed position on infant formulae to avoid confused messages in future events. This would enable water companies to provide correct advice to members of the public.

- The water requirements of farms and farm animals need to be included within the planning process. In the case of dairy cattle these are significant volumes with a requirement of between 70 - 90 litres of water per head per day.

The National Farmers' Union (NFU) noted that as animals are removed from pastures and housed in barns to escape the floods they need to be provided with water. During the summer 2007 floods farmers felt that they were not provided for by the water company or emergency responders and were left to source their own water supplies.

Recommendation 8:
Through Water UK water companies should review with drinking water regulators and public health organisations the likely scale of consumers' requirements for water during emergency events and how this requirement may change throughout an event. We recommend that plans for the provision of emergency drinking water supplies should take as their starting point that each person should be supplied with a minimum of 20 litres a day (i.e. twice the current assumption).

Recommendation 9:
Water companies should review the efficacy of their emergency supply assets to cope with such minimum levels (e.g. bottled water supply chain, numbers and location of bowsers and tankers). Dialogue with supermarkets and other bulk providers of bottled water should also ensue to determine how best to ensure adequate supplies during emergency events.

Recommendation 10:
Water companies should ensure that they maintain a full and up to date register of key stakeholders and contact lists for organisations responsible for vulnerable consumers, and of any special communication requirements they may have. These registers should be accessible by overflow call centres and emergency response teams.
The registers should highlight in advance the appropriate actions required for each group of vulnerable consumers, including farm animals. Companies should also consider any limitations of bottled water supplies for consumers with specific medical needs or infants.

2.3.3 – Communications. Good communications are the cornerstone of maintaining public health during a flooding event, both to advise consumers what they can and cannot drink, and how to deal with sewage flooding, and where to find potable water if necessary. Without such adequate, timely and appropriate communication the risk of public health breakdown is heightened. In none of the incidents this summer was there any breakdown in public health.

There was much national media coverage of the events highlighting drinking water issues, as well as local news stories through local radio and TV. As the events unfolded, the use of these media outlets by the water companies to inform consumers became central to their response. Traditional methods were also used by water companies - loud hailers, direct mail shots, call-centres, and company websites - but some of these were overwhelmed by the sheer and unanticipated weight of enquiries put on them.

When water consumers were interviewed after the event (0, 0) the main response was that they did not feel adequately informed of what was going on or what they should be doing. This was particularly evident in the case of the restoration of piped water supplies that involved the provision of non-potable water for sanitation purposes ahead of the provision of potable water.

The research carried out by the Consumer Council for Water noted that whilst local radio and Sky Television were particularly good sources of information they tended to focus more on victims of the flooding rather than the issues around the loss of water supply. In their response to the Review Group CC Water noted:

- Consumers felt that communication had been patchy.

- Consumers would have preferred to have been supplied with water 'not fit for drinking' so that they could flush the toilet.
- Not all consumers knew when water was back on or whether it was safe to drink.

To clarify this situation CC Water recommends (and we agree) that there should be:

"A clearly defined communications process in the case of a declared disaster involving national as well as local media, with clear bulletins and information being given at specific time slots." (CC Water, 0)

Recommendation 11:
Water companies should undertake to review their communication strategies for addressing customers, the wider public, other agencies and the media during emergencies to ensure that they are suitable for widespread service failure taking account of:
- *The possibility of widespread disruption to channels of communication;*
- *The need for clear, simple and up to date information and advice;*
- *The likely weight of calls on web-sites and call centres;*
- *The demands on senior management time;*
- *The roles of the emergency services and other agencies;*
- *The particular needs of vulnerable consumers.*

Where water supplies are interrupted, keeping consumers advised of the availability and location of alternative water supplies is obviously paramount. This can be done by radio, TV, post, door to door leaflets, word of mouth, adverts in papers and websites. Technological solutions, for example, using a web-based application are being developed and deployed by a number of companies and show great promise. However, high-tech solutions on their own may not be the complete answer - for example, where consumers do not have access to the web or where an electricity supply breakdown disables electronic equipment. There is clearly a need for diversity in communication channels but consistency in the messages going through them.

Recommendation 12:
Water companies should consider developing proforma standardised text and vocabulary to ensure that messages to consumers are consistent across and between water companies and that this consistency is maintained in the event of the use of emergency or overflow call centres.

Water companies have a system in place whereby when their normal call centres become overwhelmed by the density of calls being made to them, an emergency call centre can be brought on line. The establishment of these centres will involve staff who do not normally have daily dealings with the water industry and its customers and, as such, it is imperative that full briefings are provided to ensure that customers can be told with justifiable confidence what is going on and what it means to that individual.

2.3.4 - Mutual Aid Scheme. Water companies have plans for deployment of alternative water supplies in the event of supply interruptions under the Security and Emergency Measures Direction (SEMD). To assist with this a Mutual Aid Scheme exists. The Mutual Aid Scheme is the means by which the resources held by the water industry as a whole can be made available to a specific water company in an emergency at any time. All the water companies are expected to play their part.

This was fully activated during the Gloucestershire incident and the logistics involved are well documented in the STW report (0). Bowsers and tankers were supplied to STW from 18 water companies and a range of food, logistics and supply chain companies.

It was felt by many respondents, especially those at the delivery end of the incident, that there were areas where improvement to the system of the Mutual Aid Scheme could be made. These include improvements in the compatibility of bowsers and other equipment between companies; the availability and provision of mid-sized tankers; and the manner in which the provision of personnel is handled under the Scheme.

2.3.4.1 Compatibility of bowsers and other equipment

It became apparent that as the Mutual Aid Scheme brought in tankers and bowsers from across the UK, there were problems of incompatibility amongst the range of different makes, components, and ages of the equipment supplied, particularly of bowsers. In particular, there is no standard specification for bowser and tanker couplings such as fittings, level indicators and security mechanisms. This resulted in problems with deploying and filling bowsers.

Recommendation 13:
Water UK should use its existing emergency planning and security network to review the state of preparedness of the industry for future events; in particular the industry's Mutual Aid Scheme should be reviewed with a view to ensuring:
- *The technical compatibility of assets;*
- *The number and readiness of such assets;*
- *The means of deploying and managing staff made available under the Scheme;*
- *The resilience of the scheme to cater for simultaneous events.*

Recommendation 14:
Through Water UK the water industry should address the standardisation of emergency supply equipment to ensure that in the event of an incident equipment from other companies or organisations is compatible.

2.3.4.2 Provision of appropriate sized tankers

It also quickly became clear that there is a shortage of appropriate tankers, particularly mid-sized tankers, available to the industry. These are particularly useful for filling smaller static bowsers in urban areas and accessing sites which can only be accessed by narrow or restricted roads. Bowser locations are planned by water companies based on distance from consumers. Water companies need to review their intended location for bowsers and ensure that suitable tankers are available to allow replenishment. Operational planning needs to ensure that only appropriately sized tankers are deployed to certain locations.

2.3.4.3 Provision of personnel

The Mutual Aid Scheme also needs extension to include a protocol for enabling and managing the provision of personnel from supporting water companies as well as equipment. This should cover operational staff and supervisors, call centre staff, communications and media staff; as well as technicians and tanker drivers, in order to clarify chains of command, communication links, and to whom such staff report.

Recommendation 15:
Water companies should rehearse emergency plans on a regular basis. This should include physically moving equipment within individual company areas and ensuring with other companies that provisions under the conditions of Mutual Aid Scheme are available. These rehearsals should include the emergency response organisations. Such rehearsals should include scenarios allowing for disruptions to access to sites and locations due to flooded roads and facilities.

2.3.5 – Public health and recovery. Mains water supplies were lost to 350,000 consumers in the Gloucester floods, their requirements being met by a combination of bottled water and water from tankers and bowsers. When it became clear that the Mythe WTW would be inundated by the ingress of flood water, STW managed the shut down of the plant in an orderly fashion to facilitate the eventual restoration of supplies. When the plant came back on stream, STW had a choice of advising consumers that supplies were initially suitable for sanitation purposes only or of waiting until potable standard had been re-established.

Opinions on the matter were divided amongst the relevant agencies (health authorities and drinking water regulators) and much to-ing and fro-ing took place before it was decided that the best approach was to advise consumers that non-potable mains water was again available in the first instance for sanitation purposes and then some days later restored to drinking quality.

We agree this was the best approach. However, various issues of conformity with drinking water regulations arise, and the views of local and national public health authorities need to be factored in. This requires resolving and formalising amongst all the relevant parties, together with the associated communication needs, so that, next time, there is an agreed approach and process available.

In their submission to the Review Group the HPA state that:

> *"There should be a clear statement of the role of DWI and the water companies in emergencies. In particular there must be clarity about roles and responsibilities in deciding on, agreeing and implementing a re-connection strategy. The risk assessment for this should be informed by appropriate public health advice but all agencies should be aware of where the statutory responsibilities lie. These issues should be clarified and rehearsed within Local Resilience Fora as well as centrally." (HPA response to Review Group)*

Recommendation 16:
Through Water UK the water industry should establish a standard approach to the temporary use of non-potable water to restore sanitation supplies, clearly outlining the conditions and situations in which it should be considered as an option. To deliver this standard the industry will need to work at a national level with government, statutory consumer organisations, public health bodies and drinking water regulators.

Whilst in most cases the physical clean up of damage caused by flooding (with the exception of sewer flooding) will not be the responsibility of the water company considerations such as availability of resources and positive public relations could influence decisions to engage with the clean up process.

Recommendation 17:
Water companies should give post-event clean up operations further consideration as an opportunity to recover service to customers including in situations where responsibility is not directly attributable.

2.3.6 - Emergency plans. As stated during the introduction to this report the water industry is used to handling floods and other emergency events. Each company has in place plans for dealing with floods or other emergencies (as part of the requirement of the Security and Emergency Measures Direction (1998)). The issues to which we have drawn attention in the preceding paragraphs all have implications for the viability of those plans in real, foreseeable emergencies.

All water companies need to revisit their emergency plans to take account of the lessons learned from the 2007 events and to which we have drawn attention above. In particular, water companies need to reflect on the potential scale, severity and frequency of future floods, to enhance their preparations and planning accordingly, and to rehearse those plans regularly, preferably as part of the local gold or silver command structures.

The need for such a review is the basis of the Review Group's key recommendation which we repeat here for emphasis.

Water companies urgently put in hand a thorough review of their emergency response and contingency plans on the assumption that the sheer scale and severity of future floods may make current plans inadequate.

The above paragraphs and recommendations cover those areas where the Review Group feels that immediate lessons can be drawn from the events of the summer of 2007 and the responses required by water companies individually and the water industry as a whole.

There are, however, a number of areas where the Review Group believes more long term consideration needs to be given. These will be addressed in more detail in the Phase 2 report. The discussion below introduces these areas and outlines some of the key issues.

2.3.7 – Drainage responsibility and sustainable options. The responsibility for drainage of water is a complicated issue. The summary below highlights the legal responsibilities for drainage of water (in England and Wales):

- Under the Building Act 1984 and the Building Regulations 2000, local authorities have powers to require the satisfactory drainage of buildings by their owners – with ministerial responsibility being with Communities and Local Government (CLG).
- Under the Water Industry Act 1991 (the Act) sewerage undertakers have a general duty to provide, maintain and operate systems of public sewers and works for the purpose of draining their areas. This is enforceable only by Ofwat, with ministerial responsibility being with Defra. Under Section 106 of

the Act owners and occupiers of buildings have an automatic right to connect to the sewer system.

- Under the Highways Act 1980, highway authorities have responsibilities for the drainage of highways – with ministerial responsibility being with the Department of Transport (DoT).
- Under Part IV of the Water Resources Act 1991 and the Land Drainage Act 1991, powers to protect land and property from inundation by water is divided between the Environment Agency, Internal Drainage Boards and local authorities – with ministerial responsibility being with Defra.
- Under the Town & Country Planning Act 1990, planning authorities may impose conditions and enter into agreements relating to the drainage of developments – with ministerial responsibility being with CLG.

Sewerage undertakers are only responsible for drainage from customers' properties, not surface water in general. Many sewers are still in private ownership and not the responsibility of the sewerage undertakers. Other responsibilities lie with the environmental agencies and local authorities as well as highway authorities, land owners and, in rural areas, internal drainage boards (Figure 2).

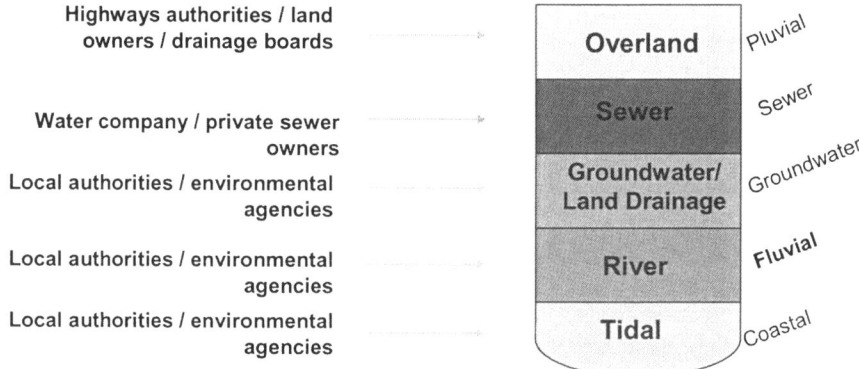

Figure 2 *Responsibility for components of urban drainage*

Almost all responders to the Review Group noted that there needs to be some consolidation and rationalisation in the roles and responsibilities of organisations in relation to flooding. For example, Gloucester County Council in their Scrutiny Report notes:

> *"...there should be a single agency with overall responsibility for ensuring the maintenance of watercourses, as the current system is not effective, and therefore recommends legislative change to create a single agency with overall responsibility for ensuring the maintenance of watercourses. The new system must include clear signposting for members of the public on how to report problems and on who is responsible for addressing those problems."*
> *(Gloucester County Council, 0)*

The EA also notes that there is

> *"...the need for clarity of responsibility and a strategic overview of all inland flood risk, and improved coordination and action on management of risk from all sources of flooding. ......We need a strategic framework to understand, mitigate and manage urban flood risk. The Environment Agency is well placed to deliver that strategic role." (Environment Agency response to Review Group)*

The Review Group will consider the range of opinions on and options for allocating strategic responsibility for flooding in Phase 2 of the review.

In the event of a flood householders and businesses are not concerned with whose water it is, but with getting the water away and minimising future risks.

The water industry cannot deal with the impact of flooding of sewerage and waste water services on its own and will therefore need to involve other stakeholders to resolve issues associated with sewers, drainage and flooding. These will include the environmental regulators, highways authorities, local authorities and drainage boards. There are currently a number of projects underway to further understand and implement sustainable drainage including Defra's Integrated Urban Drainage pilots and Making Space for Water programme.

In addition UK Water Industry Research (UKWIR) is currently completing its 21st Century Sewer Design research project which reviews the ability of sewers to cope with climate change impacts.

Key to delivering a long-term drainage strategy is the concept of sustainable drainage. This focuses on slowing down the transfer of water, particularly in urban areas where impervious surfaces and constrained channels move water quickly from where it falls into the drainage system. Whilst sustainable drainage alone would not have offered much assistance for events of the magnitude of summer 2007, the inclusion of sustainable drainage solutions into modern planning should be encouraged.

The water industry supports sustainable drainage initiatives and is involved in all 15 Defra Integrated Urban Drainage pilots. The Sustainable Urban Drainage System initiative (SUDS) is seen as a valuable tool to improve the manner in which the sector deals with surface water. However, SUDS are not flood defences. They can be used to ensure there is no increase in downstream flood risk but their primary purpose is providing surface water treatment. They are part of the solution but not sufficient on their own.

However, there are currently significant barriers to the design and implementation of SUDS:

- The right of water companies to refuse a design that is not appropriate;
- Health and safety issues associated with open water features that would be owned by private companies in residential areas;
- Many techniques are relatively new and the long term costs of maintenance are not well understood;
- Water industry responsibility for SUDS should be limited to those SUDS features that have connectivity with surface water sewerage systems (porous pavements, isolated infiltration systems, soakaways and swales should remain in private/local authority ownership).

The water industry is keen to resolve issues preventing the implementation of sustainable drainage systems and, through Water UK, is lobbying Government to adopt legislation which reflects the existing Code of Practice, developed by the national SUDS working group. In this it is particularly important that the following points are addressed by legislation:

- Sewerage undertakers being involved in the planning process;
- Ownership and maintenance and funding of SUDS;
- Definition of works capable of constituting SUDS;
- Design standards;
- Restrictions on rights of drainage to SUDS and connection to surface water drainage;
- Rights of discharge from SUDS to watercourses;
- Protection of groundwater, and surface water quality, in accordance with the Water Resources Act 1991 and EC Directives.

Central to the issue of maintaining sewers is to ensure that they are used only for the purpose and products for which they are designed. Misuse of the sewer network through the disposal of inappropriate materials such as cooking oils, sanitary products and other waste can drastically reduce capacity or cause internal blockages that could lead to backing up and flooding in the event of heavy rainfall.

The Review Group proposes to carry out further discussions with the other parties involved around the issues of sewers and sustainable drainage for the Phase 2 report.

2.3.8 – Resilience. The flooding in the summer highlighted two very different issues – pluvial flooding associated with heavy rainfall and exceeding drainage capacity, and fluvial flooding caused by rivers exceeding their discharge capacity. In Yorkshire and Humberside sewer flooding and drainage overflows due to sheer volumes of rainfall were key to the problem whilst in the Midlands river flooding overwhelmed the Mythe WTW, electricity sub stations and local rail and road networks.

Both incidents serve to highlight that parts of the water industry's infrastructure are susceptible to failure in the event of severe floods. In many cases if an asset fails there is sufficient back up available (for example when the Bransholme pump failed in Hull temporary pumping arrangements were put in place once the pumping station itself was overwhelmed). However there are areas of the infrastructure network where water companies recognise potential single points of failure, where there is no contingency to cover a failure of the asset. These can range from a number of customers at the end of a single piped supply through, to a water treatment works supplying a dedicated area.

The latter was the case with the Mythe WTW. This incident highlights the consequences of failure of single points of supply. In most cases there are engineering solutions that could secure the isolation of these assets: however, these will require (often significant) capital investment to accomplish. Gloucester County Council's Scrutiny Inquiry recommended that:

> *"...Severn Trent Water secure a secondary piped water supply for Gloucestershire, possibly via the Strensham Water Treatment Works, in order to ensure that they are able to maintain a piped water supply to Gloucestershire in the event of any future loss of the Mythe Water Treatment Works." (Gloucester County Council, 0)*

Discussions need to be held with the economic regulators to determine how such capital programmes should be assessed to ensure proper and appropriate funding. This will need to take the form of a risk based approach.

There are no generic standards for the resilience of the water industry's networks and guidance on this from the environmental regulators and government departments would be well received. Water companies already carry out flood risk reviews on assets, particularly above ground assets, though the understanding of the ability of sewer networks to cope with excess water is less well assessed or understood. We agree that all water companies should revisit the identification of major assets whose loss would seriously imperil maintenance of water services to large numbers of customers, and should evaluate what options may exist to improve the resilience of the system.

However, the existing layout of water networks means that it is unlikely to be practical to provide backup supplies to all properties and, given that the investments required to provide duplicate means of delivering water services to even the majority of consumers are likely to be significant, it is far from clear that there would be an economic justification for a wholesale upgrading of infrastructure resilience or whether customers would be willing to foot the bill.

A co-ordinated approach by and with the water industry is also needed, not least because the effectiveness of these various networks is only likely to be as good as the weakest link. With such a co-ordinated approach, it should be possible to produce plans that will address vulnerable sites or the needs of major blocks of consumers on a prioritised basis thereby maximising the value of available capital.

Clearly any improvements to the resilience of infrastructure will require not only that significant capital investment, but stakeholder consultation and political involvement. The Review Group plans to study this issue further through discussions with key stakeholders for the Phase 2 report.

2. 4 Timetable and actions

Given the severity of the events in June and July 2007, the risk of similar events happening again in the near future and the timely nature of the report in the light of the water industry price review cycle, it is important that the water industry reacts promptly to address the recommendations made in this report, as well as those made by other reports both published and underway.

Water UK has an extensive network of working groups comprised of members of water companies with expertise in specific areas such as emergency planning; regulation; sewers and sewerage and asset management. Through these working groups Water UK can call upon knowledge and experience that will be essential in delivering the recommendations outlined in both this report and in those other reviews of the flooding that have taken place since June 2007.

Recommendation 18:
The Review Group's final recommendation is that an appropriate group is established to oversee the actions on these recommendations, and those of other reviews, and to identify an appropriate method of reporting progress. This group should be governed by Water UK and should take its membership from water companies and other organisations with the appropriate knowledge and influence to complete the actions required.

2.5 Phase 2 – Longer Term Issues

The Review Group plan to follow up the Phase 1 report with a series of visits and discussions with key stakeholders to understand fully some of the key issues. As noted in part 1 of this report the aim is to publish the Phase 2 report in Spring 2008.

The scope of the report will remain flexible but will cover the questions around long-term policy and investment issues highlighted so far. This will include but not be limited to:

- Climate change impacts and implications for investment;

- Resilience of water infrastructure and assets;

- The allocation of responsibilities for flood prevention and remediation;

- Sewers – automatic right of connection, sewer flooding, ownership / private sewers, suitability of design standards in a changing climate;

- Impact of EU Flooding Directive;

- Public expectations in the event of future flooding.

Key References

1 The Cabinet Office, December 2007, "Learning Lessons from the 2007 Floods"
2 Hull Independent Review Body, November 2007, "The June 2007 floods in Hull. Final Report by the Independent Review Body 21st November 2007"
3 Yorkshire Water Services Ltd, December 2007, "Initial review of 'The June 2007 floods in Hull', Final Report by the Independent Review Body, 21st November 2007", Confidential report.
4 Severn Trent Water, September 2007, "Gloucestershire 2007. The Impact of the July Floods on the Water Infrastructure and Customer Service – Final Report"
5 Gloucestershire County Council Overview and Scrutiny Management Committee, November 2007, "Scrutiny Inquiry into the Summer Emergency 2007".
6 Environment Agency, December 2007, "Review of Summer Floods".
7 Association of British Insurers, November 2007, "Summer Floods 2007: Learning the Lessons"
8   Consumer Council for Water, September 2007, "Response to Loss of Water Supply"

Appendix 1 – Membership of Review Group on Flooding

Sir John Baker - Independent Chair
Jim Marshall – Secretary to Review Group
Jonson Cox – Group Chief Executive, Anglian Water
John Cuthbert – Managing Director, Northumbrian Water
Ronnie Mercer – Chairman, Scottish Water
Colin Skellett – Chairman, Wessex Water

Appendix 2 – Organisations providing substantial responses

Anglian Water
Northern Ireland Water
Northumbrian Water
Scottish Water
Severn Trent Water
South East Water
South Staffordshire Water
South West Water
Southern Water
Sutton and East Surrey Water
Thames Water
Three Valleys Water
United Utilities
Wessex Water

Consumer Council Northern Ireland
Consumer Council for Water
Environment Agency
Natural England
Water Services Regulation Authority
(Ofwat)

Defra
Department for Regional Development
Dumfries and Galloway Council
Highways Agency
Local Government Association
Enterprise Plc
Flood Protection Agency
Gloucestershire County Council
Gloucestershire Police
Grampian Strategic Coordinating Group
Strathclyde Police
Water Direct
Water UK Emergency Planners Network

Association of British Insurers
Energy Networks Association
National Farmers Union
Food Standards Agency
Health Protection Agency
The Met Office
UKCIP

RISK ASSESSMENT METHODOLOGY FOR WATER UTILITIES -RAM-W™
LESSONS LEARNED

J. J. Danneels

Sandia National Laboratories, P.O. Box 5800,Albuquerque, NM 87185-0719

1 INTRODUCTION

In partnership with the Awwa Research Foundation (AwwaRF) and Sandia National Laboratories (Sandia), shortly after the events of September 11, 2001, the Environmental Protection Agency (EPA) undertook a program to improve security at water utilities across the United States. During the Clinton administration, the importance of our critical infrastructures was highlighted by the National Security Council in Presidential Decision Directive 63 (PDD 63). PDD 63 was superseded by President Bush when he signed Homeland Security Presidential Directive 7[1] (HSPD-7) into law. HSPD-7, like its predecessor PDD 63, establishes a national policy for Federal departments and agencies to identify and prioritize critical infrastructure and key resources within the United States and to protect them from terrorist attack. HSPD-7 encourages Federal departments and agencies to form public and private partnerships to pursue the goal of lowering risk to our national assets due to malevolent events.

This paper describes the lessons learned from the effort to assess and mitigate the vulnerabilities of drinking water utilities.

1.1 Law Requires Vulnerability Assessments

The Bioterrorism Preparedness and Response Act of 2002 (PL 107-188) required community water systems that serve populations of greater than 3,300 persons to conduct vulnerability assessments. According to EPA statistics, approximately 4,800 water utilities fit this category[2]. When combined, these water utilities serve over 256 million people.

2 VULNERABILITY ASSESSMENT PROCESS

In cooperation with the EPA and AwwaRF, Sandia National Laboratories (Sandia) created the Risk Assessment Methodology for Water Utilities[3] known as RAM-W™. RAM-W™ was the most widely used methodology to assess security risks at large water utilities. Several thousand water utility owners/operators, regulators, and water industry consultants have been trained in the use of RAM-W™. Figure 1 illustrates the process followed in RAM-W™ and demonstrates the iterative nature of the methodology. This methodology was developed through decades of security research and development at Sandia. Ideally, the level of threat chosen drives the entire analysis. In many high-security applications, the threat is determined by a federal entity (e.g., the Department of Energy or the Nuclear Regulatory Commission) and a designated security analyst then evaluates the effectiveness of the security system. Most high-security applications also employ an on-site guard force, usually armed and well trained, to respond to malevolent incidents. Managers of the majority of civilian infrastructures do not employ a dedicated response force and operate geographically distributed assets, the majority of which reside in the public realm.

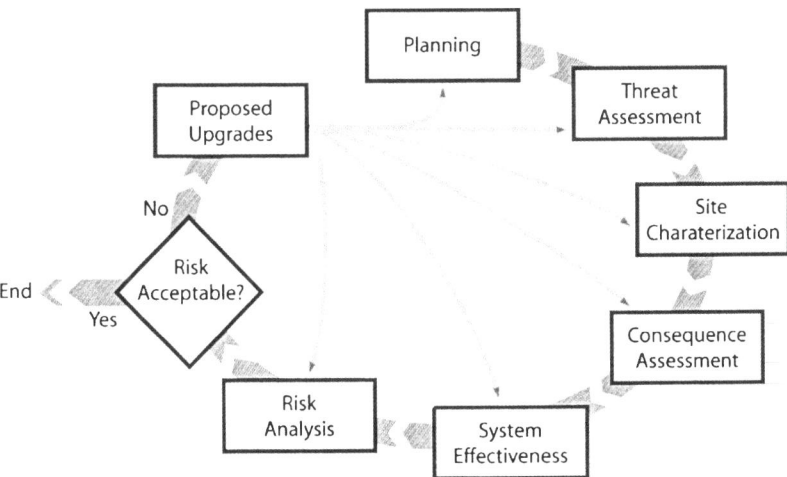

Figure 1 *RAM-W™ Process*

3 RESULTS

Sandia conducted several vulnerability assessments during the development and validation of RAM-W ™. A team of security professionals would visit a site, work with the water utility to evaluate their assets, and then take what was learned to tailor Sandia's core vulnerability assessment methodology for water utilities. The updated methodology and the results would then be shared with the water utility. Feedback from the water utility and from an AwwaRF Advisory Panel would then be incorporated after each assessment. After a series of assessments, purposefully including a wide variety of water utilities (surface water only, combined surface and ground water, ground water only, multiple treatment processes) the methodology was rolled out for use. Water utility owners/operators and consultants then applied the methodology at several hundred locations.

After the large water utility vulnerability assessments were complete, AwwaRF and Sandia teamed to collect information to better understand (1) how well the process worked, (2) areas of concern, and (3) what further developmental efforts to pursue. A report entitled *"Results from the Water Utility Vulnerability Assessment Lessons Learned Study*[4]*"* was then issued. The rest of this paper describes the challenges encountered during the performance of their vulnerability assessments as related by the water utility operators and water utility consultants.

3.1 Defining the Threat to Water Utilities

Although encouraged to contact local law enforcement and other authorities, most water utilities found it difficult to obtain relevant threat data. As stated earlier, the specified threat drives the risk analysis. Therefore, water utilities were faced with a high degree of ambiguity about what the actual threats might be while having to undertake risk reduction programs that may cost millions of dollars. Much of the utilities' infrastructure resides in the public realm, is broadly distributed, and would be very difficult to protect.

The Federal government has not defined a threat that can be used as the basis of a security design for the water infrastructure, nor is there agreement in the water community about what threats to consider. Therefore, the water utilities analyzed a multitude of threats and threat levels. Neighboring water utilities often used significantly different threats during their risk assessment. The recommended approach is to evaluate a threat spectrum to gain a better understanding of what the consequences might be from a range of potential attacks. The water utility will then need to determine the security level they would like to attain for their system.

3.2 Contamination of Water Supplies

One of the least understood threats to the drinking water industry is contamination, particularly in the water distribution system. At the beginning of the program to assess the vulnerabilities of water utilities, very little was known about malevolent water contamination and even fewer analytical tools were available to help understand and analyze the problem. Since 9/11, several groups, including AwwaRF, the EPA, and the Centers for Disease Control, have collaborated to collect and characterize information on contaminants that may pose a significant health threat in drinking water systems. Prioritizing contaminants, developing methods to rapidly detect them, developing a full understanding of contaminant fate and transport, developing estimates for contamination risks to water distribution systems, creating programs for isolating and treating contaminants, and final restoration of clean water supplies are all making significant progress, but still require further development.

3.3 Agreement on Critical Assets

One of the most important steps to take before getting very deep into the vulnerability assessment involves bringing folks from across the various divisions within the water utility together and following a prioritization process to come to agreement on the priority of critical assets. RAM-W™ provides what appears to be a simple, pairwise comparison process for performing this analysis. Turns out that it takes quite a bit of time and several meetings to accomplish the goal of developing a true priority scheme for the utility's assets. The lessons learned feedback indicated that the final list was quite a surprise to most of the participants and could not have been anticipated. Highly visible assets are often thought to be the most critical, but over and over the water utilities found assets that weren't very visible ended up being the most crucial to their operations.

3.4 Security Culture

This is an issue that not only plagues organizations new to security, but high-security organizations struggle with it as well. There are no short-term solutions. Convincing everyone in the organization that how well they perform their security responsibilities could have a tremendous impact on the health of their customers, that staying vigilant constantly for something that in all likelihood will never materialize, and being committed to procedures that make their jobs more difficult, is not an easy task. To be successful at improving the security culture requires:

- Training
- Management leadership
- Commitment
- Practice

3.5 Total System Operation

During the on-site assessments, the security team noted that very few of the water utility participants understood the total operation of their system. In fact, at a couple of the large metropolitan water utilities, there were only 1 or 2 employees that could explain why the system operated the way it did and how parts of it were designed. This made it very challenging to determine where to collect additional data and how important certain operations might be to the whole. Knowledge preservation and how it's effectively communicated during a time when most water utilities are being downsized is a challenging task, but one that must be undertaken.

3.6 Contingency Plans

Contingency plans are a critical component of any security program. Planning, documenting, and practicing the response to major system disruptions should be an on-going effort. Many water utilities had made the effort to write a plan, but few had gone beyond tabletop exercises to determine the efficacy of those plans. Fewer still had actually run large-scale exercises with emergency responders and other stakeholders. In the middle of an emergency is not the time to find out how well your plan is thought out and how likely you are to recover your operations.

4 RECOMMENDATIONS

The water distribution system has long been known to represent one of the greatest security vulnerabilities. Current challenges include a lack of clear understanding of the fate and transport and consequences of potential contaminants within a water distribution system coupled with generally easy access into the system. To minimize the potential risks from a malevolent contamination attack, it is first necessary to develop computational tools that can predict the fate and transport of contaminants within distribution systems, or more generally, how contaminants might move in a hydraulically complex pipe network.

Methodologies for conducting vulnerability assessments should include a framework for cleanup and recovery. The tools to estimate the fate and transport of contaminants within a water distribution system could also play a significant role in developing a methodology for recovery after such an event and could serve as the instrument to integrate both components for the protection of drinking water systems.

Mitigating threats may require the development of new approaches that greatly enhance the time an adversary needs to complete a malevolent act. Threats can be countered by storing high-consequence assets underground, limiting the paths an adversary might exploit and thereby creating long task times. For example, pumping stations could be protected better by installing them below ground in protected shelters.

In testimony to the United States House of Representatives Committee on Science entitled "*H.R. 3178 and the Development of Anti-Terrorism Tools for Water Infrastructure*," Sandian Jeffrey J. Danneels suggested several alternatives that might provide the improved security desired at a much lower cost than the physical security approaches currently in use. Research dollars should be made available to study alternatives that put final treatment of the water supply closer to the consumer, consider much of the present potable water system as non-potable to decentralize the impact of a potential event, and evaluate the efficacy of creating municipal bottling facilities and other novel approaches that provide the level of security demanded by the water consumer and which may not be achievable through any other means.

5 CONCLUSIONS

Understanding and analyzing the vulnerabilities within the water infrastructure is a very important, yet challenging, undertaking. The efforts completed to date have highlighted several vulnerabilities that will require significant amounts of effort to correct. A large investment will be required to provide even minimal levels of security for this important resource. "When is enough enough?" will be a difficult question to answer and will be debated for years to come.

Acknowledgements

Sandia is a multiprogram laboratory operated by Sandia Corporation, a Lockheed Martin Company, for the United States Department of Energy under Contract DE-AC04-94AL85000.

References

1 *Homeland Security Presidential Directive/HSPD-7;* Office of the Press Secretary; Issued by President George W. Bush, Washington, D.C., December 2003.

2 *Risk Assessment Methodology for Water Utilities (RAM-W^{TM}),* Second Edition, Awwa Research Foundation and Sandia National Laboratories, Published by Awwa Research Foundation, Denver, Colorado, 2002.

3 *FACTOIDS: Drinking Water and Ground Water Statistics for 2001*; Environmental Protection Agency; http://www.ngwa.org/pdf/01factoids.pdf

4 *Results from the Water Utility Vulnerability Assessment Lessons Learned Study*, Awwa Research Foundation and Sandia National Laboratories, published by Awwa Research Foundation, Denver, Colorado, 2004.

RISK-BASED APPROACHES TO WATER QUALITY MANAGEMENT: INTEGRATING PUBLIC HEALTH METRICS IN WATER SAFETY PLANNING

G. Howard

DFID, Abercrombie House, Eaglesham Road, East Kilbride, Glasgow G75 8EA

1 INTRODUCTION

Despite the public health basis for water supply improvement, there remains little systematic attempt to quantify or monitor the public health benefits expected from investments. Attention over many years has primarily been given to monitoring the quality of water for indicator bacteria and chemicals rather than considering what health risks is posed by waters of different quality and the public health basis for setting drinking-water quality objectives.

In the 3rd edition of the WHO Guidelines for Drinking Water Quality (1) a renewed emphasis was placed on defining public health benefits as the basis for water quality improvement. In particular, it was recommended that health-based targets be set, supported by water safety plans. Health-based targets can take a number of forms, ranging from expected disease reduction to specified technology targets. In all cases they need to be based on quantifiable health gains that have been demonstrated to occur.

Assessing public health benefits of water has to take into account the wide range of organisms and chemicals now known to exert an adverse health effect. Thus trying to assess overall impacts on health needs to consider a wide range of possible effects, some of which may be difficult to detect in the short-term.

Interventions to address one particular water quality problem may inadvertently lead to the introduction of other hazards – or risk substitution. This may be found with the formation of harmful by-products by disinfection, but may also occur when new forms of water supply are introduced to reduce exposure to a particular hazard. This happened, for instance, in Bangladesh where sinking of shallow tubewells to reduce microbial contamination of drinking water led to increased exposure to arsenic (2). Where risk substitution may occur, the health benefits from the intervention need to be assessed and compared to the newly introduced risk. Because this often involves comparing dissimilar hazards with very different impacts on public health, a common metric must be found.

2 USING DISEASE BURDEN APPROACHES

Havelaar and Melse (3) proposed using a disease burden approach offered a more effective way of assessing the safety of drinking water than traditional methods. These authors recommended the use of Disability-Adjusted Life Years (DALYs) as a common metric. The use of DALYs permits the comparison of microbial and chemical threats, thus allowing trade-offs between different types of contaminant that may derive from the use of different water sources or use of different forms of disinfection could be quantified. Disease burden approaches include the full spectrum of health effects from individual contaminants allowing a more meaningful assessment of the disease burden represented by particular levels of water quality (3).

Using the work by Havelaar and Melse, WHO in the 3rd edition of the guidelines for drinking water quality, adopted a disease burden approach for drinking water quality (1). The GDWQ went as far as establishing a 'reference level of risk' of 10^{-6} DALYs, a level that is broadly equivalent to the 10^{-5} cancer risk that has been used for many years when establishing guideline values for chemicals. By establishing such reference levels of risk, there is an assumption that there is some mechanism by which the disease burden can be quantified. WHO suggest that this could be through epidemiological surveillance or through the use of quantitative health risk assessments (QHRA).

2.1 Quantitative health risk assessment (QHRA)

QHRA can be applied in both developed and developing countries, for differing reasons but with the same objective of improving policy and practice with regard to improving water supplies and wastewater treatment and disposal. QHRA offers a cost-effective way of reaching decisions that are supported by public health evidence, provided the assumptions used and constraints of the study are made transparent. In this regard, the use of QHRA provides points of comparison to expected public health gains; assessment against targets that are not easily quantified through epidemiological study; or, between different technological options whose delivery has been too limited to make a thorough assessment of public health impact.

In industrialised countries, the use of quantitative risk assessment is a means of estimating of what level of residual health burden may result from delivery of drinking water in situations where expected numbers of excess cases of illness are too low for routine epidemiological surveillance to identify (1). It can also used as a means of understand risks posed by water meeting numerical standards for indicator bacteria, which have been shown in some circumstances to still be a significant amount of gastro-intestinal disease (4,5).

In developing countries the use of QHRA allows an assessment of different technologies. This can be done whilst the numbers deployed remains reasonably small, thus reducing the subsequent cost of replacing those that have proved ineffective or too high a risk. QHRA has a particular value in settings where there is significant chemical contamination as well as microbial contamination, for instance in those regions affected by arsenic and fluoride.

3 CASE STUDIES

The two case studies below provide a picture of the type of studies undertaken to date to assess disease burdens associated with drinking water in developing countries. The assessment undertaken in Kampala, Uganda, was an attempt to quantify the likely public risk posed by different water sources used by an urban population. The case study in Bangladesh is quite different. It is a rural study and was specifically designed to assess the level of risk associated with moving to different alternative water supplies installed to mitigate the risks of arsenic contamination of shallow groundwater.

3.1 Uganda

In Uganda, a QHRA was undertaken as part of a wider project to develop water safety plans (6). A specific objective within this project was to test whether the simplified approach to QHRA outlined by WHO in the Guidelines for Drinking Water Quality was workable in environments with limited data. The approach adopted is shown in table 1.

Table 1 *Simplified procedure for calculating disease burden (adapted from WHO, 2004)*

Raw water quality, organisms per litre (C_R)	Will probably be calculated from concentrations in standard volumes (e.g. 100ml) and may not be directly for pathogens
Treatment effect (PT)	Estimated or calculated removal of pathogens
Drinking water Quality (C_D)	$C_R \times (1-PT)$
Consumption of unheated drinking water (V)	Estimated or calculated
Exposure by drinking water, organisms per litre (E)	$C_D \times V$
Dose-response(r)	From literature
Risk of infection per day ($P_{inf,d}$)	E x r
Risk of infection per years ($P_{inf,y}$)	$1-(1-P_{inf,d})^{365}$
Risk of diarrhoeal disease given infection ($P_{ill|inf}$)	From literature
Risk of diarrhoeal disease (P_{ill})	$P_{inf,y} \times P_{ill|inf}$
Maximum disease burden (mdb)	Calculated from available data and from data reported in literature
Susceptible fraction (fs)	From literature
Disease Burden (DB)	$P_{ill} \times mdb \times fs$

An assessment was made of the utility provided piped water supply in the capital city, Kampala, and comparisons made with the principal alternative water supply – protected springs – commonly used by the poor for at least part of their water supply. Large numbers of people used multiple sources of water for drinking, often switching between poorly protected springs and tap water depending on their financial situation of a daily basis (8). The assessment focused solely on microbial hazards as there was no evidence of significant chemical hazards in the source or finished waters at concentrations likely to lead to widespread health effects.

The utility piped water supply had two treatment works – Gaba 1 and Gaba 2 – which served 871 kilometres of pipe, passing through a number of service reservoirs. Both plants employed conventional treatment, although there was no coagulation-flocculation-settling step in Gaba 1. The piped network supplied 20% of the population through household connections. Of the unserved population, 68% used tap water as their sole, primary or secondary source of water. The remaining population relied on protected springs and a small percentage on other alternative supplies. As households collecting water from a communal tap had to store water at home and the risks of re-contamination high, an assessment was also made of the risk posed by this water.

The assessment derived expected disease burden estimates for three reference pathogens – *E.coli* O157, rotavirus and *Cryptosporidium parvum* – as recommended by WHO (1). Dose-response relationships were derived from the literature and allowance was made for the impact of HIV status on susceptibility to infection, particularly for *Cryptosporidium,* which has been noted as a significant pathogen for adults and a cause of persistent diarrhoea in children, and for which water is a well established route of transmission in developing countries (7). An assumption of 1 litre per day on unboiled water was made taken from the WHO Guidelines for Drinking-Water Quality and allowances were made for the use of other water sources. A discounting factor was used to try to make explicit within the model that many people, particularly the poor, used multiple sources of water in Kampala (8).

Data were collected on a range of indicator organisms to estimate the likely presence of pathogens, For *E.coli* O157, thermotolerant coliforms were used as an indicator. Based on the literature, it was assumed that 95% of all thermotolerant coliforms were *E.coli* (1) and that 8% of *E.coli* were pathogenic (9). For the sake of the assessment the figure for pathogenic *E.coli* were assumed to be O157, although this would be unlikely in reality it provided a useful means of capturing the health impact of the full range of pathogenic bacteria. Given limited data on dose-response for *E.coli* O157, the established dose-response for shigella was used given similarities between the two organisms (9). Coliphage was used as a surrogate for rotavirus and sulphite-reducing clostridia were used as a surrogate for *Cryptosporidum parvum*. In both cases, the surrogates were treated as if they were equivalent to one pathogenic organism.

Water quality data collected through two specific assessments at each treatment works and through the distribution system, and through accessing routine monitoring data. Data on the quality of protected springs and household water from a routine surveillance programme were used to assess risks from these sources of water. The results of the assessment are shown in table 2 below.

Table 2 *Results of Uganda QHRA*

	E. coli	Rotavirus	C. parvum
Gaba 1 works (2002)	2.9^{-6}	7.88^{-3}	6.11^{-5}
Gaba 2 works (2002)	1.63^{-6}	7.10^{-3}	1.02^{-4}
Distribution system (1998)	5.26-^{4}		
Distribution system (1999)	2.92^{-4}		
Protected springs	8.67^{-2}		
Household water	2.82^{-2}		

Table 2 shows that the treatment works produced water of generally very good quality and that the public health risks posed from water coming from the plants was consistent with the WHO reference level of risk. Water within the distribution system, however, was much poorer and represented a significant increase in risk, being two orders of magnitude higher than the WHO reference level of risk. This remain, however, much lower than the risk posed by the protected springs and indeed for recontaminated household water.

3.2 Bangladesh arsenic mitigation

This study was designed to assess the potential health risk posed by alternative water supplies installed to mitigate the effect of arsenic (10,11). The data on disease were more extensive than those for Uganda, but the assessment had to consider the risk from both arsenic and microbial hazards.

Bangladesh had reached 97% water supply coverage in rural areas in the early 1990s through the use of shallow tubewells fitted with handpumps. The installation of nearly 11 million such tubewells was an important contributor to a decline in diarrhoeal disease. However, in 1993 arsenic was first detected in the northwest of the country and a national survey in the late 1990s indicated that 29% of tubewells exceeded the national standard of 50µg/l and that 46% exceeded the WHO Guidelines Value of 10µg/l (12).

As a consequence a large-scale mitigation programme was developed and in 2004 the Government published a National Policy for Arsenic Mitigation (13). This was supported by an implementation plan. A number of technologies were identified in the implementation plan for use in arsenic mitigation including dug wells, rainwater harvesting, pond sand filters and deep tubewells. Priority was given to the first three of these and deep tubewells were only to be used when others were not considered viable.

Unfortunately, no consideration was given to the potential for these technologies to re-introduce microbial hazards and the risk substitution potential was considered to be high (10). As a consequence, a study was set up to assess the risks associated both with each technology from microbial hazards and arsenic. A cluster survey approach was adopted to identify a statistically representative sample of each technology. A total of 156 water supplies were included (36 each for dug wells and deep tubewells, and 42 each for rainwater harvesters and pond sand filters). Each water source was then sampled in the dry season and monsoon and water analysed for thermotolerant coliforms, with limited confirmatory *E.coli* and arsenic.

A deterministic point risk model was developed based on calculated median risks that could be run on generally available spreadsheet packages. The point risk estimates were compared to those that were found through a probabalistic model and found to be consistent. For arsenic, measured concentrations was used as the input data and applied directly to the dose-response relationship. Only those health effects for arsenic that are most strongly supported and for which disease burden estimates were available were used. Disease burden estimates were taken from Havelaar and Melse (3) and Murray and Lopez (14). The dose-response model of Yu et al (15) was used, with the volume of water consumed taken from Watanabe et al (16).

For microbial hazards, three reference pathogens were selected – *a* composite bacterium using traits from *Salmonella*, *Shigella dysenteriae* and *E.coli* O157, rotavirus and *Cryptosporidium parvum*. The inputs to the model were thermotolerant coliforms and it was assumed that 85% would be *E.coli* based on a log-normal distribution. In order to derive likely pathogen concentrations represented by the *E.coli* in the water, relationships were established based on long-term sewage monitoring in Australia. The median ratios for *E.coli* to the pathogenic bacterium and for rotavirus were both 10^5, with the median ratio of

E.coli to protozoa 10^6 (11). A variety of dose-response models were used, as described in Ahmed et al (17) and Howard et al (11).

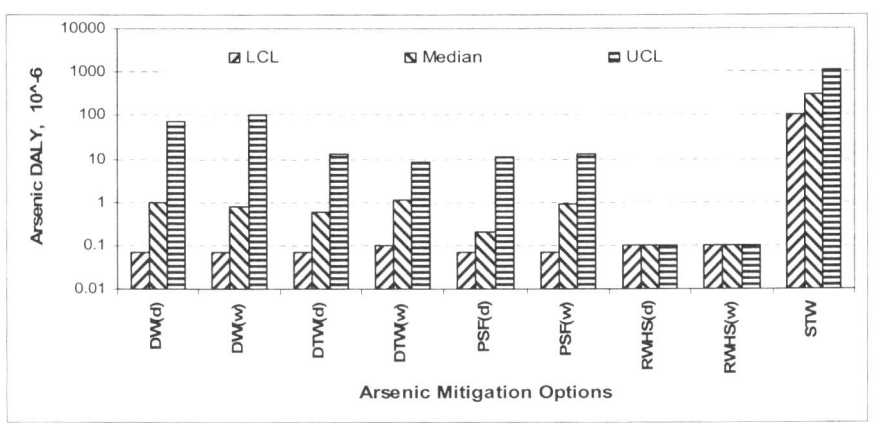

Figure 1 *Arsenic disease burdens in Bangladesh*

The results of the modelling are shown in Figure 1 above and Table 3 below. This shows that in general the microbial risks associated with these technologies is much higher than the arsenic risk, although there were a large number of dug wells in particular where the national standard was exceeded. This is not surprising as the level of risk posed by the Bangladesh standard for arsenic (185^{-6}) is the same as for 1.4 thermotolerant coliforms cfu/100ml. The risks from viral contamination dominated the microbial disease burden until high levels of contamination were reached and viral infection has reached saturation point. Protozoal risks were largely negligible.

Table 3 *Upper 95%ile microbial disease burden in Bangladesh*

Technology	Dry season	Monsoon
Dug well	1.11^{-2}	1.34^{-2}
Deep tubewell	6.98^{-5}	1.26^{-3}
Pond sand filter	1.07^{-2}	1.3^{-2}
Rainwater harvesting	6.48^{-3}	3.73^{-3}

Of the technologies assessed, only deep tubewells approached the WHO reference level of risk for microbes and this only in the dry season. In the monsoon, the risk increases substantially. Rainwater harvesters were significantly above the WHO reference level of risk, although the estimate for the dry season requires further consideration because this depends on the source of the thermotolerant coliforms found. If these are derived from washing-in from the occasional storms that occur during the dry season, the estimate is likely to be valid. If, however, more of the thermotolerant coliforms were environmental and were present as a result of re-growth, then the estimate is likely to significantly over-state the risks to health because there is no evidence that re-growth of bacteria are a significant cause of gastrointestinal disease (18).

Both pond sand filters and dug wells showed a high level of risks in both dry and monsoon seasons and are four orders of magnitude above the WHO reference level of risk for microbial hazards. The only realistic ways of controlling these risks would be through chlorination, although experience with community-level chlorination has not generally been positive.

4 DISCUSSION

The case studies demonstrate that the approach can be used to deliver useful and important answers to key questions, where these are rigorously defined and the design of the assessment is clear about the assumptions made. QHRA is more flexible and lower-cost than epidemiological studies or routine epidemiological study, although the quality of outputs relies heavily on the reliability of local health data.

Both case studies show that QHRA need to be designed to answer specific questions and to accommodate a range of different problems in the assessment design. The most important lesson that is emerging is that there are no blueprints that can be automatically applied, particularly in developing countries where context and data vary enormously, and that without a clear understanding of the local context such risk assessments can be difficult to interpret.

In Uganda, the study showed that the piped water supply represented a potentially higher risk than expected and emphasised the need for implementation of the water safety plan that was being designed at the time (19). It demonstrated that overall, greater efforts needed to be made to extend access to the piped water system which was of much lower risk than the alternative supplies. The study also demonstrated that even with limited data, quantitative risk assessments could be implemented that gave reasonable findings and which could be used to direct future investment in operation, maintenance and capital investments.

In the Bangladesh case, the outcome of the study was widely used by policy-makers and project implementers in assessing the technology options available for arsenic mitigation. It provided clear guidance that when dug wells and pond sand filters were used, chlorination was essential, at least in the monsoon, to keep risks within an acceptable limit. The study also demonstrated that the use of tubewells sunk into older, deeper aquifers that are arsenic-free represent the best form of water supply from a public health perspective.

This was reinforced by evidence collected in this survey and others that households preference is for deep tubewells than other water sources and that these were the type of supply most likely to remain functional (20). The Bangladesh study demonstrated the value of using data from similar environments when establishing the relationship between indicators and pathogens and the value of accessing local disease records.

There remain weaknesses as shown in both case studies. In all cases the assumptions are quite significant and although most potential sources of error were minimised as far as possible, in each case key sources of error remained. In both countries there are uncertainties introduced by the assumed relationships between pathogens and indicators, although the Bangladesh case study was a clear improvement in reducing these uncertainties. Without direct analysis of pathogens it is unlikely such uncertainties can be reduced, but the reality is that such analyses are unlikely to be routinely available in developing countries for the foreseeable future. In both case studies, however, uncertainty would be reduced if analysis had been for *E.coli* rather than thermotolerant coliforms.

In Uganda, the estimates for the non-bacterial pathogens are very uncertain and cannot be taken as representing more than an initial formative assessment of the likely risks. For the bacterial pathogens, the estimates appear realistic but are heavily influenced by the discounting used to try to capture the reality of multiple source use. The final figures are risks for the whole population of Kampala. Thus risks to communities and household relying heavily on the protected springs is likely to be significantly under-estimated. The quantity of water consumed was also set relatively low for people living in hot climates and a higher consumption of water would result in significant increases in risks.

In Bangladesh, It was assumed that the water consumed reported by Watanabe et al (16) was unboiled, although this was not explicitly stated. Had this referred to overall water consumption and unboiled water consumption was lower, this would have had a significant impact on microbial risk, although not on the risk from arsenic. It should also be noted that a number of arsenic health effects were not considered, including those at early onset, because there is no consensus on the severity weighting these should receive or because the evidence of the role of arsenic in causing these effects is still emerging. Nonetheless, it is worth noting that inclusion of other end-points would have significantly increased the arsenic-related disease burden.

5 CONCLUSION

The use of QHRA to estimate public health burdens posed by water supplies is a useful tool in directing both policy and practice. Both case studies show how the outputs of QHRA can be used to effective change in urban and rural water supply which have a sound public health basis. The case studies demonstrate that QHRA can be performed using relatively simple models, using existing analytical methods and do not rely on expensive additional water quality analysis. The results, however, must always be interpreted carefully and care should be taken to ensure that the assumptions made are always clearly set out and uncertainties acknowledged and where possible quantified.

6 REFERENCES

(1) *World Health Organization, Guidelines for Drinking-Water Quality*, 3rd edition, 2004, World Health Organization, Geneva, Switzerland.
(2) M.F. Ahmed, *Arsenic Contamination: Bangladesh Perspective*. ITN-Bangladesh, Dhaka, 2003.
(3) A.H. Havelaar, J.M. Melse, *Quantifying public health risks in the WHO Guidelines for Drinking-Water Quality: a burden of disease approach.* 2003, RIVM, Bilthoven, Netherlands (www.who.int/water_sanitation_health) .
(4) P. Payment, L. Richardson, J.Siemiatycki , R.Dewar R, M.Edwardes, E.Franco, *Am. J Pub. Health* 1991 **81**(1991) 703-708.
(5) P.Payment, P.R.Hunter, in: **L**.Fewtrell, J.Bartram J (eds.) *Water quality: guidelines, standards and health,* International Water Association, London, 2001, 61-88
(6) G. Howard, S.Pedley, S.Tibatemwa,. 2006 Quantitative microbial risk assessment to estimate health risks attributable to drinking-water supply: Can the technique be applied in developing countries with limited data? *J. Wat. Health, 2006 4 (2006), 49–56.*

(7) P.Kelley, K.Sri Babooi, P.Nduban, M.Nchito, N.P.Okeowo, N.P.Luo, R.Feldman, M.J.G. Farthing, *J. Infect. Dis. 176(1997), 1120–1123.*

(8) G.Howard, J.Teuton, P.Luyima, R.Odongo Int. *J. Environ. Health Res. 12 (2002), 63-73.*

(9) C.N. Haas, J.B. Rose, C.P. Gerba, 1999 *Quantitative Microbial Risk Assessment.* John Wiley, New York, 1999.

(10) G. Howard, M.F. Ahmed, A.J. Shamsuddin, S.G. Mahmud and D. Deere, *J. Health, Popul. & Nutr. 24 (2006) 346–355.*

(11) G. Howard, M.F. Ahmed, P. Teunis, S.G. Mahmud, A. Davison, D. Deere, *J. Wat. & Health 5(2007), 67–81.*

(12) BGS, DPHE Arsenic contamination of groundwater in Bangladesh. *BGS Technical Report WC/001/19, Volume 1: Main findings. British Geological Survey, Keyworth, 2001.*

(13) Government of Bangladesh, *National Policy for Arsenic Mitigation and Implementation Plan for Arsenic Mitigation in Bangladesh,* Government of Bangladesh, Dhaka, 2004.

(14) C.J.L Murray, A.D. Lopez, *The Global Burden of Disease: A Comprehensive Assessment of Mortality and Disability from Diseases, Injuries and Risk Factors in 1990 and Projected to 2020.* Harvard School of Public Health, Harvard, Massachusetts, 1996.

(15) W. Yu, C.M. Harvey, C.F. Harvey, *Wat. Resour. Res. 39(2003), 1146–1163.*

(16) C. Watanabe, A. Kawata, N. Sudo, M. Sekiyama, T. Inaoka, M. Bae, R. Ohtsuka, *Toxicol. Appl. Pharmacol. 198(2004), 272–282.*

(17) M.F. Ahmed, S.A.J. Shamsuddin, S.G. Mahmud, H. Rashid, D. Deere, G. Howard, *Risk Assessment of Arsenic Mitigation Options (RAAMO).* Arsenic Policy Support Unit, Govt. of Bangladesh, Dhaka, Bangladesh, 2005

(18) P.R. Hunter, in: J. Bartram, J. Cotruvo J, M. Exner, C. Fricker, A. Glasmacher (eds), *Heterotrophic plate counts and drinking-water safety.* London: IWA Publishing, 2003,119-36.

(19) G. Howard, S. Godfrey, S. Tibatemwa, C. Niwagaba, *Urban Water Journal 2(2005), 161–170.*

(20) A. Kabir, G. Howard, *Internat.J. Environ. Health 17(2007), 1–17.*

HOW STANDARDS CAN ASSIST THE ASSESSMENT OF, RECOVERY AND PREVENTION OF FUTURE EMERGENCIES

R. Greaves

British Standards Institution, 389 Chiswick High Road, Chiswick, London, W4 4AL

1 INTRODUCTION

The British Standards Institution (BSI) was founded in 1901 to co-ordinate the development of national standards in the United Kingdom and was incorporated by Royal Charter in 1929. BSI is also a founder member of the European standards organisations (CEN and CENELEC) and the international standards organisations (ISO and IEC). BSI was formally recognised by the British Government as the UK National Standards Body under a Memorandum of Understanding signed in 1982, but had in effect filled that role since 1901.

BSI works with government and industry to put into place the standards, systems assessment and registration, product testing and certification, and training that enable companies to compete on equal commercial terms in the European and international market places.

BSI is a not-for-profit organisation. It comprises three divisions:

- BSI British Standards (NSB)
- BSI Management Systems
- BSI Product Services

As well as being the UK National Standards Body, the organization operates in over 110 countries, has 56 offices globally and employees 2,250 staff worldwide.

2 REASONS FOR STANDARDIZATION

Standards are a powerful tool for organizations of all sizes, supporting innovation and increasing productivity. Effective standardization promotes forceful competition and enhances profitability, enabling a business to take a leading role in shaping the industry itself.

Standards can be a very useful tool for organisations to achieve their strategic objectives and there are many reasons why they decide to develop new standards or to become involved in the maintenance of existing standards. Some of the reasons underpinning the business case for involvement in standards include:

- Brand protection
- Customer expectations for consistency in delivered products and services
- Minimising risk in a volatile business market
- Effective internal communications
- Effective internal co-ordination
- Effective internal control
- Assist in complying with legislation
- Pre-empt government regulation by introducing good practice
- Clarity of specifications with suppliers
- Elimination of unacceptable practices
- Dissemination of good practices – internal and external
- Demonstrate and control global leadership
- Recognition for leadership
- Codification of tacit knowledge
- Integration of diverse operations/business units
- Introduce continuous improvement
- Flexible to rapid change
- Ease of use in complex environments
- Develop measurement techniques

2.1   Participation in standardization can be a powerful marketing tool

Manufacturing products and/or supplying services to appropriate standards maximises companies' compatibility with those manufactured or offered by others, thereby increasing potential sales and widespread acceptance. Standards are an important enabler for this to occur. As companies look to grow their markets as well as their own market share, being at the forefront of standardization is crucial. It gives companies the ability to help ensure the quality of the technical content in standards (and ultimately the performance of products and services), and helps demonstrate and solidify their position as industry leaders.

As consumers become increasingly informed about their choices, conformity to recognised standards becomes pivotal. An example is the international standard for environmental management (ISO 14001), increasingly used by businesses to demonstrate environmental responsibility. Those organisations involved in the development of those environmental standards have made significant gains because of customers' recognition of their expertise and leadership in this field.

2.2   Standards provide the basis for certification and quality marking

Standards can set out certifiable requirements that can be independently verified and provide assurances to customers that the products and services they are purchasing operate, function and perform in accordance with good practice. Furthermore, they can provide the basis for quality marking schemes, allowing customers to easily recognise those products and services that comply with quality standards, Users often look for the independent verification that technical standards provide. Quality marks earned by businesses whose products and practices consistently stand up to rigorous examination are instantly recognizable and act as respected badges of quality, safety and performance.

2.3   Standards improving business

In modern business, effective communication along the supply chain and with legislative bodies, clients and customers is imperative. Standardization can deliver measurable benefits when applied within the infrastructure of a company itself. Business costs and risks can be minimised, internal processes streamlined and communication improved. Standardization promotes interoperability, providing a competitive edge necessary for the effective worldwide trading of products and services.

3 OVERVIEW OF STANDARDIZATION

What are standards?

Put at its simplest, a standard is an agreed, repeatable way of doing something. It is a published document that contains a technical specification or other precise criteria designed to be used consistently as a rule, guideline, or definition. Standards help to make life simpler and to increase the reliability and the effectiveness of many goods and services we use. They are intended to be aspirational - a summary of good and best practice rather than general practice. Standards are created by bringing together the experience and expertise of all interested parties such as the producers, sellers, buyers, users and regulators of a particular material, product, process or service. They are also developed voluntarily.

Standards are designed for voluntary use and do not impose any regulations. However, laws and regulations may refer to certain standards and make compliance with them compulsory.

One common misconception is that standards are obligatory. This is not true. All standards are voluntary, unless when taken up into legislation (which is very rare). Furthermore, they can take many different forms such as guidelines, methods, codes of practice, requirement specifications and vocabulary.

There are specific rules for drafting standards that must be adhered to. These are designed to ensure that standards meet their aim of providing, for common and repeated use, rules guidelines or characteristics for activities. They are founded on usability, verifiability and commonality.

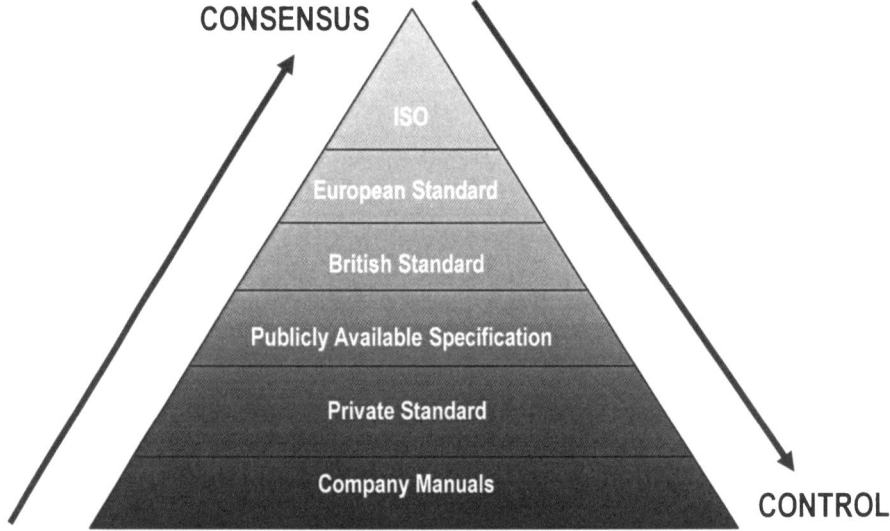

Figure 1 *Standards Pyramid – a representation of the structure of standardization models*

Lead times for standards vary from a matter of months to several years. British Standards are usually developed within 12–15 months, whilst international standards take around 3 years. Commissioned standards such as PAS and PS can be developed within months to meet industry requirements.

4 BESPOKE AND FAST TRACK STANDARDIZATION

In order to develop standards and services which are responsive to particular industry needs, BSI developed its Standards Solutions department. This department was founded in 2001 as BSI Professional Services and is an operating unit of BSI British Standards.
This department of BSI retains BSI's core values of Independence, Integrity and Innovation but develops standards and services in respond to specific sector needs. A core service is the ability to develop Publicly Available Specifications (PASs).

PASs are developed using a similar process to that of British Standards and are commissioned by an external sponsor, subject to stakeholders consultation and do not require full consensus. These differences between a PAS and formal consensus British Standards enable PASs to be developed and published faster in order to satisfy urgent business needs particularly in areas of technology, innovation or service industries. Some PASs may evolve into formal consensus standards (e.g. PAS 56 - Business Continuity, published in 2003 - was replaced by a British Standard (BS25999) in 2006), demonstrating BSI British Standards' pioneering involvement in business continuity ahead of any other standards maker.

Fast-Track standards are:

- Tailor-made for each individual sector, developed to their needs
- Fast, flexible and effective

- Sector sponsored
- Endorsed by BSI and co-branded with sector stakeholders
- Developed in 6 - 9 months
- Guided and supported by a dedicated Project Manager
- Developed through a rigorous consensus-based process
- Enabled for the development of certification schemes
- An opportunity to lead an industry at national and international levels

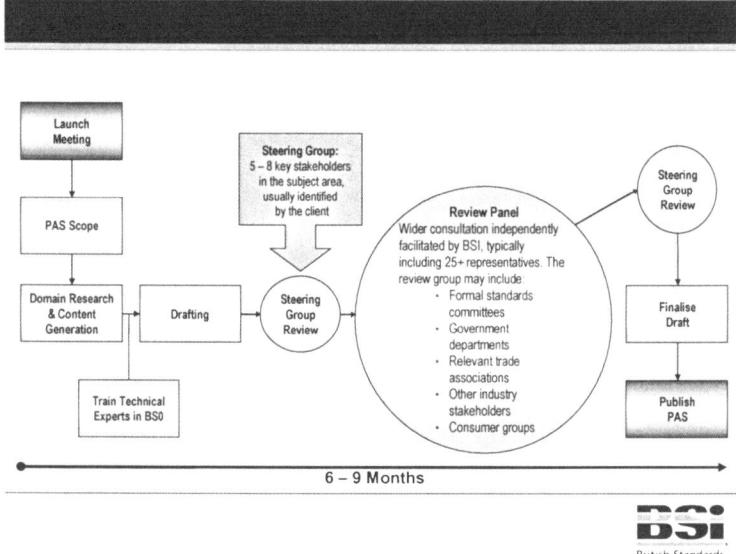

Figure 2 *The fast track standardization model*

5 STANDARDS FOR WATER EMERGENCIES

Standardization activities in the environment and water quality

BSI has been developing environmental and in particular water quality standards now for well over 20 years. The majority of the water quality ones are published as sections of BS 6068 Water quality, which is split up into six parts as follows:

Part 1: Terminology
Part 2: Physical, chemical & biochemical methods
Part 3: Radiological methods
Part 4: Microbiological methods
Part 5: Biological methods
Part 6: Sampling

These British Standards are identical to their international counterparts published as International Standards and many of these also European counterparts. The UK work is co-ordinated by BSI technical committee EH/3 and its subcommittees. Much of the current work is to support the Water Framework Directive, especially on the biological methods side. BS 6068 now contains over 200 British Standards ranging from the more traditional classical methods to up to date instrumental methods.

The EH/3 committee is expanding its purely national work programme at present, which includes revision of BS7592 on sampling of Legionella and BS1427 Guide to on site test methods for the analysis of waters. This latter standard could be particularly valuable as a tool in emergency situations when there is insufficient time to wait for results from samples sent to laboratories. Both will be available during Quarter 4 2008 in completely revised and updated forms.

Standardization activities in risk:

BS25999: Business Continuity (first developed as PAS 56)

BS 31100: Code of practice for risk management (under development)
   - Ensures an organization achieves its objectives
   - Ensures risks are proactively managed in specific areas or activities
   - Oversees risk management in an organization
   - Provides assurance on an organization's risk management
   - Reports to stakeholders, e.g. through disclosures in annual financial statements, corporate governance reports or corporate social responsibility reports

BS ISO 28000:2007 Specification for security management systems for the supply chain

Specifies the requirements for a security management system, including those aspects critical to security assurance of the supply chain. Security management is linked to many other aspects of business management. Aspects include all activities controlled or influenced by organizations that impact on supply chain security. These other aspects should be considered directly, where and when they have an impact on security management, including transporting these goods along the supply chain.

This International Standard is applicable to all sizes of organizations, from small to multinational, in manufacturing, service, storage or transportation at any stage of the production or supply chain that wishes to:
   a) establish, implement, maintain and improve a security management system;
   b) assure conformance with stated security management policy;
   c) demonstrate such conformance to others;
   d) seek certification/registration of its security management system by an Accredited third party Certification Body; or
   e) make a self-determination and self-declaration of conformance with this International Standard.

There are legislative and regulatory codes that address some of the requirements in this International Standard. It is not the intention of this International Standard to require duplicative demonstration of conformance. Organizations that choose third party certification can further demonstrate that they are contributing significantly to supply chain security.

BS 7578:2006 Security screening of individuals employed in a security environment – Code of practice

This British Standard gives recommendations for the security screening of individuals to be employed in an environment where the security and safety of people, goods or property is a requirement of the employing organization's operations and/or where such security screening is in the public interest.

BS 7499:2007 Static site guarding and mobile patrol service – Code of practice

This British Standard gives recommendations for the management, staffing and operation of an organization providing security guarding (see Annex A) services on a static site and/or mobile patrol basis.

Under development
- BS 8484 Security protection of lone workers
- BS 7984 Key holding and response service

Fast track standards activity

Case Study – PAS 96
This PAS was sponsored by the Centre for the Protection of National Infrastructure (CPNI)
And provides an approach to the threat of malicious attack on the food and drink industry
The PAS seeks to:
- Inform all those involved in the food and drink industry of the nature of the threat
- Suggest ways of deterring attack
- Recommend approaches that will mitigate the effect of an attack should it happen

The PAS identifies three generic threats to food and drink:
- Malicious contamination with toxic materials causing ill-health and even death;
- Sabotage of the supply chain leading to food shortage;
- Misuse of food and drink materials for terrorist or criminal purposes.

These threats could be carried out by a number of individuals or groups, including:
- People with no connection to the organization;
- Those with a contractual relationship such as suppliers and contractors
- Alienated or disaffected staff

Applicability of the PAS

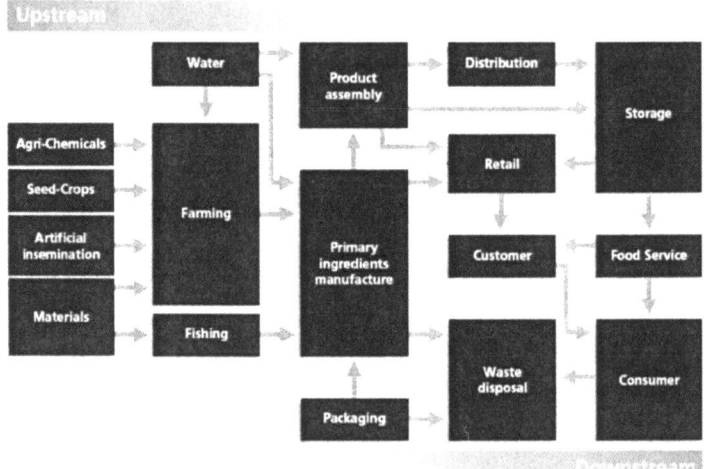

Figure 3 *The applicability of PAS 96*

Areas covered within the PAS include:

- Response levels
- Protocol for threat assessment and mitigation
    - Assessing the threat
    - Assuring personnel security
    - Pre-employment prudence
    - Systems for control of temporary staff
    - Building employee inclusiveness
- Controlling access to premises
- Secure mail-handling
- Controlling access to services
- Secure storage of transport vehicles
- Controlling access to materials and processes
- Assurance that sources of material are reliable and threat aware
- Product security - Tamper-evident consignments
- Reception arrangements for materials
- Quality control arrangements
- Contingency planning for recovery from attack
- Audit and review of food defence procedures
- Organization of the key sources of advice and information
- Guidance for specific parts of the food and drink supply chain
- Defending Food: A food and drink defence checklist

Case Study – PAS 68 & 69

This work was sponsored by the Centre for the Protection of National Infrastructure (CPNI) and is aimed at providing assurance that vehicle security barriers will provide the level of impact resistance that they seek.

PAS 68 specifies a classification system for the performance of vehicle security barriers and their supporting foundations when subjected to a single horizontal impact.
PAS 69 complements PAS 68 and PAS 69 provides guidance on the selection, installation and use of vehicle security barriers to ensure that they are selected and placed as effectively as possible.

6 CURRENT ACTIVITY

- PAS 68 is being taken to Europe and developed as a CEN Workshop Agreement (PAS is being revised and will be submitted later)
- PAS for Crisis Management due to start
- PAS for Mail Screening sponsored by the CPNI
- PAS 1188 – Flood protection products specification
- PAS 64 – Professional water damage mitigation and restoration of domestic buildings
- PAS for Fire Restoration for domestic dwellings

7 CONCLUSION

BSI is in a unique position to be able to help the water industry in its capacity as the National Standards Body of the UK. It is specifically proposed that there are a number of existing methods in which BSI could work with the industry to address the current issues.

- Develop new sector sponsored co-branded standards for the industry rapidly (6-9 months)
- Adapt existing standards (e.g. risk management) to meet the specific needs of the industry's stakeholders
- Identify and disseminate compliance requirements
- Set up an industry-wide consortium
    - Share knowledge
    - Develop consensus-based standards
    - Share costs
    - Deliver services the sector needs
    - Broad-based consultation

In addition, it is important that industries are aware of the routes into BSI and its activities. Part of the development of standards requires input from a broader base of interested parties. There are a number of routes into BSI.

1. BSS (British Standards Society)
The British Standards Society (BSS) is made up of and run by people who use standards and are interested in the developments and issues surrounding standards. The online Society enables members to network and share information and take part in special interest discussions. A close relationship with BSI British Standards allows an

exchange of key views from members. Membership to the Society is free and discussions take place in a secure website. If you would like to be kept up to date with standards and related developments in the UK and internationally, register today. If you would like to come together with other standards users joining is easy and free – simply fill in the online registration form in the About Us section of the BSS online meeting place http://ecommittees.bsi-global.com/BSSMP/

2. On-line DPC tool

BSI is pleased to announce the launch of its online Draft Review system for national Drafts for Public Comment (DPCs). A selected number of drafts for national standards (BS only) will be available via our new easy to use website. You will be able to read the full draft and leave comments and suggestions about its content which we may use in the final British Standard. Please visit our site at http://drafts.bsigroup.com.

3. Consumer and Public Interest Unit

All standards affect the public directly or indirectly, even though most are produced to serve the immediate needs of business and industry. Many, though, have a direct and beneficial impact on the general public. These include 'traditional' consumer related standards such as those for domestic appliances, or signs and symbols, as well as those newer types of standard for sustainability, social responsibility or services.

BSI is committed to trying to ensure that representation on its technical committees and access to the standards-making process is as wide as possible and maintains a Consumer and Public Interest Unit (CPIU), responsible for co-ordinating the participation of those stakeholders who would not otherwise normally be involved at a day-to-day level, e.g. consumers and individual specialists in subjects such as child safety or ergonomics.

The objective is to influence the content of standards to reflect the needs and proper expectations of the general public with regard to factors such safety and security, labelling, accessibility, fairness and redress.

Very many subjects are covered by the CPIU, but current priorities are in the areas of sustainability, security, accessibility, and the new and expanding field of services standardization. We also continue to cover important ongoing work on product safety, child safety and symbols, where our expertise and contribution are crucial.

Representatives are recruited and supported by BSI. They come from diverse backgrounds and have a range of high quality expertise and experience. Those who are unfamiliar with standardization are given suitable training and guidance in the standardization process, including specific skills required for researching, reporting and attendance at meetings both here and abroad.

If you are interested in being involved with this work, please contact consumer@bsigroup.com.

4. BSI Education

BSI supports those involved in education from information aimed at 7 year olds through to higher education. http://www.bsieducation.org/Education/default.php

References

1. BSI   Website: http://www.bsi-global.com/en/Standards-and-Publications/About-standards/
2. BSI Website: http://www.bsieducation.org/Education/default.php

THE XX EDITION OF THE TORINO OLYMPIC GAMES EXPERIENCE: PLANNING FOR AND RESPONDING TO DRINKING WATER CONTAMINATION THREATS AND INCIDENTS

R. Binetti[1], P. Olivier[1], L. Meucci [1], L. Cappuccio [1]

[1]SMAT S.p.A. the Turin Metropolitan Water Utility, Torino, Italy

1 INTRODUCTION

SMAT S.p.A. (Società Metropolitana Acque Torino) is the Turin Water Company and deals with both drinking and wastewater services for Turin town and all the Metropolitan area. The average annual production of drinking water is of about 260 million cubic metres; the inhabitants supplied are more than 1.6 million widespread over a territory of 4,000 km^2 and divided in about 200 municipalities. During the Olympic Games Event SMAT was designed to be in charge of the internal management of all the Olympic Venues for both drinking and wastewater services, as well as for the emergency management of the Olympic Games sites and the Olympic Municipalities. Just to give few numbers, 17 days of competitions took place in 15 Olympic sites and 2,500 competitors coming from 85 nations were involved; 500,000 L/days of drinking water was the amount of water supplied.

The organisations involved in the security activities were numerous and of different categories: - the Prefettura (the representative office of the Home Affair Minister), the police bodies (Polizia and Carabinieri), the Military Geografic Institute, the CONI for the sports events, the local bodies such as Civil Protection, Region, Province, the Services Companies (that deal with motorways and transport, electricity, gas, airport, telecommunication and water). During the Olympic Games Events beside the above governmental bodies and the local agencies, another bureau, TOROC (Torino Organising Committee) specifically in charge for these sport events organisation was created, so that a city command centre was established as a reference point for the communication among all the bodies/agencies and the Olympic municipalities involved. In the TOROC building, ready for prompt intervention, a representative of each of the service companies (SMAT included) was present in order to monitor the situation and to be informed in advance if an incident occurred. Also SMAT adopted a Crisis Operative Center and beside the "normal" on call group, another enlarged group was created as it was reckoned important to sort out the correct flow of information (what to share and to whom to report data to).

Within these preparations for the Olympic Games Event, SMAT also implemented a Response Plan for Drinking water Contamination Threats and Incidents (PRDWC) in order to deal with possible chemical, microbiological and radiological contamination of the water sources and of drinking water supplied. The PRDWC is divided into three steps which can occur either sequentially or simultaneously. The first step regards the threat evaluation, the second step consider the water quality verification management, the third step is related to the action response which can range from the simple communication of the event to the bodies/agencies involved to the interruption of the water supplied. In this paper, the second step will be discussed.

2   RESPONSE PLAN FOR DRINKING WATER CONTAMINATION

After a vulnerability assessment[1] of the entire water supply system some security intervention has been implemented: roughly 10,500 drains and manholes were welded, about 90 sewers were secured, many other tanks were modified in order to present a better security challenge and many other alarm systems were installed (the vulnerability assessment details are beyond the scope of this paper).

The first activity of the second step of the Response Plan for Drinking Water Contamination Threats and Incidents (PRDWC) was the identification of the possible contaminants including organic and inorganic chemical contaminants (i.e. hydrocarbons, pesticides, pharmaceuticals chemical weapon substances, toxic metals, cyanides, etc.), microbiological contaminants (viruses, bacteria and parasites), biotoxins and radionuclides. The possible contaminants were determined from consulting information and lists provided by the Istituto Superiore di Sanità[2] (Italian Superior Health Institute) and by other International Organisations (WHO, EPA,CDC, EMEA, etc)[3].

The water quality verification section of the Response Plan for Drinking Water Contamination Threats and Incidents (PRDWC), described in this paper, is an extension of an old plan implemented since the 1990s to cover accidental pollution of the water resources; it was developed because of the use of vulnerable sources for drinking water production for the Turin municipality. The analytical activities planned should lead to the presumptive identification of the pollutant/s in order to express a global judgment about the risk and about the extent of the eventual pollution. In particular, in the PRDWC plan, intentional contamination has been included beside accidental contamination, for which the old plan was developed for. In any case, during any incident, it is well known the existing shift in terms of time between the demand and the answer curves. To reduce as much as possible the specific vulnerability the "time" aspect is then the key factor.

As far as time factor is concerned the early warning systems are the best tools to rapidly detect accidental or intentional contamination of the water. The main advantage from the on-line monitoring equipment is the availability of data really representative of the water quality analyzed (24 hours per day). At present, to obtain a good protection system, many different type of sensors/analyzer have to be used at the same time[4]. In the Turin municipality, three on-line monitoring stations equipped with more than 20 monitors are installed on the most vulnerable water intakes of the drinking water treatment plants for the measurement of the following parameters: - toxicity (fish test and Musselmonitor®, http://www.mermayde.nl/), gamma radioactivity, algae, TOC, nitrate, UV-Vis spectra, ammonia, pH temperature, dissolved oxygen, conductivity, redox potential, turbidity. In order to be able to trust the data produced by any analytical instrument, either on-line or produced in laboratory, a good quality assurance system should be adopted[5]. With the aim of obtaining a quality baseline as a reference point, it is very important to handle the huge amount of data collected by the online equipment. For a normal distribution, the values that are outside the range of ± 3 times sigma represent abnormal values. The probability that the warning signal is a false-positive is four in ten thousand. The availability of global water parameters baseline makes it easy to put in evidence of any deviation from the "normal" water quality or from the normal deviation due to seasonal changing or to possible contamination. The results of measurements obtained in different episodes are not shown, but are available elsewhere[5-7].

Beside these early-warning systems, that monitor more than 20% of the drinking water produced for Turin Municipality, some *screening analytical protocols* have been developed by SMAT internal laboratories personnel consulting "state of the art" reported literature on the subject[8]. These protocols can be applied to samples collected in any point of the drinking water supply system and in particular to control the distribution system which is well known to be the most vulnerable part because of its accessibility to those it serves and its geographic span.

3 ANALYTICAL ANSWER/APPROACH TO DRINKING WATER CONTAMINATION EVENTS

The analytical approach developed for emergency water samples in response to a specific contamination threat includes : -
-   the event evaluation
-   the screening
-   the confirmation

As part of the event evaluation the site characterization has to be carried out; with respect to this, some access security guidelines have been developed[9]. The evaluation of the site hazards is useful to establish the necessity to implement eventual special handling techniques or in the worst case to involve special teams like Hazmat or special trained personnel. In addition the sampling guidelines, the sample transport and the sample storage guidelines have been defined.

As part of the site characterization the personnel who are charged with carrying out the site characterization perform the so called "*field screening*" (Table 1). The field screening is aimed at: -
-   defining the appropriate precautions required during the sampling according to the hazard tentatively identified;
-   giving additional information for the evaluation of the threat credibility;
-   providing tentative identification that will be analytically confirmed in laboratory.

The field screening includes the physico-chemical and non-specific parameters such as pH, conductivity, temperature, turbidity, dissolved oxygen and the disinfectant residual. To analyse these parameters, standardized methods are used. In addition, a toxicological screening is included using the ECLOX® system. This is a rapid broad spectrum toxicity test (a chemiluminescence test). The test utilizes a plant enzyme which when mixed with other reagent produces light. Certain pollutants when present in water interfere with the chemical reaction and reduce the amount of luminescence. The extent of inhibition of chemiluminescence is assumed to be proportional to the concentration of contaminant.

It takes about 40 minutes to obtain all the results of the field screening parameters on the sample collected. The protocols described have been developed to take the minimum possible time to perform as this, as previously mentioned, is a critical factor in emergency analysis.

Table 1   *Field screening*

TYPE	CLASS	PARAMETERS/ MEASURING PROCEDURES	TIME	
CHEMICAL	PHYSICO-CHEMICAL	•pH •Conductivity •Turbidity •Dissolved Oxygen	20 min	40 min
	NON-SPECIFIC	•Disinfectant residual	40 min	
TOXICOLOGICAL	AGENTS INHIBITING CHEMILUMINESCENCE	•Inhibition of chemiluminescence (ECLOX)	20 min	

Other than the field screening three other analytical screenings have been defined:
- The *basic screening* (Table 2) aimed at the evaluation of the "BASELINE" variation;
- The *extended screening* (Table 3) aimed at the detection of the major number of analytes using a contained number of techniques;
- The *complete screening* (Table 4) that includes the confirmatory chemical analysis, explorative screening for toxins and specific bacteriological contamination, specific determination such as radionuclides, anions, protozoa, and so on.

The *basic screening (2-3 hours)* includes, in addition to the field screening parameters, also UV 254 absorbance, TOC, ammonia, nitrite, heavy metals, cyanide, ATP, and toxicological screenings made in laboratory (i.e. Microtox®). Meanwhile, in order to take advantage, in terms of time, the microbiological indicators required by the regulation are also performed. These last parameters will be taken into account as confirmatory analysis whenever the ATP test result is positive.

For bacteriological and toxicity screening, some result interpretation has to be made. With respect to non-specific bacteriological contamination, where there is some controversy on the interpretation of results, there is generally a certain correlation between cellular ATP measurement and the number of viable bacteria present. The choice of adopting ATP measurements has been made for the following reasons: the method is rapid, simple, field applicable and cost effective[10]. The detection limit assumed was 10^2 CFU/ml (even if lower number of bacteria were detected) and the alarm threshold is > 300 RLU (Relative Luminescence Units). As regards acute toxicity screening both the metabolic inhibition test that utilizes luminescent bacteria and the rapid chemiluminescence test detect similar concentration of a variety of toxins. Neither assay provides consistently reproducible results. Screening results obtained during an emergency or during a period of heightened security can be interpreted by plotting the values on a control table constructed using background data. In agreement with other the opinion of other authors[11] negative results do not guarantee the safety of the water; positive results suggest the possible presence of a toxic substance and indicate the need for more specific analysis.

Table 2   *Basic screening*

TYPE	CLASS	PARAMETERS/ MEASURING PROCEDURES	TIME	
CHEMICAL	PHYSICO-CHEMICAL	•pH •Conductivity •Turbidity •Dissolved oxygen	20 min	2-3 hours
	NON-SPECIFIC	•Disinfectant residual •UV254 nm •TOC	40 min	
	INDICATORS	•Ammonia •Nitrite	40 min	
	METALS	•Fe, Mn, Cu, Ni, Pb, Cr, Al, Cd, Zn	90 min	
	IONS	•Cyanide	120 min	
BIOLOGICAL	MICROBIOLOGICAL PARAMETERS (required by law)	•Coliform bacteria •Escherichia coli, •Enterococci •Clostridium perfringens •Colony count 37°C •Colony count 22°C	4 days	
	NON-SPECIFIC BACTERIOLOGICAL CONTAMINATION	•ATP	30 min	
TOXICOLOGICAL	TOXIC	•Inhibition of Vibrio Fisheri Bioluminescence (screening test, comparison and/or final)	60-180 min	1 hour
	AGENTS INIBITING CHEMILUMINESCENCE	•Inhibition of chemiluminescence (ECLOX)	20 min	

The *extended screening (6 - 8 hours)* which includes all the parameters above mentioned plus a screening for the presumptive identification of a wide range of organic compounds at sub-acute contamination level (parametric values, guidelines or exposure levels etc.). The pollutant classes screened were the following: organohalides (organic solvents containing halogens, mainly as chlorine), solvents, aromatic hydrocarbons, polynuclear aromatic hydrocarbons (PHA), phenols and chlorphenols, pesticides and herbicides. The instrumentation used was a gas chromatograph and a mass spectrometer (ion trap); the pre-concentration was carried out using solid phase microextraction (SPME). The minimum detectable concentration for each class, excluding false negative/positive results that can be reported is: -

- aromatic hydrocarbons range 1 - 8 ppb (QL 0.1 - 0.6 ppb);
- organohalides range 8 - 40 ppb (QL 4 ppb);
- pesticides range 0,5 - 40 ppb (QL 0.1 - 0.5 ppb) ;

- phenols range 1 - 50 ppb (QL 1 - 2 ppb) ;
- PHA range 0.5 - 50 ppb (QL 1 ppb).

The above reported concentrations have to be compared to the human acute LD_{50} concentrations of these classes of compounds that are normally tens, hundreds and even thousands of mg/Kg. It can be said that the sensitivity obtained with this screening explorative method can be considered more than sufficient because, as already mentioned earlier the objective of this plan was to detect substances at acute toxicity concentration levels.

Another presumptive identification using HPLC with UV photodiode array detector (PDA) coupled with fluorescence detector, can detect large molecular weight substances (e.g., biotoxins) and/or thermally unstable compounds, polar hydrophilic compounds which cannot be easily extracted from an aqueous sample and are not amenable to gas chromatographic procedures. If these compounds are present in amounts greater than background can sometimes be analyzed by direct aqueous injection of the sample without extraction or pre-concentration. Otherwise SPME or SPE may be utilized for compounds that can be extracted.

If the ATP screening result is positive, in addition to the microbiological indicators, both aerobic and anaerobic on non-specific culture media are carried out followed by biochemical identification (obviously these investigations will take more time).

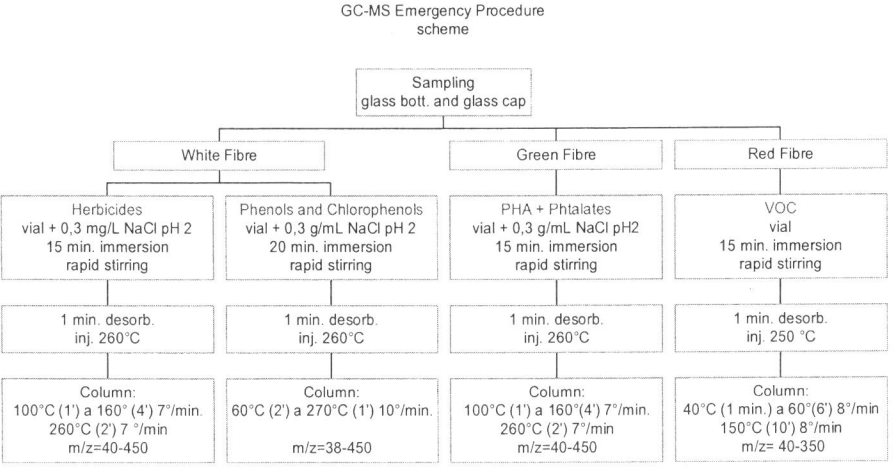

Figure 1   *Screening procedure for volatile and semi-volatile organic compounds*

Table 3 *Extended screening*

TYPE	CLASS	PARAMETERS/ MEASURING PROCEDURES	TIME	
CHEMICAL	PHISICAL- CHEMICAL	•pH •Conductivity •Turbidity •Dissolved oxygen	20 min	6-8 hours
	NON-SPECIFIC	•Disinfectant residual •UV254 nm •TOC	40 min	
	INDICATORS	•Ammonia •Nitrite	40 min	
	METALS	•Fe, Mn, Cu, Ni, Pb, Cr, Al, Cd, Zn	90 min	
	IONS	•Cyanide	120 min	
	ORGANICS VOLATILE	•Purge & Trap GC FID/ECD	40 min	
	ORGANICS SEMI- VOLATILE	•SPME GC-MS	60 min	
		•SPE HPLC-PDA /Fluorescence	360 min	
	ORGANICS NON VOLATILE	•SPE HPLC-PDA /Fluorescence	360 min	
		•Direct injection HPLC- PDA / Fluorescence	90 min	
BIOLOGICAL	MICROBIOLOGICAL	•Coliform bacteria •Escherichia coli, •Enterococci •Clostridium perfringens •Colony count 37°C •Colony count 22°C •Salmonella spp •Staphylococci pathogens •Pseudomonas aeruginosa •Fungi	4 days	
	NON-SPECIFIC BACTERIOLOGICAL CONTAMINATION	•ATP	30 min	
		•Aerobic and anaerobic culture using non-specific media and possible biochemical identification	2-5 days	
TOXICOLOGICAL	TOXIC	•Inhibition of Vibrio fisheri Bioluminescence (screening test, comparison and/or final)	60-180 min	Min 1 hour - Max 3 hours
	AGENTS INIBITING CHEMILUMINESCENCE	•Inhibition of chemiluminescence (ECLOX)	20 min	

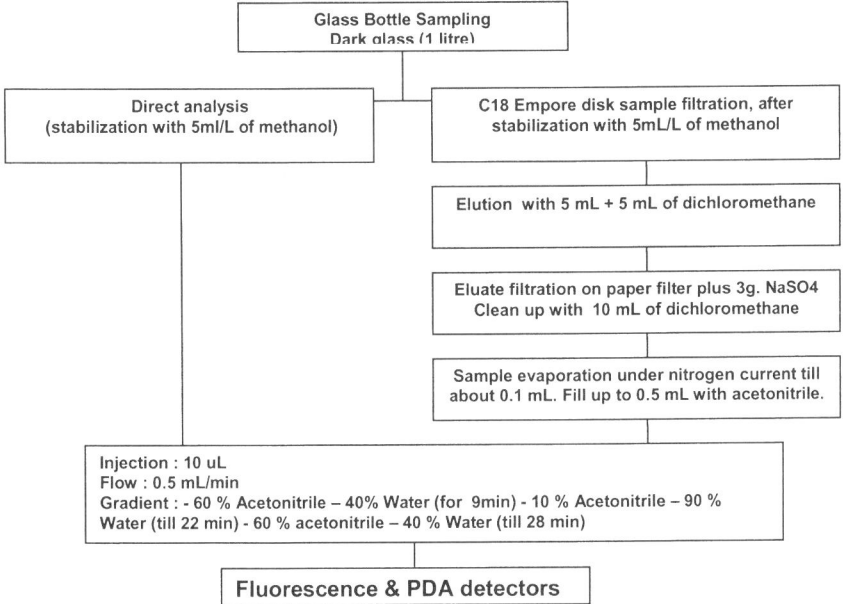

Figure 2 *Non-volatile organic compounds screening procedure*

The *complete screening* includes explorative screening for radionuclides which is carried out on the entire gamma spectrum (20-2000 Kev). In the case of detection of presumptive radiation levels above the "normal" values, more detailed analysis will be performed by ARPA (Local Environmental Protection Agency) laboratory. There is a written agreement between the two organisations. Biotoxin detection screening is carried out in an explorative manner (in case of positive results, the sample is sent to a specialized laboratory). The chemical weapon agent screening is carried out exclusively by military NBC teams.

Once the type of contamination is addressed, some confirmation analysis and other specific tests have to be performed including the detection of radionuclides, toxins, viruses, protozoa and so on. The decision of performing one screening protocol rather than another (see Figure 3) is taken by an appropriate competent person from the relevant laboratories, on the basis of the information gathered from the entire threat evaluation chain and from the warning procedure code in act (alert or normal situation).

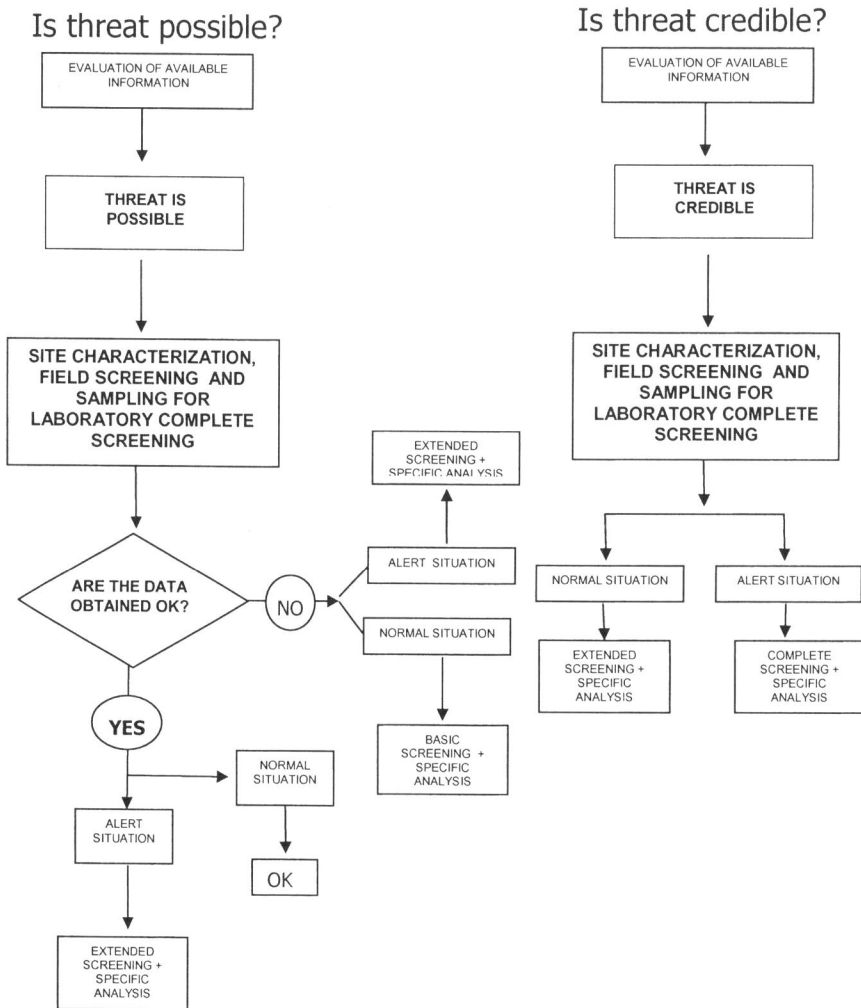

Figure 3   *General considerations for the definition of analytical approach*

The confirmatory analysis with standardized methods is carried out to evaluate the extent of the contamination and the effective decontamination action to be applied as the response to the event. The identification of the methods to be used has to take into account a balanced approach relating to the quantification of existing techniques to provide reliable and quality assured analytical data.

Table 4 *Complete screening tests in addition to that ones included in the extended screening*

TYPE	CLASS	PARAMETERS/MEASURING PROCEDURES	TIME	
CHEMICAL **+** **BIOLOGICAL** **+** **TOXICOLOGICAL**	CHEMICAL CONFIRMATION	Ions, Volatile Organics, Semi-volatile organics	120 min	2 hours
	SPECIFIC INTENTIONAL BACTERIOLOGICAL CONTAMINATION	Anthrax, Cholera, Plague, Tularemia (rapid immunoassays)	30 min	
	TOXINS	Ricin toxin, botulinum toxin, staphylococcal enterotoxin B, microcystin (Rapid enzyme test)	30 min	
OPTIONAL	RADIONUCLIDES	Gamma Radioactivity (NaI)	60 min	
		ICP-MS	120 min	
	INORGANIC CATIONS	Ion Chromatography	120 min	
	PROTOZOA	Giardia (Cysts), Cryptosporidium (oocysts)	1-2 days	
	VIRUS	Enterovirus and citophatic virus for kidney cells BGM	7 days	
	LEGIONELLA	Legionella spp (PCR), Legionella pneumophila (PCR)	2-3 days	

3   CONCLUSION AND LESSONS LEARNED

The most important aspect pinpointed during the planning activities of the PRDWC and the related mock tests carried out was that to be able to detect any contamination of drinking water, it is necessary to have a lot of background and screening data. As this was carried out before the Olympic Games Events, periodically some test exercises should be performed in order to maintain, check and revise the procedures. A general educational programme has to be carried out; security is everyone's responsibility. As for the analytical issues, any implemented activity has to be constantly revised; security is not an end point but a goal that can be achieved only through continuous efforts to assess and upgrade the system.

With respect to communication it is very important to clarify who is who within the public administration; in this case a communication protocol and a number of forms to fill in were created in order to facilitate the choice of the information that has to be forwarded. It must be pointed out that is quite hard to give exact and prompt information when dealing with possible contamination of unknown substances, specially during alert situation occurring in such prestigious very large events that are in the spotlight of all over the world.

References
1   Association of State Drinking Water Administrators, National Rural Water Association, *Security Vulnerability Self-Assessment Guide for Small Drinking Water System Serving Population Between 3,300 and 10,000*, 11/2002

2   Ottaviani, M., Drusiani, R., Lucentini, L., Ferretti, E., Bonadonna, L. Sicurezza dei sistemi acquedottistici (*Security of water system*), Rapporti ISTISAN 05/4, Istituto Superiore di Sanità, Roma, 2005

3   WHO: www.who.int/search/en/, WHO's "*Public health response to biological and chemical weapons*" www.who.int/csr/delibepidemics/biochemguide/en/index.html, CDC Emergency Preparedness and Response www.bt.cdc.gov, US EPA Water Contamination Information Tool (WCIT), USEPA List of Drinking Water Contaminants & MCLs http://www.epa.gov/safewater/mcl.html#mcls

4   Byer, D., Carlson, K.H., Real time detection of intentional chemical contamination in the distribution system, Journal *AWWA*, 97:7, 130-133, 7/2005

5   Meucci, L., Binetti, R., Bocina, G., Fabrizio, F., Quality Assurance Management of on-line monitoring equipment, *3rd International Conference on Automation in Water Quality Monitoring, AutMoNet 2007*, Ghent, Belgium, 09/2007

6   Giacosa, D., Meucci, L., Badino, G., On line Po river mussel monitoring, *Proceedings International IWA Conference on Automation in Water Quality Monitoring*, Vienna Austria, 05/2002

7   Binetti, R., Magnoni, M., Meucci, L., Bertino, S., On line radioactivity surveillance on River Po, *Proceedings International IWA Conference on Automation in Water Quality Monitoring*, Vienna Austria, 05/2002

8   Office of Ground Water and Drinking Water, Water Security Division, Response Protocol Toolbox: Planning for and Responding to Drinking water Contamination threats and Incidents - *Module 4: Analytical Guide, US EPA 817-D-03-004* www.epa.gov/safewater/security 12/2003

9   Office of Ground Water and Drinking Water, Water Security Division, Response Protocol Toolbox: Planning for and Responding to Drinking water Contamination threats and Incidents - *Module 3: Site Characterization and Sampling Guide, US EPA* 817-D-03-003 www.epa.gov/safewater/security 12/ 2003

10   Giacosa, D., Meucci, L., Buffa, E., Pignata, C., Fea, E., Gilli, G., Screening microbiologico e tossicologico in caso di sospetta contaminazione dell'acqua destinata al consumo umano: un esempio applicativo, *Workshop proceedings on Ecotossicologia, Chimica, Microbiologia: dal laboratorio alle applicazioni on-line*, Edizioni CNR-ISE, Milano, 11/2007

11   States, S., Newberry, J., Wichterman, J., Kuchta, J., Scheuring, M., Casson, L., Rapid Analytical Techniques for Drinking Water Security Investigations, *Journal AWWA, 96:1, 52-64, 1/2004*

SENSITIVE, SELECTIVE AND SIMPLE UV-SPECTROMETRY FOR CONTAMINANT ALARM SYSTEMS

J. van den Broeke

scan Messtechnik GmbH, Brigittagasse 22 - 24, 1200 Vienna, Austria. E-mail: jvandenbroeke@s-can.at

1 INTRODUCTION

When monitoring drinking water, either at the source or in the distribution system, identification of low probability/high impact events that might compromise water quality, and as a consequence public health, is an important goal. Early identification is the prerequisite for an effective response that reduces or entirely prevents the adverse impacts of such a contamination. Systems that facilitate such early detection and identification are referred to as early warning monitoring systems or alarm systems. Typical causes of high impact contaminations are natural events (e.g. flooding, anoxia, algal blooms), accidental anthropogenic events (e.g. inadvertent discharges, spills) or intentional discharges (e.g. vandalism, terrorism). As the number of potentially harmful compounds is enormous, only a fraction of these can be monitored, as the development of specific sampling strategies for all contaminants would be an impossible task.

The development of new variables, such as alarm parameters, that allow for an integrated assessment of changes in water quality using surrogate or aggregate variables, instead of searching for all specific contaminants individually, can be a useful strategy to cover a much broader range of potential threats than possible with conventional monitoring solutions.[1] As spectral data, and their evolution over time, are very rich in information, the loss in compound specific information is compensated for by a much better overall picture of change in water quality and the possibility to detect changes that will not be picked up by conventional, single contaminant directed, monitoring programmes. This contribution describes all facets of the use of online UV/Vis spectroscopy for monitoring of drinking water quality.

2 METHODOLOGY

2.1 The Instrument

For the purpose of describing the principles behind and capabilities of online spectrometers, the spectro::lyser™ (Figure 1) of the Austrian company s::can Messtechnik will be used as the reference instrument, as this is the most widely used of this type of instrument. This spectrophotometer records light attenuation in the wavelength region between 200 – 735 nm. The measurement is performed in-situ, without sampling or sample pre-treatment, thus preventing errors due to sampling, sample transport and storage, etc. A measurement cycle takes between 20 and 60 seconds, making possible a high measuring frequency and detection of rapid changes. For long term stability of the signal, a split light beam design is used; one beam passes through the sample while the other travels along a parallel pathway inside the instrument and thus acts as an internal reference. This second beam is used to cancel out fluctuations in light source energy and instrumental fluctuations due to environmental conditions.

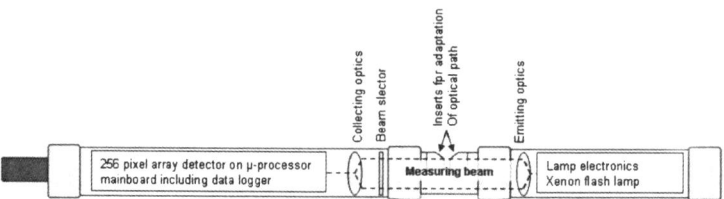

Figure 1   *Schematic representation of the spectro::lyser™ submersible spectrometer probe*

2.2 Fingerprints

The spectra, referred to as fingerprints (Figure 2), obtained with such on-line spectrometers are used for the characterisation of the sampled water. Within these, sometimes apparently featureless, fingerprints one can find a huge amount of information about the composition of the water analysed. Analysis of the general shape of the spectrum or absorption at specific wavelength yields information about turbidity, nitrate concentration, and sum parameters such as SAC_{254} (Spectral Absorbance at 254 nm), TOC (Total Organic Carbon), which are commonplace in water analysis.[2,3] Turbidity due to suspended substances causes light scattering and shading, thus influencing the absorption over the entire fingerprint. This is an important factor that influences in-situ measurements and requires compensation in order to obtain reliable and reproducible readings. The resulting compensation allows the analysis of dissolved components such as DOC (Dissolved Organic Carbon) and the colour.[2,3]

The wavelengths used for determining all these parameters have been selected using principal component analysis (PCA) and partial least square regression (PLS) on hundreds of datasets containing both UV/Vis spectra and reference values of these parameters[2], the latter being determined using established laboratory techniques. Characteristic and quantitative relationships between the parameters described above and the absorption at certain wavelengths were thus established.

3 SPECTRAL ALARM PARAMETERS

There are several ways to evaluate the information provided by the fingerprints, including the qualitative interpretation of spectral deviations from a site specific reference spectrum (e.g. peaks, shoulders, gradients, analysis of the derivative spectra, etc.), the comparison of spectral differences between measuring points in a measurement network, i.e. spatially-resolved delta spectrometry, and the evaluation of changes of the spectral features over

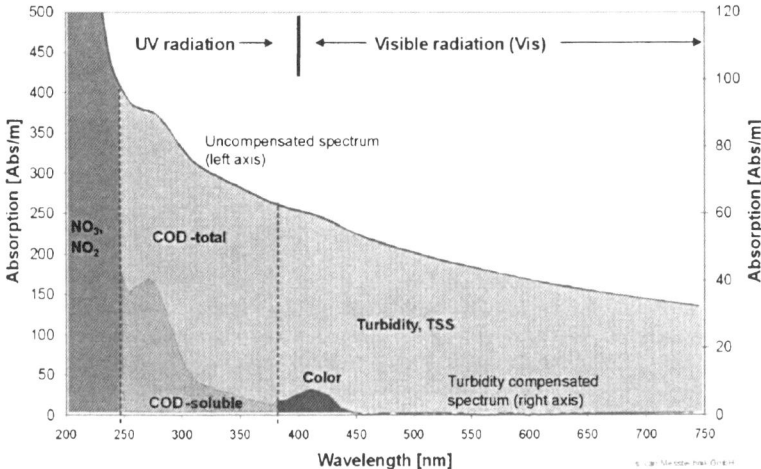

Figure 2   *UV-Vis spectrum, and examples of parameters derived out of this spectrum together with their characteristic absorbance profiles.*

time, i.e. time-resolved delta spectrometry.

Time-resolved delta spectrometry[4] tracks changes in the shape of the fingerprint over time, as well as the speed with which changes occur. As changes due to extreme natural events and anthropogenic changes are typically faster than gradually occurring natural changes, such as seasonal changes in water composition, it is possible to identify unusual water compositions solely on the basis of changes in time. This means that contaminants that do not provide very distinct signals can still be detected, because they still cause a (non-distinct) change in the absorption spectrum. It also means that in cases where no water body can serve as a reference for unimpaired water quality, for example due to continuous fluctuations, the use of UV/VIS spectrometry nevertheless provides the possibility to detect irregularities on the basis of the size and speed of changes in the fingerprint. The multi-dimensional information provided in UV/VIS spectra offers a much greater information potential on top of the "normal" baseline compared to single parameters like turbidity or DOC (Dissolved Organic Carbon), for which time resolved data analysis will not allow such an identification of abnormal changes.

For definition of spectral alarm parameters, the following procedure has to be followed:

- *Learning period*: During the learning period (usually some months) spectra are collected. After this period the baseline spectrum (shape and features) is established.
- *Abnormality definition*: Absorption spectra or their 1st- and 2nd-order derivatives can be used to identify deviations from "normal" spectral features (Figure 3). Detection of deviations is possible even with a continuously changing baseline, as in this case the trigger values for abnormal changes will be set outside the limits of the normal deviations or at wavelengths little affected by the normal fluctuations.
- *Alarm definition*: Alarm definition is based on the concept of virtual contaminants. Since one never knows in advance which contaminant may be the next to enter the

system, several groups of "virtual" contaminants are defined. These virtual contaminants are designed in such a way that by monitoring this limited set of parameters, the whole range of contaminants visible in the UV/VIS spectrum is covered.

- *Sensitivity definition*: Sensitivity of the virtual contaminants to changes in the water matrix has to be adjusted individually with respect to risks involved and acceptable false alarm levels. This is based on empirical and statistical evaluation.
- *Definition of actions to be taken in case of alarm*: Examples for possible actions in the case of an alarm would be: *Alarm level 1*: Slight deviation from normal situation: Automatic sampling triggered, waiting for laboratory analysis, no additional action needed. *Alarm level 2*: Strong abnormalities: Automatic sampling plus urgent action at site needed.

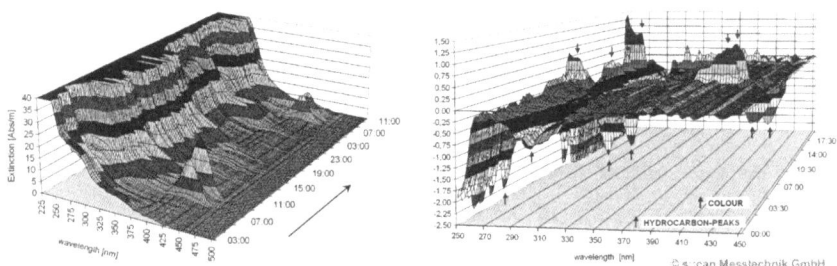

Figure 3   *Example of the enhanced visibility of spectral events (left) when using the 1st derivative of the spectrum (right)*

4 APPLICATIONS

The following paragraphs describe a number of applications of online spectrometer probes for drinking water monitoring. The described cases vary from measurement of the standard spectral parameters, such as turbidity and TOC, to the use of spectral alarm parameters and development of site specific process parameters.

4.1 Vienna Waterworks

Mountain springs in geological regions consisting of karst formations are the main source for drinking water supply of the city of Vienna, Austria. The water quality from these springs is normally very good, but it does show temporary instabilities due to natural events such as intense rainfalls. Rapid fluctuations in water quality can occur in such cases, which can lead to increases in turbidity, dissolved organic substances as well as bacteria numbers in an almost unpredictable way during storm weather. To be able to ensure drinking water quality and security at all times in this system, it is vital to monitoring the composition of the raw water continuously; the raw water quality is obviously one of the most important factors determining the final quality of the drinking water, and the rapid changes of raw water quality can limit the efficient efficacy of the treatment procedures during drinking water production.

Before 2001, only a few very fundamental springs were being monitored online. Complete buildings equipped with power supply, pipe installations and special foundations had to be built for housing the then available instrumentation for monitoring the spectral absorption coefficient at 254nm ("UV254"), TOC, turbidity, electrical conductivity and pH. Both the actual costs and the situation of the springs avoided the setting up of a broader monitoring network covering all essential springs. Sample pre-treatment such as a membrane filtration had to be installed prior to some measuring devices, thus the operation of the monitoring stations required regular maintenance increasing the total costs of ownership. The availability of the spectro::lyser™ allowed Vienna Waterworks, since 2001, to monitor the water quality even of springs without power supply by means of online measurement instrumentation. As no pipe installations are to be provided and no pumps, filters, membranes and reagents are needed the total costs of ownership are much reduced despite increasing the number of measurement sites.

Before starting to establish a monitoring network based on the spectrometer probes, a period of long-term validation took place: In order to verify the equivalence of spectral and conventional measurements, the spectrometer probes were operated in parallel to existing TOC cabinet analysers. It was also possible to verify the equivalence of spectral and conventional measurements both for turbidity and SAC ("UV254") for periods of several months. For verification of the parameter specifications and the inter-instrumental comparability, twelve instruments were installed at the same spring for a period of six months. In total readings of more than 18 months were analysed and have proven the equivalence, measuring performance and the long-term stability of the spectrometer probes.

At present Vienna Waterworks operates s::can monitoring systems in order to monitor turbidity, SAC254, Nitrate, TOC, DOC, temperature and electric conductivity at 22 locations (springs, catchments of various springs and essential points in the transportation network). The results of these monitoring systems are transferred in real time to an early warning system that can be accessed from three central stations. This early warning system manages the raw water sources 24 hours a day: Whenever the readings exceed limits that are specific to each parameter, the water of the spring of concern will not be used for drinking water production but drained off. In this way the proper quality of raw water used for the drinking water supply is ensured, and no other water than of perfect quality can enter the transportation network. It has been observed that the quality of the raw water can vary within very short intervals and online monitoring is indeed necessary to provide water of constant composition to the drinking water production. Monitoring just one parameter would not meet the needs of efficient and safe drinking water supply, as different types of events cause different changes in the composition of the spring water.

4.2 Testing the sensitivity in drinking water

A number of studies have been performed where the sensitivity of the spectrometer probe and the spectral alarm parameters was tested towards the presence of possible contaminants in drinking water. During these studies, real contaminants were spiked in various drinking water at concentrations between 0 mg/L and the LD50 concentrations of these substances, where the LD50 does is calculated based on a typical consumption of 2L of drinking water per day. The detection limits were defined as the point where the increase in absorption due to a contaminant exceeds the standard deviation in the measurements due to instrumental noise by a factor of 3.[5] The group of substances tested includes mainly pesticides, chemical warfare agents, and bio toxins. The detection limits observed during these studies are presented in Table 1.[6,7] It must be noted that the detection limits vary significantly amongst the substances, as the chemical structure of the substances determines how much light they absorb. Furthermore, it was observed that the detection limits can vary by as much as a factor of 2 depending on the composition of the drinking water, with drinking water with higher TOC contents generally showing higher detection limits, i.e. less sensitive monitoring.

compound	detection limit
Linuron	0.01 mg/L
Saxitoxin	0.04 mg/L
Isoproturon	0.05 mg/L
Oxamyl	0.1 mg/L
Mevinphos	0.1 mg/L
Aflatoxin	0.2 mg/L
LSD	0.2 mg/L
Aldicarb	0.8 mg/L
Metamidophos	1 mg/L
Azinphos – methyl	2 mg/L
Fenamiphos	1.5 mg/L
VX	4.5 mg/L
Ricin	5.9 mg/L

Table 1   *Detection limits observed for potential contaminants in drinking waters.*

4.3 Monitoring for the Superbowl

The city of Glendale (AZ) in the USA hosted the 2008 Superbowl, which being the final of the American Football season is one of the biggest sporting events of the year in the US. In a multi-year project to improve its drinking water supply system, and partly to ensure the highest possible water quality during this event, the municipality of Glendale decided to expand its online monitoring efforts. As part of these efforts, multiple monitoring stations have been erected throughout the supply network, each of these stations combining a spectro::lyser™ UV-probe with additional sensors such as conductivity and residual chlorine sensors. All these data are collected and transmitted to a central database in real time.

The main advantage of employing these monitoring stations is a much improved insight into the normal dynamics of the water (quality) in the network. Using the spectral information it is directly possible to assess daily water quality variations, variations in water composition due to changing water sources, measurement of retention times of water in the network, optimise regeneration of Activated Carbon filters as well as detection of operational events (switching of pumps, power outages in treatment steps, ...). One valuable results was the detection of higher than expected corrosion in a new cast iron pipe system. The presence of elevated levels of iron oxide in the water was directly discernable in the fingerprint of the water, whereas traditional sensors did not show any response. As a result of these measurements, the problem could be identified much earlier as would have been possible otherwise, and measures were taken to stabilise the water and reduce the corrosivity.

For the Superbowl event itself the spectro::lyser™ probes were equipped with the spectral alarm parameters, which were trained on the normal water quality in the weeks leading up to the event. Then during the Superbowl event, a continuous watch was kept on all water quality data, allowing immediate response in case any unexpected changes would occur. No significant water quality deviations were detected during the entire weekend (figure 4).

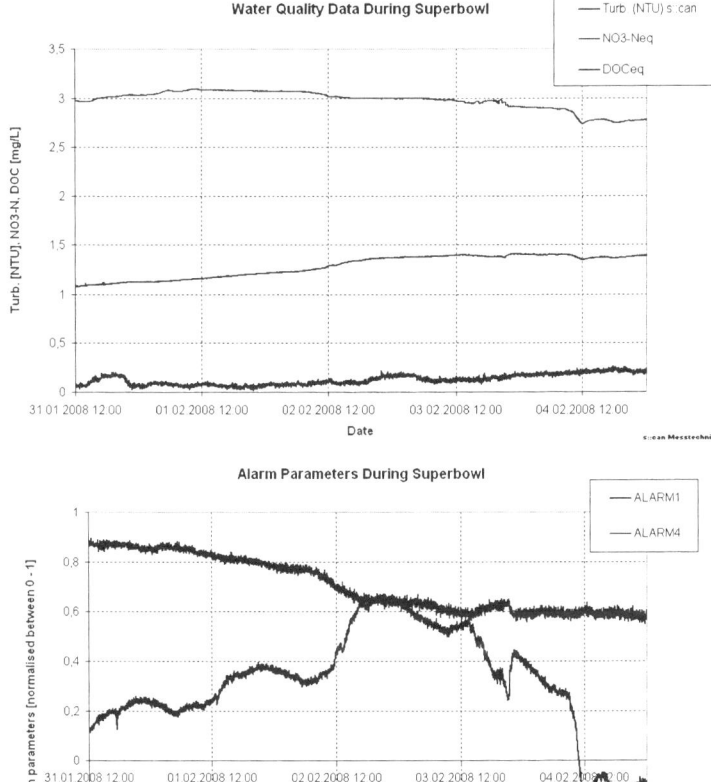

Figure 4 *Representation of the variation in the parameters measured at the stadium of the Superbowl during the weekend of the event. The upper graph shows the classical parameters turbidity, NO3 and DOC, which show no significant changes. The lower graph shows two spectral alarm parameters, of which ALARM1 shows a significant change during the day of the Superbowl, February 3, which however remained within the normally observed variation at this site (between -0.25 and + 1.0) and was no cause for an alarm.*

4.4 The Power of Spatially Resolved Delta Spectroscopy

A final example of the use of UV/Vis spectroscopy in drinking water quality monitoring of the use of delta spectroscopy in control of drinking water treatment steps. Delta spectroscopy is the calculation of the change in spectrum over time or over space. The resulting spectrum, which quantifies the change in the signal, the so-called delta spectrum, can be used to characterise changes, such as those caused by water treatment processes. Spatially Resolved Delta spectroscopy was used to measure changes that occurred while the water underwent several subsequent treatment steps at a pilot plant of Amsterdam Waterworks, i.e. the effect of ozonation, pellet softening and carbon filtration on the spectrum was investigated. A powerful example of which can be achieved with this technique is the monitoring of Assimilable Organic Carbon (AOC) formation and removal in these treatment steps.

During oxidation the high molecular weight organic molecules are converted into smaller organic molecules. The total amount of organics in the water, however, does not change significantly. As a result, classical spectroscopically measurable sum parameters such as DOC are not very well suited to assess AOC concentrations. For measurement of AOC a new spectral algorithm was required. The approach used was the following; to predict the change in AOC concentrations for the treatment steps in the pilot plant both the differential concentrations ($\Delta AOC = [AOC_{before}] - [AOC_{after}]$), as well as the differential spectra ($\Delta Abs = Abs_{\lambda,before} - Abs_{\lambda,after}$) were calculated (figure 5). The resulting spectral algorithm shows a strong overall correlation with the ΔAOC for all three treatment steps considered in this study. Furthermore, it also shows significant to strong correlation when applied to the single treatment steps (figure 5).[8]

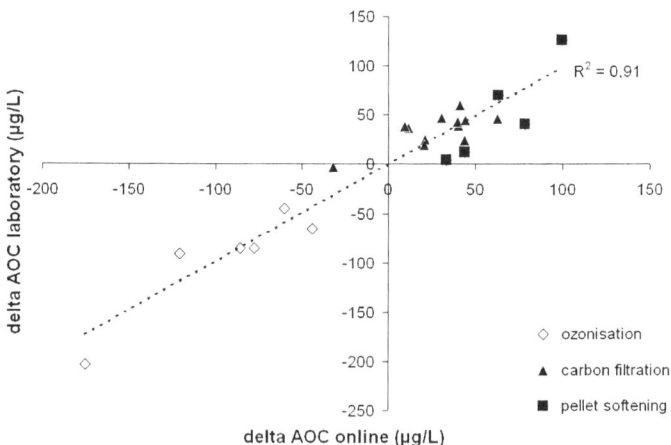

Figure 5   *Correlation between AOC changes over individual process steps and predicted AOC change based on differential spectra.*

The AOC calibration obtained in this way is a surrogate parameter, in the sense that the AOC concentrations in the water are far below the concentrations that can be distinguished using a UV/Vis spectrophotometer without sample pre-concentration.

Therefore, the correlation observed is between not between the AOC itself and the spectrum, but between other properties of the spectrum that are linearly related with AOC.

The results of the AOC measurements were all obtained at a pilot plant of Amsterdam Waterworks (NL), and validation of the results at different sites is currently underway.

5 CONCLUSIONS

The use of UV/Vis spectrometry for monitoring of drinking water quality and security has proven to be a highly valuable addition to online monitoring capabilities. The measurement of parameters such as turbidity, nitrate and sum organics can be performed simultaneously with a single instrument, which is robust, easy to operate and cost effective compared to classical analysers. The use of the spectrometer probe in this type of applications has been highly successful as evidenced by the example of the source water monitoring system operated by the waterworks of Vienna.

Additional capabilities of the spectrometer probe include the detection of accidental and intentional contamination of drinking water. Many substances can be detected with very high sensitivity, i.e. down to 10 µg/L concentration range, although the detection limits are directly dependent on the chemical structure of the target substances. Using the described spectral alarm system, the presence of non-natural pollutant substances can be detected. Further analysis using e.g. gas or liquid chromatography with mass spectrometry detection can then be used to unambiguously identify the detected contaminant.

Finally, it has been shown that the use of delta spectroscopy opens up an even broader field of applications. Using changes in the spectrum that occur between measurement sites separated over space, it is possible to calculate change in the concentrations of substances, which can then be used for process control and optimisation. In the presented example, the change in AOC could be monitored, although AOC itself is not directly visible in the spectra collected by a single spectrometer without the use of delta spectroscopy.

References

1. T.H.M. Noij and I. Bobeldijk, *Water Science and Technology, 2007, 47(2), 181.*
2. G. Langergraber, N. Fleischmann and F. Hofstädter, *Water Science and Technology, 2003, 47(2), 63.*
3. L. Rieger, G. Langergraber, M. Thomann, N. Fleischmann and H. Siegrist, *Water Science and Technology*, 2004, **50**(11), 143.
4. G. Langergraber, A. Weingartner and N. Fleischmann, *Water Science and Technology 2004, 50(11), 13.*
5. V. Thomsen, D. Schatzlein, D. Mercuro, *Spectroscopy, 2003, 18(12), 112.*
6. J. van den Broeke, A. Brandt, F. Hofstädter and A. Weingartner, in '*Security of Water Supply Systems: From Source to Tap'*. J. Pollert and B. Dedus (Eds.), Springer Verlag, Dordrecht, The Netherlands, p.19 (2006).
7. B.H. Tangena, D. de Zwart, J. van den Broeke, A. Brandt, M.J. van der Schans, Dutch Institute of Public Health and Environment (RIVM), *Report 609121.001, 2006.*
8. J. van den Broeke, P.S. Ross, A.W.C. van der Helm, E.T. Baars, L.C. Rietveld, *Water Science and Technology, 2008, accepted for publication.*

FULLY AUTOMATED INSTRUMENTATION FOR NUCLEIC ACID TESTING IN THE FIELD

D.J. Squirrell[1], M.A. Lee[1] and P. Wakeley[2]

[1]Enigma Diagnostics, Tetricus Science Park, Dstl Porton Down, Wiltshire SP4 0JQ, UK
[2]Veterinary Laboratories Agency (Weybridge), Addlestone, Surrey KT15 3NB, UK

1 INTRODUCTION

Testing for microorganisms has been revolutionised since the invention of the polymerase chain reaction (PCR)[1,2] and by its development into quantitative "real-time"-PCR (rt-PCR).[3,4] rt-PCR is becoming the method of choice for diagnosing infections, especially where traditional culture-based detection is of limited use. It requires instrumentation that can heat and cool the reaction mix whilst at the same time measuring the increase in concentration of target DNA by deploying, typically, fluorescent reporter molecules (see Figure 1). In addition to the rt-PCR itself, there is also a requirement, in most cases, for sample pre-treatment to remove PCR inhibitors and, for almost every instrument, a computer for control and results presentation. The method is therefore very much a laboratory-based technique carried out by skilled operators.

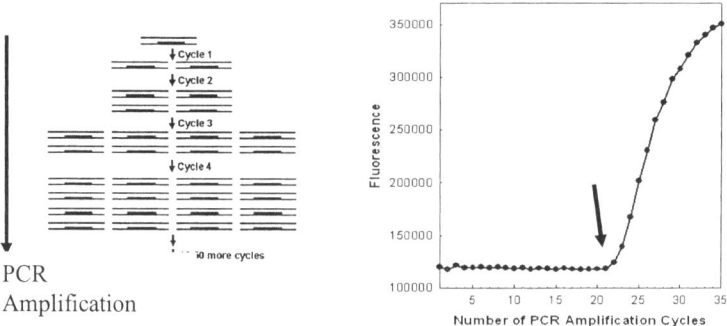

Figure 1 *In real time-PCR, the cycle-by-cycle doubling of target DNA sequences (shown as double black lines) is probed using a reporter sequence (red line) that fluoresces when it binds to the specific target of the assay. Following a number of cycles when the signal is below the detection threshold of the fluorimeter, amplification can be monitored and the take-off point ("Ct value", arrowed) allows the concentration of target in the initial sample to be determined.*

This paper reports the development and field testing of a fully integrated rtPCR system where all of the steps are carried out automatically in a single, compact, portable instrument.

2   FULLY AUTOMATED INSTRUMENTATION

2.1 Specification

The system described here was designed for use by UK Armed Forces to confirm detection events reported by on-line sensors. The list of requirements for the instrument for this rôle included:
- fast results (less than 1 hour essential, less than 30 minutes desirable)
- fully automated (no manual steps other than sample addition)
- no computer (onboard display and traffic-light result output)
- man-portable and ruggedised
- to be operated from a vehicle (without environmental control)
- pre-packaged reagents (stable for 72hrs without refrigeration)
- single sample analysis for a known target (confirmation) for bacterial and viral agents
- usable by an operator wearing protective clothing
- full test data storage with GPS for date/time stamping (for evidential purposes)
- minimal training needed.

2.2 The Enigma FL instrument

The instrument developed to fulfil the requirements above is shown in Figure 2. It has been designed for wall mounting in a vehicle with access and connections at the front. Sample processing is carried out in the upper half with thermal cycling and fluorescent monitoring in the bottom part.

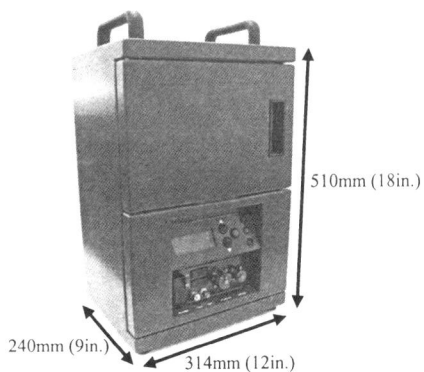

510mm (18in.)

240mm (9in.)

314mm (12in.)

Figure 1   *The Enigma FL fieldable real-time PCR instrument. Sample processing is carried out by a robotic system in the upper half. The lower part comprises the thermal cycler and fluorimeter. The front panel has an 80-character alpha-numeric display, traffic-light LED's, buttons to control operation and various connectors. The steel casework has a chemical agent resistance coating.*

The instrument is ready to run within 5 seconds of being switched on. The required test is selected from a menu on the display panel. The sample is added to a single use cartridge containing all of the necessary reagents for the test which the operator places in the processing chamber. After the door is closed, the operator presses the green button to start the analysis. The progress of the assay is reported on the display ending with a report on the result which is accompanied by an (optional) audible signal and a traffic light output (with red for a positive, green for a negative, and amber indicating that a problem with the test has been detected - through the in-built assay controls - and that a re-test is required).

2.2 Sample processing and reagents

Sample processing is carried out by a robotic system that works with the cartridge. This is in a carousel format and contains not only reagents, but also all of the consumable parts needed for extracting and purifying nucleic acid from the sample. The cartridge and its component parts are shown in Figure 3. The Electrically Conducting Polymer (ECP) capillary is detachable and there is a gap in the cartridge through which it can be lowered.

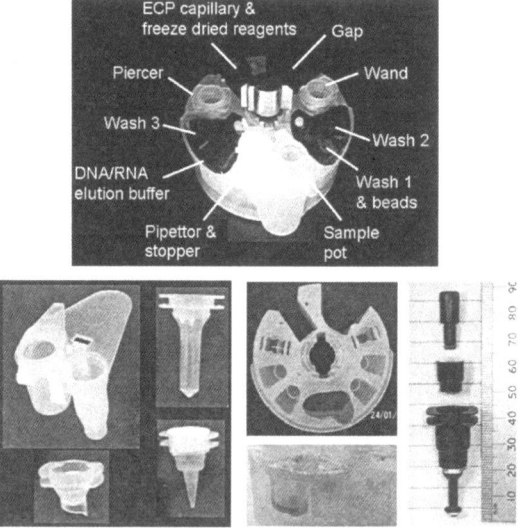

Figure 3   *The picture at the top shows an assembled assay cartridge. It is approximately 8cm in diameter and 3.5 cm high, the scale being chosen to facilitate handling and to allow for the processing of relatively large sample volumes.The component parts are shown individually beneath (first panel, clockwise from top left: sample pot, wand, pipettor and piercer; second panel: empty carousel top, detail showing reagent wells bottom; third panel: stopper top, double-foiled pot containing freeze dried PCR reagents middle and ECP capillary bottom).*

Sample is added to the sample pot. This contains a guanidium salt which acts as a chaotropic agent for lysis of cells and virus particles to release nucleic acids. The chaotrope is provided as granules to which up to 0.8 ml of aqueous sample may be added. Using the wand, into which a magnet can be inserted and withdrawn, silica-coated magnetic beads are transferred from the first wash solution and released into the lysed sample. Released nucleic acids bind to the silica surface of the beads which are then transferred through the three wash buffers where compounds from the sample that might inhibit PCR are removed. The nucleic acids are released into a small volume of elution buffer. The magnetic beads are removed and discarded and then, using the pipettor, 25 microlitres of the eluted nucleic acid are added to freeze-dried reagents that are contained in a chamber above the ECP capillary.

Centrifugation of the cartridge transfers the PCR reaction mix into the ECP capillary. The final stage in preparation for the PCR is stoppering the capillary and transferring it out of the cartridge and downwards into the real-time thermal cycler. Sample preparation times vary from 10 to 15 minutes depending upon the nature of the sample.

Throughout the sample preparation sequence, the use of dried reagents and magnetic bead technology allows dilution of the target material to be avoided to facilitate the detection of low concentrations of the target species.

2.3 PCR

The ECP capillary, (Figure 4), is lowered into a chamber where it makes electrical contact for a heating current to be applied and is located so that it is: (1) in the air path for a cooling fan; (2) in the field of view of a thermopile which makes infra-red temperature measurements from the shaft of the capillary; and (3) in a position for the fluorimeter to make measurements through the window at the bottom. The optics are configured to be compatible with any of the standard fluorescence reporting chemistries used for PCR.

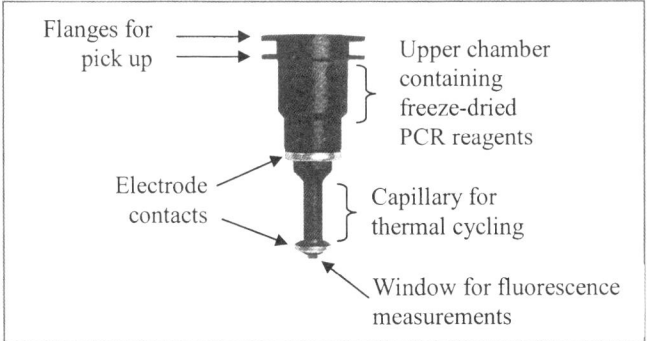

Figure 4   *Details of the ECP capillary are shown above. It is moulded from a carbon-filled, electrically conducting, nylon polymer to form a tube for the PCR. This is heated directly by passing a current through the portion containing the reaction mix. Rapid heating is possible because of the low thermal mass.*

If the nucleic acid to be detected has an RNA, rather than a DNA genome, as is the case for many disease-causing viruses such as Influenza and Foot-and-Mouth Disease Viruses, this has to be converted to DNA before testing by PCR. Conversion takes 2 minutes at 48°C using a polymerase enzyme that can accomplish both this reverse transcriptase reaction and, subsequently, be employed for the PCR amplification steps.

The PCR phase is carried out using thermal cycling parameters that are programmed for a given assay. Heating and cooling of the reaction mix is controlled through a feedback circuit governed by the sample temperature as computed from the thermopile signals using a numerical model of heat flows within the system.

2.3 Fluorescent reporting and results analysis

Hybridisation probes are used to detect the formation of specific products. The probes are generally in the form of labelled sequences whose binding to the target can be detected using a FRET (fluorescent resonant energy transfer) process. Amplification can be monitored as soon as the concentration of bound probe reaches the limit of detection of the fluorimeter and then followed until the reagents become exhausted and further amplification is inhibited by reaction products.

An algorithm is used to analyse the fluorescent data cycle by cycle and a positive result can be called as soon as the characteristic pattern of exponential growth has been established. This means that high concentration positive samples can be detected fastest, (see Figure 5). Quantitative results are available from the algorithm as the initial target concentration can be determined from the amplification curve. The level of information can be limited to a simple yes/no answer if appropriate. Negative test results require a full 40 or 50 cycles of amplification before the result can be called, when PCR reaction controls which are added at a low level are measured. If the controls are not amplified then a re-test is advised. Limits of detection are typically between 20 and 200 bacterial cells or virus particles per millilitre of sample and the dynamic range covers over 8 logs.

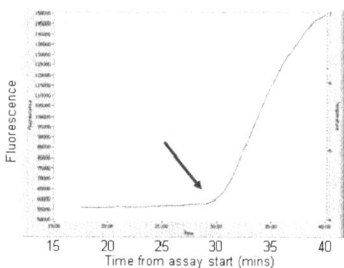

Figure 5   *A trace from a test run with 10^6 spores ml^{-1} of* Bacillus atropheus, *(a simulant for anthrax), shows the total assay time needed for higher levels of target agent. Sample preparation program are longest for spores, but the assay result above was apparent around 30 minutes from the start of the test. Another 2-3 minutes is required for a result to be called for every factor of 10 lower in concentration of target in the sample.*

The fluorimeter has two LED excitation sources and six emission channels to allow multiplexed assays. If required, post-amplification melt anaysis[5] can be used to provide additional multiplexing and/or another level of confidence in the test result.

It should be noted that with the real-time PCR approach, the reaction tube can be kept permanently sealed to prevent the problem of cross-contamination with amplification products which can cause of false positive results. The ECP capillary, made from a carbon-fibre filled polymer, is very robust which means that it is particularly resistant to physical damage.

3 FIELD TRIALS

The system was tested in field trials to assess practical deployment issues and to prove that fully automated testing, as required, was feasible. A veterinary pathogen, Bovine Viral Diarrhoea Virus, (BVDV), was chosen as the target pathogen for the trials. This is an endemic disease with a prevalence of around 1%, but affecting around 90% of UK dairy herds. Calves infected *in utero* develop poorly and sicken and die at the age of around 12-18 months. Laboratory-based real-time PCR testing of blood samples is the current standard method used to confirm diagnosis of infection.

Following an initial laboratory trial on known positive and negative archive samples, which showed comparable results with the laboratory assay could be obtained, tests were carried out on three farms in Southern England. Two positive blood samples were detected on the third farm, with the tests performed by the vet, unaided, after only about 10 minutes training. The positive and negative results that were obtained were confirmed by tests in the reference laboratory, (see Figure 6). One of the positives was on a pooled sample.

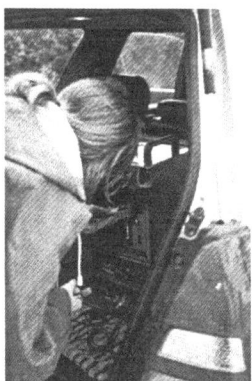

Sample	Automated pen-side results	Confirmatory laboratory results
Calf 1 (on farm 3)	+	+
Calf 2 (on farm 3)	-	-
Calf 3 (on farm 3)	-	-
Pool 1 (on farm 3)	+	+/-/-
Pool 2 (on farm 3)	-	-/-/-
Pool 3 (back at lab)	-	-/-/-
Pool 4 (back at lab	-	-/-/-
Pool 5 (back at lab)	-	-/-/-

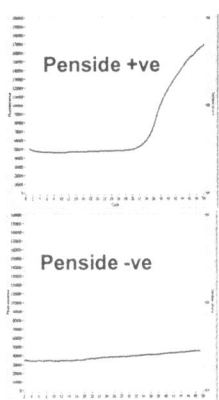

Figure 6   *It was found that tests can be carried out most conveniently with the instrument in a vehicle, (here in the boot of an estate car). The results from the farm where animals tested positive are given in the table and PCR traces from positive and negative blood samples are shown.*

As well as proving the feasibility of pen-side testing, the trials also enabled deployment options to be investigated. It was considered that biosecurity could best be maintained by addressing the handling of samples and by keeping the instrument in the car used to visit the farm, keeping it parked at the farm entrance to avoid contamination of both machine and vehicle.

4   CONCLUSIONS

We have clearly demonstrated the feasibility of definitive PCR-based testing for the detection of pathogens
- by non-technical operators
- in complex samples
- in the field
- with full automation
- in robust, compact equipment

Potential uses of the system include contamination testing on water samples or swab samples, the diagnosis of infection and confirmation of detection events following alarms from on-line detectors.

We have learnt that operation from a vehicular platform is preferable and that the instrument can be deployed very rapidly. Definitive results can be obtained within an hour and there is a low training requirement for operators. Procedures for maintaining biosecurity are feasible and pooled sample analysis works which may be useful in survey applications.

References

1   R. K. Saiki, S. Scharf, F. A. Faloona, K. B. Mullis, G. T. Horn, H. A. Erlich and N. Arnheim, *Science*, 1985, **230**, 1350.
2   K. B. Mullis and F. A. Faloona, *Methods Enzymol.*, 1987, **155**, 335.
3   R. Higuchi, G. Dollinger, P. S. Walsh and R. Griffith, *Biotechnology*, 1992, **10**, 413.
4   R. Higuchi, C. Fockler, G. Dollinger and R. Watson, *Biotechnology*, **11**, 1026.
5   K. M. Ririe, P. R. Rasmussen and C. T. Wittwer, *Anal. Biochem.*, 1997, **245**, 154.

OPTIMISATION OF NMR METHODOLOGY FOR NON-TARGETED DETECTION OF WATER CONTAMINANTS

A. J. Charlton[1], J. A. Donarski[1], B. D. May[1] and K. C. Thompson[2]

[1]Department for Environment, Food and Rural Affairs, Central Science Laboratory, Sand Hutton, York, YO41 1LZ, UK
[2]ALcontrol Laboratories, Templeborough House, Mill Close, Rotherham, South Yorkshire, S60 1BZ. UK

1   INTRODUCTION

In 1997 the Laboratory Environmental Analysis Proficiency (LEAP) Emergency Scheme[1] was initiated. The primary function of this scheme is to establish the reliability of water testing laboratories to determine the contents of a simulated potable water contamination sample. In the event of a potable water contamination incident, water testing laboratories may be called upon to rapidly identify contaminants that are present in the water with little information about the likely source of contamination.[2,3] Samples are analysed for inorganic and organic chemical contamination. Here we present the analysis of three consecutive LEAP organic contamination emergency samples using cryoprobe nuclear magnetic resonance (NMR) spectroscopy and demonstrate the complementarity of the approach with existing analysis methods. We demonstrate that a wide range of compounds can be simultaneously detected and characterised at μgL^{-1} levels with minimal sample pre-treatment prior to NMR analysis.

Determination of organic contamination of the potable water supply is usually performed using techniques based on the application of high performance liquid chromatography (HPLC) or gas chromatography (GC). The choice of chromatographic method is determined by the suite of analytes that is to be detected and therefore these approaches are inherently targeted to detect specific compounds. Both liquid chromatography and gas chromatography require that the target analytes are retained on the chromatographic column and this requires careful selection of the analytical approach to be used. The choice of approach is particularly difficult when analysing for unknowns and the method chosen will always lead to some exclusions. Therefore, the detection method that is used to monitor elution from the chromatographic column requires consideration. The most frequent detection method for GC in water laboratories is mass spectrometry (MS), whilst for HPLC it is diode array detection (DAD) and increasingly mass spectrometry (MS or MS/MS). Mass spectrometry can also be used to determine the mass of the detected molecules and further ionisation fragments, potentially facilitating the identification of unknown compounds. However, ionisation efficiency is very variable especially for HPLC-MS and therefore some difficult to ionise substances such as chlorpyriphos-ethyl, chlorpyriphos-methyl, dicamba, dichlobenil, fenchlorphos, fenitrothion, parathion-ethyl, parathion-methyl and pentachlorophenol are poorly detected and can be under reported by orders of magnitude for non-targeted analysis when measured against inappropriate internal standards. Additionally, chemical derivatisation is also often required for semi-polar and polar substances analysed by GC-MS, thus limiting the range of applicability of the GC methods. In emergency incident situations, the derivatisation step is often omitted leading to poor extraction and chromatographic performance. Both of these can lead to significant under-reporting of polar or semi-polar substances. The approximate concentration levels reported (measured against non-polar internal standards) could be up to two or three orders of magnitude too low.

NMR spectroscopy is commonly perceived to be a relatively insensitive technique, however developments in the field of NMR are constantly leading to improved sensitivity and include improvements such as increases in the maximum available magnetic field strength[4] and cyroprobes[5]. The NMR signal arises from the population difference between aligned nuclei within a magnetic field. This population difference is directly correlated to magnetic field strength, therefore higher measurement sensitivity is achieved when using higher field strength magnets. Cryoprobes decrease the electrical noise of recorded NMR signals as the major electrical components of the probe are cooled to 25 K, therefore increasing the signal to noise ratio and hence the measurement sensitivity. In the absence of high salt concentrations, the sensitivity obtained by using the cryoprobe is approxiamtely 3 to 4 fold greater than that obtained by using the equivalent standard probe and this corresponds to a reduction of 9 to 16 fold in experimental time[5]. Further improvements in the signal to noise ratio of NMR measurements can be made by effective instrument set-up. This includes the choice of pulse sequence, correct calibration of required pulses and effective choice of probehead[6]. Sample acquisition times can also be adjusted to obtain the desired sensitivity. Recent advances in the sensitivity of NMR spectroscopy have facilitated the determination and characterisation of molecules present in solution at much lower concentrations than prior to these developments.

NMR spectroscopy is a non-selective technique and when using solution state ^{1}H NMR spectroscopy, any soluble molecule that contains non-exchangeable ^{1}H atoms will give an observable resonance. These signals are presented on the chemical shift (δ, in ppm) scale which is machine independent. Therefore, spectra acquired on one spectrometer can be directly compared to spectra generated on another, even at different magnetic field strengths. Chemical shift is dependant on the chemical structure and the local chemical environment of the molecule under observation. Further information about the chemical structure is inherent in the NMR spectrum as *J*-couplings. The NMR measurement is therefore highly specific and well suited to discriminating between similar compounds (including isomers). Peak area is directly correlated to ^{1}H concentration and therefore can be used to determine analyte concentration. Solution state NMR spectra can be generated in a wide range of liquids, including but not limited to water, methanol, chloroform, acetone and dimethyl sulphoxide. Therefore, spectra from pre-concentration of a sample using rotary evaporation, SPE columns or liquid / liquid extractions can generally be acquired. However, one of the principal advantages of NMR spectroscopy for the detection of unknown contaminants in potable water is that the analysis can be performed without significant sample preparation. In ^{1}H NMR spectroscopy this often involves just adding a small amount of deuterated water, which may contain an internal standard if quantitative measurements are being performed.

2 EXPERIMENTAL

Three water samples were used in this study. They were independently prepared at CSL for the LEAP emergency exercises 10, 11 and 12. The concentrations of the organic contaminants in these three samples are listed in Table 1. A blank water sample was analysed with each test sample to determine the presence of naturally occurring contaminants.

Samples were prepared by adding 10% (v/v) deuterium oxide (^2H_2O) containing 1 mM 3-(trimethylsilyl)- propionic acid-d4, sodium salt (TSP) and a K_2HPO_4/KH_2PO_4 buffer (150 mM, pH 7.0). Pre-concentrated samples were prepared using rotary evaporation at a final vacuum of *ca* 800 Pa and a temperature of 60°C for a period of 2.5 hours, 500 mL of sample was concentrated to a volume of *ca* 10 mL giving an approximate concentration factor of 50.

All NMR experiments were carried out using a Bruker Avance 500 MHz NMR spectrometer equipped with a TCI cryoprobe. Data acquisition and processing was performed with the use of Topspin v 1.3 (Bruker, Germany). Spectra were acquired at a central frequency of 500.1323505 MHz, using on-resonance pre-saturation to suppress the intensity of the water signal, followed by a 1D NOESY pulse sequence[7] with irradiation of the residual water signal during the mixing time (200 ms). Using this sequence the intensity of the water signal was reduced to below that for the internal standard and therefore optimisation of the receiver gain (and therefore the measurement sensitivity) was not limited by the water signal. An observation pulse length of 7.14 μs and a delay between transients of 10 s were used. 65536 complex data points were acquired with a spectral width of 20.6 ppm, giving an acquisition time of 3.17 s. Four unrecorded (dummy) transients and 8192 acquisition transients were used giving a total maximum experiment time of approximately 30 hours 37 minutes. The free induction decay was processed at approximately hourly intervals during data acquisition. Examination of these data was used to determine the minimum experimental time required to observe contaminant resonances. A line-broadening coefficient of 2 Hz was applied to the free induction decay. After Fourier transformation and manual phase correction the resulting spectra were referenced to the TSP peak at 0 ppm.

The ^1H NMR spectra acquired from both the water blanks and the test samples were compared. Where significant differences between the blank and the test sample were noted, these data were compared to data obtained from the contaminants present in the samples.

A library of ^1H NMR spectra had previously been generated for the monitoring of the presence of contaminants in food, introduced by both accidental and malicious means. These data were used to determine the identity of contaminant resonances. Where compounds not present in these libraries were encountered, the structure determination power of NMR spectroscopy was utilised.

3 RESULTS AND DISCUSSION

Table 1 summarises the results obtained for the ^1H NMR analysis of the three LEAP emergency scheme potable water samples tested. It gives the substance concentrations; the ^1H chemical shifts; whether pre-concentration was used and the minimum detection time.

	Analyte	Concentration (µgL⁻¹)	^{1}H Chemical shift	Observed without pre-concentration	Observed with pre-concentration	Minimum experimental time used to observe (Hr)
LEAP Water Sample 10	Picric Acid	504	8.987	Yes	Yes	0.2[b]
	Amitrole	50	7.792	Yes	Yes	2.0[b]
	Dicamba	50	7.400 7.237 3.909	Yes	Yes	0.5[b]
	Pyridine	200	8.541 7.915 7.492	Yes	No	2.0
	Mevinphos	80	5.853 3.923 3.901 3.753 2.382	Yes	Yes	1.0[b]
LEAP Water Sample 11	Aldicarb	75	7.740 2.820 2.812 2.010 1.470	Yes	Yes	0.3[b]
	Aniline	339	7.261 6.9091[a] 6.8943[a] 6.8797[a] 6.864[a]	Yes	No	1.3
	Glyphosate	202	3.731 3.026 3.002	Yes	No	2.0
	MCPA	11	7.257 7.193 6.757 2.246	No	Yes	4.0[b]
	Phosphamidon	70	3.961 3.865 3.480 2.247 2.088 1.20	No	Yes	0.3[b]
LEAP Water Sample 12	2,4-D	75	7.533 7.311 6.899 4.571	Yes	Yes	0.3[b]
	2,6-Dibromophenol	5	N/D	No	No	>30Hr
	Pentachlorophenol	20	N/D	No	No	N/A

Table 1   *Summary of results obtained from ^{1}H NMR spectroscopic analysis of potable water and the concentration of the organic contaminants present in the three LEAP water samples tested.*

---

[a] Two overlapping multiplets.
[b] Acquisition time used after pre-concentration.

The data generated from LEAP water sample 10 (see Table 1) shows that five compounds were detected (picric acid (δ = 8.987 ppm), pyridine (δ = 8.541, 7.915 and 7.492 ppm), amitrole (δ = 7.792 ppm), mevinphos (δ = 5.853, 3.923, 3.901, 3.753 and 2.382 ppm) and dicamba (δ = 7.400, 7.237 and 3.909)) . NMR spectroscopy has therefore been used to detect all of the compounds present in the contaminated sample without significant sample preparation at concentrations as low as 50 μgL^{-1}. A single dataset from the contaminated sample was required to detect and identify all five of these chemically diverse compounds.

The data generated from LEAP water sample 11 (see Table 1) gave similar results with the analytes aldicarb (δ = 7.740, 2.820, 2.812, 2.010 and 1.470 ppm), aniline (δ = 7.261, 6.909, 6.894, 6.880 and 6.864 ppm) and glyphosate (δ = 3.731, 3.026 and 3.002 ppm) being detected without sample pre-concentration. To detect 2-methyl-4-chlorophenoxyacetic acid (MCPA; δ = 7.257, 7.193, 6.757 and 2.246 ppm) and phosphamidon (δ = 3.961, 3.865, 3.480, 2.247, 2.088 and 1.210 ppm) the sample was pre-concentrated by 50 fold using rotary evaporation.

The data generated from LEAP water sample 12 (see Table 1) showed that 2,4-dichlorophenoxy acetic acid (2,4-D; δ = 7.533, 7.311, 6.899 and 4.571 ppm) could be detected without sample pre-concentration. The compounds 2,6-dibromophenol and pentachlorophenol could not be detected. 2,6-dibromophenol could not be detected even after 50 times pre-concentration, this is presumably because the compound evaporated during the rotary evaporation process. Pentachlorophenol contains only exchangeable protons and therefore could not be observed using ^{1}H NMR spectroscopy.

One limitation of ^{1}H NMR spectroscopy is the need for non-exchangeable proton resonances. Recent developments to NMR spectroscopy instrumentation led to the realisation of a phenomena reported in 1953[8] called dynamic nuclear polarisation (DNP). In DNP-enhanced NMR spectroscopy, the distribution of electron energy states is transferred to the nuclear spins creating a non-Maxwell Boltzmann distribution of nuclei. This leads to an enhanced NMR signal with recent reports of a 40,000 fold increase in sensitivity over that for conventional NMR measurements. Initial results have been reported using aqueous solvents for sample dissolution[9-10]. These advances in DNP-enhanced NMR spectroscopy will enable the generation of natural abundance ^{13}C NMR spectra using trace levels of material. These spectra require only that a carbon atom be present in the molecule, which by definition includes all organic molecules. Further testing of the applications of this methodology for trace analysis is required, however initial results suggest that this technique offers a realistic solution to the problem of identifying unknown contaminants.

It was also noted that some analytes could not be concentrated using rotary evaporation. Further evaluation of simple robust sample pre-concentration technologies (e.g. SPE or liquid-liquid extraction) will facilitate the detection of these compounds by NMR spectroscopy.

4 CONCLUSIONS

Cryoprobe ^{1}H NMR spectroscopy is an excellent method for the determination of contaminants in water, which has hitherto been untested for use in emergency scenarios. The unbiased nature of the NMR measurement enables the generation of spectral libraries facilitating the identification of contaminants in water. NMR spectroscopy will also yield considerable information about the identity of the contaminating compound in the absence of standard compounds or spectral libraries. Inherent in the spectrum is chemical structural information that can be used to guide further analysis or to identify an unknown contaminant.

For contaminant levels above about 200 μgL^{-1}, no sample pre-concentration is normally required. The only sample pre-treatment is the addition of a suitable internal standard and deuterated water (buffered). As NMR measurements are non-invasive and the samples are often contained in solution within glass tubes, highly toxic and explosive compounds can be analysed with limited risk to the operator and the instrumentation.

The detection limit of the NMR measurement is not a fixed parameter and is correlated linearly with the square root of the data acquisition time. As the throughput is inversely correlated to acquisition time, doubling the acceptable detection limit results in a four-fold increase in sample throughput. A compromise therefore exists between the sensitivity (detection limit) required and the sample throughput. The data shown in Table 1 indicates that a detection limit of 200 μgL^{-1} can be achieved in approximately 2 hours, using whole water samples without the risk of analyte losses. Using pre-concentration methods this acquisition time can theoretically be reduced to approximately 1 minute, translating to a throughput increase from 12 to up to 1440 samples per day with a suitable pre-concentration step. However, in the most challenging of circumstances where the desired detection limit is close to 50 μgL^{-1} and a pre-concentration step is not used, a throughput of two samples every three days is applicable. However, this is unlikely to be the case in most emergency scenarios as few substances are acutely toxic at such low concentrations and even if the spectral acquisition time is optimised for very high sensitivity detection, the data can be read throughout the experimentation such that higher concentration contaminants would be detected in the shortest possible time.

Advances in NMR spectroscopy have resulted in the availability of a technique that is capable of detecting trace levels of contaminants in water without *a priori* knowledge of the nature of the contamination. This is particularly useful for the detection of malicious contamination and NMR spectroscopy provides an effective screening method for the presence of chemical toxins and soluble or liquid explosives in the drinking water supply.

The wider applications of this methodology include monitoring and detection of contaminants in environmental water samples (e.g. effluents) and water supplies for food and drink production processes. Further studies have demonstrated the applicability of the methodology for the detection of contaminants in bottled water and other soft beverages.

The high cost of NMR cryoprobes currently restricts the broad adpotion of this methodology. However, national laboratories such as CSL provide facilities for the application of these methods during emergency scenarios.

REFERENCES

1 B. May, in *Water Contamination Emergencies Enhancing Our Response.* Eds. K. C. Thompson and J. Gray, RSC Publications, 1st edn, 2006, pp.229-235

2 K. C. Thompson and S. Scott, in *Water Contamination Emergencies Enhancing Our Response.* Eds. K. C. Thompson and J. Gray, RSC Publications, 1st edn, 2006, pp.216-228.

3 G. O Neill, C. Ridsdale, K. C. Thompson and K.Wadhia, in *Water Contamination Emergencies: Can We Cope?* Eds. K. C. Thompson and J. Gray, RSC Publications, 1st edn, 2004, pp 100-109.

4 L. R. Dixon; W. D. Markiewicz.; W. W. Brey, and K. K. Shetty, *IEEE T. Appl. Supercon.*, 2005, **15,** 1334-1337.

5 J. Losonczi and I. Green, *Am. Lab.* 2004, **36,** 26-29.

6 W. F. Reynolds and R. G. Enríquez. *J. Nat. Prod.* 2002, **65** 221 - 244

7 A. J. Shaka, C. Bauer and R Freeman. *J Magn Reson,* 1984, **60,** 479-485

8   A. W. Overhauser, *Phys. Rev.* 1953, **92**, 411.
9   J. H. Ardenkjaer-Larsen, B. Fridlund, A. Gram, G. Hansson, L. Hansson, M. H. Lerche, R. Servin, M. Thaning, and K. Golman, *PNAS*, 2003, **100**, 10158-10163.
10 J. Wolber, F. Ellner, B. Fridlund, A. Gram, H. Johannesson, G. Hansson, L. H. Hansson, M. H. Lerche, S. Mansson, R. Servin, M. Thaning, K. Golman and J. H. Ardenkjaer-Larsen. *Nucl Instrum Methods Phys Res A.* 2004, **526** 173 – 181.

H. Clay-Chapman,

Three Valleys Water, Bishops Rise, Hatfield, Hertfordshire AL10 9HL

1   INTRODUCTION

The explosion at the Buncefield Oil depot not only caused an immediate and highly visible image that is still etched in people minds but it has also left a legacy that continues for many different groups of people. This paper aims to provide an insight into one of those legacy issues- the impact on water quality. It will cover the types of information and data management issues that were prevalent in the early days of the explosion when a multi-agency team was convened. Those involved faced the challenge of managing a potential water quality situation concerning a range of chemical compounds of which very few had expertise in managing. It will also cover the mechanisms that were put in place to build on the certainties, cope with the uncertainties and the key lessons that have been learnt.

2   THE INCIDENT

The explosion at the Buncefield Oil terminal happened at 06.01 hours on Sunday 11th December 2005. A national response was initiated involving both category 1 and 2 responders and the catastrophe was reported by the media during the ensuing hours, days and weeks. In response to the incident and considering the global significance of the event a major incident investigation board was convened under the chair of Lord Newton. To date, it has published seven reports[1] on the cause of and response to the incident which are recommended as background information to the overall incident and impacts.

2.1 Water Quality Concerns

The explosion and fire posed an immediate set of risks for Three Valleys Water Company. An emergency response team was quickly established on the Sunday morning to assess the risks and put in place mitigation controls. Of paramount importance for water quality was the requirement to ensure that both the drinking water supply would not be affected by any contamination from the site and that the fire-fighting response would not deplete water supplies and cause a secondary issue within the distribution network. The Company's water resource experts were also mobilised to assess the possible impact to the public abstraction points which surround the Buncefield depot. As the day progressed, it was apparent that due to the extent and visibility of the plume of smoke, international concerns were being raised and options to extinguish the fire were being considered.

During the Sunday, the assessment of risk by the water resources experts highlighted a number of potential pathways from the site. The oil depot at Buncefield sits on clays overlying chalk, a major aquifer and the general groundwater flows are to the south east. While the Oil depot is outside Environment Agency (EA) estimated 400 day travel time, it is within the total catchment for a number of sources. The possibility of uncontained liquids to contaminate the aquifer was of concern. More importantly, however, the risk of run off down the dry river valleys toward the River Gade and Ver and effectively short-circuiting the potential contamination pathway was of significant importance to the Company. These risks are shown diagrammatically in figures 1 to 3.

During the decision-making process as to whether to let the fire burn or to extinguish it, a significant amount of lobbying of Gold command took place to ensure those responsible would be fully aware of the risks this decision would pose. Despite the potential risk to the environment, the decision was made to extinguish fire and so, as a precaution, the site at Bow Bridge which was deemed the most susceptible to surface water run off, was shut down and isolated from supply.

The ensuing days were spent responding to requirements for fire fighting water and ensuring that all possible controls that could be in place were implemented to prevent contamination of water quality within the network and at Company sources.

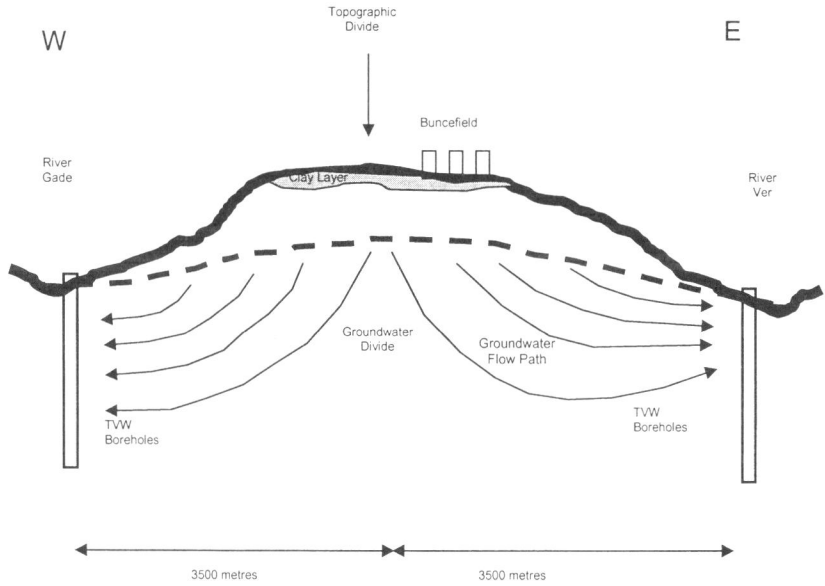

Figure 1   *This shows the general topography of the area with Buncefield depot position at the highest point*

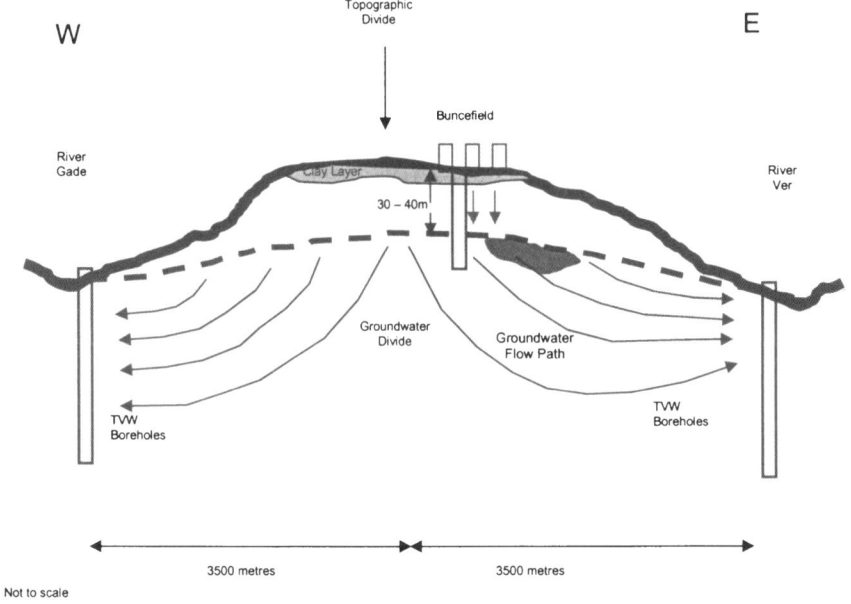

Figure 2   *This shows the impact that any contamination on site would have on the aquifer*

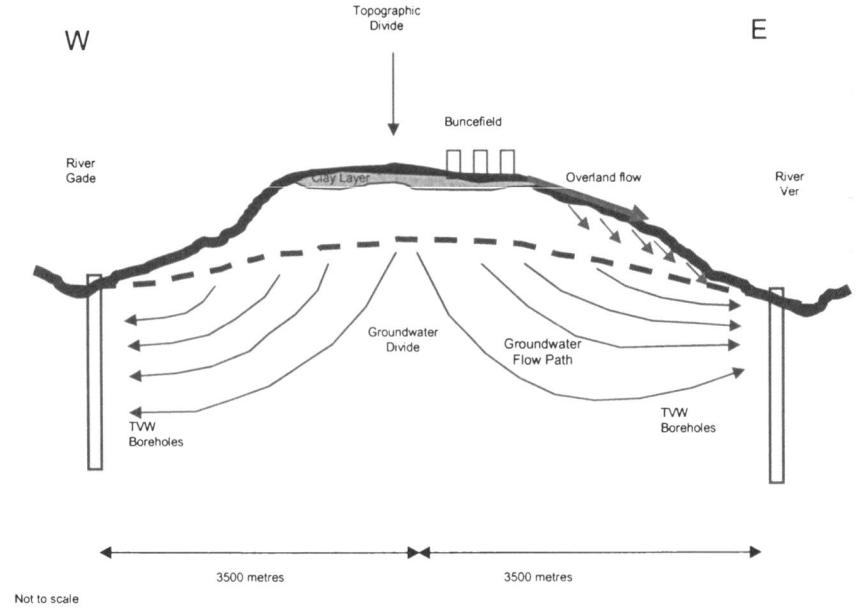

Figure 3 *This shows the impact of run-off from the site*

Five days after the explosion a conference call took place involving EA, DWI, HPA and water company personnel chaired by Prof. Virginia Murray, CHaPD, HPA. The outcome of this initial meeting of experts quickly highlighted that there was much to discuss, more questions than answers and that an expert multi – agency team needed to meet to map out the way forward. The urgency was such that the meeting took place the following day at the DWI Offices, Whitehall.

2.2 Multi-agency Concerns

At the initial multi-agency meeting those present highlighted the following key questions to be answered were:-

What are the risks?
How do we deal with them?
Will there be any secondary or tertiary impacts?

Of paramount importance was to find the answers to the following questions:-

What do we know?
What do we need to know?
What do we have to do to find information and by when?

Everyone present contributed by presenting their current position and highlighting specific concerns and issues. From this a detailed set of questions were developed. These are tabulated in Figure 4 and have been grouped into the different areas of concern.

It was quickly apparent that there were more questions than answers and recognition that there was limited expertise in dealing with the compounds present and the quantities involved. The extent of the detailed work required to provide the answers was not underestimated. To enable answers to be found the initial response team was formalised into the multi-agency liaison group and terms of reference were drawn up to clarify roles, responsibilities, outputs and enable knowledge and information sharing. The purpose of this group was to bring together all disciplines involved in protection of the environment, human health and drinking water to improve consistency of approach, apply control and attempt to restore normality.

Initially, this group met weekly to collate and consider the information that was being produced. The volume and extent of knowledge required to make the necessary objective and pragmatic decisions was significant and without exception all representatives were fully committed to the work for several months. As the situation stabilised and alternative controls were implemented for the various work components the meeting reduced to monthly and the more detailed background work merged back into most people's day job.

The site
What liquids are we dealing with? How can we stop them leaving site? What disposal containment arrangements need to be made? What possible arrangements are there to store the liquid until they can be safely treated?

The environment
What do we know about the general water quality from river, sewers, waste water treatment works, aquifers, ground waters and the distribution network in the vicinity of Buncefield? What do we need to put in place to ensure it can be measured, monitored and controlled? Do we understand all the constituents of fire fighting foam and the impacts of any bi-product formation? What are the safe levels of these substances for environment and drinking water?

Analysis
Who has analytical methods available and are they sufficient for the volume of monitoring required and the type of substrate of interest? Does the analytical capability require enhancing?

Monitoring
Are there sufficient monitoring locations available? What criteria are to be used to define or install suitable monitoring locations? What considerations need to be given to access arrangements? Who will approve/grant access on privately owned land where access is essential?

Treatment
What treatment options are available? What work needs to take place to prove effective treatment is occurring and gain the endorsement by public health and water quality and environmental regulators

Figure 4   *Key questions*

2.3 Lessons Learnt

The lessons learnt are as follows:-,

   2.3.1   The site.   Despite the best endeavours of all concerned, at the time of the explosion and during the fire-fighting activities it was very difficult to obtain accurate and consistent answers on the quantities of fuel lost from the structures and bunds and the fire-fighting liquid applied. It was very apparent that there was a substantial amount of mixed liquid which required specialist removal and treatment. Removing this liquid was an enormous logistics exercise over a period of two to three months. Fortunately, the weather was very dry during this period and so the added complication of substantial rainwater adding more liquid to the quantities to be contained and removed to safe storage together the possibility of uncontrolled run-off did not occur.

The up-to-date information provided by the Oil Company representatives about how containment was being managed and the extent of controls which were being put in place was invaluable and provided much reassurance during times of enormous uncertainty.

2.3.2 The environment and public health.   The potential environmental impact of vast quantities of fuel and firefighting liquid at the depot in partially destroyed tanks and bunds was a significant risk that required an immediate solution. The options for relocating thousands of mega litres (Ml) of contaminated and potentially carcinogenic material to a safe and secure location posed a number of challenges. It was very fortuitous that Thames Water Services had some temporary spare capacity at a nearby waste water treatment works and made this facility available until treatment options for these liquids could be developed. At the time it was not known that it would take almost two years for the clean-up of this liquid and that further issues would arise with regard to the quality of the discharge water from this site. This experience does raise the issue of when responding to a major emergency the full implications of offering assistance without compromising existing regulatory requirements should be assessed at the time. If necessary, the potential minor impacts that may occur from offering facilities to avert a major impacts, especially for the benefit of other stakeholders, should be subject to pre-agreed national guidance so that the lesser regulatory infringement is adopted in favour of the significant benefit afforded by the response action. Regulators should agree and present a balanced decision on this issue at the time.

The public health consequences were of equal concern there being no documented safe levels of the contaminants established for the environment and drinking water. Interim precautionary standards were quickly established. Subsequently, effort was devoted to a detailed evaluation of the public health risks. This culminated in the following guidance (Fig 5) being published for the components for the fire-fighting fuel {PerFluoro Octanate Sulphonate-(PFOS) and PerFluoro Octanoic Acid (PFOA) } that could be present in drinking water.

Item	Regulatory requirement	Guidance value (concentration)		Minimum action to be taken
		PFOS	PFOA	
Tier 1	Regulation 10 (Sampling: further provisions)	>0.3 µg/l	>0.3 µg/l	Consult with local health professionals Monitor levels in drinking water
Tier 2	Regulation 4 (2) (Wholesomeness)	> 1.0 µg/l	>10 µg/l	As tier 1 plus: Put in place measures to reduce concentrations to below 1.0 or /10 µg/l as soon as is practicable
Tier 3	Water Undertakers Information Direction 2004	> 9.0 µg/l	>90 µg/l	As tier 2 plus: Ensure consultation with local health professionals takes place as soon as possible Take action to reduce exposure from drinking water within 7 days.

Figure 5 *Guidance for PFOS and PFOA in drinking water*

2.3.3   Analysis.   In the early period of the incident the analytical aspects of proved to be one of the most contentious issues. Initially, there was limited analytical availability, with only one water company laboratory, the environment agency laboratories and an external laboratory being capable of the analysis. The initial numbers of samples required and the urgent requirements for rapid turn-round of data places additional burdens on the laboratory staff. As the methodology for the parameters of concern was still under development and had not had the benefit of robust quality assurance and inter laboratory performance checks there were additional factors to be considered.

The early results did show unusual positive detections which were not corroborated on re-analysis, quite quickly deficiencies were highlighted in the sampling and analytical methodologies. A continuous programme of improvement was instigated to prevent and eliminate non-representative results.

In addition, the limit of detection for the PFOS analytical method was <0.1 µg/l at the time. As time has progressed further improvements have been made and more water companies had developed their own methodologies. The limit of detection for PFOS is now <0.013ug/l thus aligning the data with the revised DWI guidance.

Further work continues on the analytical side as the results of inter-laboratory comparison exercise highlight the difficulties of analysing for a substance that has been designed to form a protective waterproof layer on any material it is in contact with.

2.3.4   Monitoring.   Monitoring of ground contamination requires access to the aquifer by deep boreholes. Initially, all available existing boreholes were used but the depth and type of these did not provide a comprehensive set of sentinel wells with which to monitor the protection the wider aquifer and groundwater. To ensure a multi-layered approach to managing risk, sets of new observation boreholes were identified, designed and installed. Due to having to reach legal agreement on access arrangements with private land owners the time periods for installation of the observation boreholes, in some instances, were protracted. The importance of providing a facility that will be relied upon to give reassurance to the experts and general public and be sustainable as the long term situation develops was critical. Thus, the time, effort and resources expended in this activity subsequently proved to be very beneficial.

To ensure comparability of data, consistent sampling and monitoring methodologies were developed and established. These have been adopted by all parties involved in the ongoing monitoring of the groundwater.

2.3.5   Treatment.   At the time of the incident the acceptable treatment options for the component of the fire-fighting foam, (PFOS) were not well known or internationally accepted. A number of R&D studies were initiated which proved that low levels of PFOS could be removed by carbon treatment and that the higher levels of PFOS required more complex treatment such as reverse osmosis.

2.3.6   The future.   Many of the impacts arising from the explosion at Buncefield and the subsequent fire fighting and incident management activities have been addressed and are available for others to learn from. For the company the outstanding matters that remain are:-

Ensuring the impacts of Buncefield are controlled and remediated so that normality can be restored.

Achieving international acceptance and understanding of analytical and treatment issues.

Managing customer perceptions of the consequences of the incident.

3   CONCLUSION

The explosion at the Buncefield and the subsequent fire fighting and incident management activities have provided a wealth of new information from which others can learn. As a water company and a category 2 responder, the importance of establishing multi-agency expert teams should not be under-estimated. Nor should the over-riding drive to work together to help others.

By working together, the impacts from the Buncefield explosion were successfully managed. As ever, there is more that needs to be done to aid multi-agency working and the recent flooding incidents have highlighted the areas that still require addressing [2,3].

There is substantial evidence from experiences at Buncefield that a collective approach delivers results in a timely and effective manner but nationally, resources do need to be dedicated to enhance the current levels of understanding between agencies and stakeholders that will bring a common and consistent approach for multi-agency response.

REFERENCES

1.Buncefield Major Incident Investigation Reports:
  i)   Progress Report published 21 February 2006
  ii)   Second Progress Report published 11 April 2006
  iii)   Third Progress Report published 9 May 2006
  iv)   Initial Report published 13 July 2006
  v)   Recommendations on the design and operation of fuel storage sites published 29 March 2007
  vi)   Recommendations on the emergency preparedness for, response to and recovery from incidents published 17 July 2007
  vii)   Explosion Mechanism Advisory Group report published 16 August 2007
All are available from www.buncefieldinvestigation.gov.uk

2. HM Government: *Emergency Response and Recovery*

3. The Pitt Review: *Lessons Learnt from the 2007 Floods*

POTENTIAL SOURCES OF MAN MADE RADIOCHEMICAL CONTAMINATION OF WATER RESOURCES WITH SPECIAL EMPHASIS ON THE NUCLEAR FUEL CYCLE

N. R. Pacey and J. Cobb

AMEC Nuclear Ltd. 601, Faraday Street, Birchwood Park, Warrington. WA3 6GN

1 INTRODUCTION

The nuclear fuel cycle involves the mining and isotopic enrichment of uranium and its fabrication into fuel rods and assemblies for nuclear fission reactors. After 'burn-up' in the reactors, there is a period of storage or reprocessing of the spent fuel and management of all the wastes associated with these activities (including decommissioning of the reactors at the end of their useful lives). Specialised facilities also exist for the production of isotope sources of many types used in medicine, industry and research. In addition, in many countries there are associated activities related to military applications of nuclear power.

The current paper provides a summary of these activities, with special reference to the UK and a range of actual and theoretical scenarios where nuclides might enter the front end of the drinking water cycle. The paper is based on information available in the public domain.

2 THE UK's CIVIL NUCLEAR POWER PROGRAMME

Most of the uranium used in the UK's nuclear programmes comes from Canada, Australia and the United States. Whilst impacts on drinking water sources from this first stage of the nuclear cycle in the UK are negligible, tailings dams and other residues have given rise to legacy issues abroad which are being addressed through various national remediation programmes.

Processing of imported refined uranium ore (yellow cake) in the UK is currently carried out at Springfields in Lancashire. Subsequent isotopic enrichment of the uranium (in the form of 'Hex') is carried out at the URENCO plant at Capenhurst in Cheshire. There are only very small authorised discharges of mainly naturally occurring nuclides from these plants into surface waters.

The three main types of civil reactors used in the UK are the Magnox stations (now mostly undergoing decommissioning) the Advanced Gas Cooled Reactors (AGRs) and a single Pressurised Reactor (PWR) at Sizewell in Suffolk. Future reactors in a "New Build" programme are likely to be PWRs of European (mainly joint French/German) or American design.

In all these reactors, the fuel is 'burnt-up' to produce useful heat. During this process, a significant new inventory of several hundred radioactive elements appear in the fuel, eventually exhausting it and making it necessary to remove it from the reactor into station cooling ponds. Table 1 shows the main elements that appear, but without a breakdown into specific nuclides (for example, there will be upward of 3-4 radioactive isotopes of iodine with different half-lives such as I-131, I-132 etc).

Table 1 *Main elements making up in reactor fuel at discharge in terms of mass*

Element	grams/tonne	Comments
Uranium	970,000	Refractory oxide, residual nuclear fuel from the initial charge of uranium.
Plutonium	5,000	Refractory oxide, but forms many oxidation states with varying solubility's.
Xenon	4,000	Inert gas
Neodymium	3,400	Refractory, low solubility
Zirconium	3,000	Refractory metal, probably in the form of oxide
Cerium	2,000	Low volatility
Caesium	1,900	Volatile & soluble in water
Ruthenium	1,700	Transition metal, with variable oxidation states and forms numerous complexes. Usually remains in fuel as a refractory, although can form volatile oxides
Samarium	700	Refractory oxide, low solubility
Technetium	650	In fuel as oxide, but most stable as the soluble pertechnetate anion.
Strontium	550	Intermediate volatility. Soluble in water
Americium	460	Oxides and hydroxides of Am(III) are relatively insoluble.
Yttrium	430	Oxide of low volatility and solubility
Krypton	300	Inert gas
Iodine	150	Volatile, inorganic and organic forms
Neptunium	150	Refractory oxide
Europium	100	Refractory oxide, low solubility
Curium	5	Refractory oxide , low solubility
Niobium	0.01	Stable oxide, low solubility

Metal cladding on the fuel and the overall reactor design (the so called multi-barrier approach) is meant to contain these radionuclides as far as possible, but in the reactor and cooling ponds, very small amounts of can escape and enter the gaseous or liquid coolants. In addition, some components inside the reactor (notably the moderator) become 'activated' by the neutron flux, producing further radioactive isotopes, for example, cobalt-60 and tritium.

2.1 Reactor Safety Cases

The operation of reactors in all parts of the world is subject to national regulatory control and, increasingly, international scrutiny. Careful assessments (Safety Cases) of plant performance during normal operation and a range of possible fault scenarios must be prepared by plant operators to demonstrate to the regulators that risks to the public and to workers are within acceptable internationally based norms at all times and are also ALARP (As low As Reasonably Practicable).

Assessments for public safety consider the most exposed groups of individuals, called 'critical groups' who are identified using habit surveys of the local population. In the UK, most routine discharges from nuclear power stations (regulated by the Environment Agency or in Scotland by SEPA) are made to the cooling water systems that then discharge to sea. As a result, most 'critical groups' for which the impacts of these discharges are made are fishermen or individuals with a diet rich in sea food. During postulated faults, most discharges would consist of the gaseous and volatile species (inert gases, iodine and caesium isotopes) and take place via an aerial route. Critical groups at these times tend to be those eating locally produced foodstuffs. During both normal operation and faults, the routes into drinking water sources are more tortuous and indirect, allowing radioactive decay, dilution and hold-up in natural systems. As a result, those drinking water from potable sources around the UK's reactor sites are not considered as especially exposed critical groups and are not normally required to be assessed by the regulators in the site Safety Cases.

2.2   Reactor Accidents and their Impacts on Drinking Water Supplies

Prior to the development of detailed safety cases and operating experience, there were some serious reactor accidents leading to release of nuclides from reactors. The first in the UK was the fire in the Windscale (now Sellafield) Pile 1 which was part of a military, not civil, programme. Accidents then occurred in civil power reactors at Three Mile Island (a PWR in the US) and at Chernobyl (a specific Russian design). Following the incident at Windscale, the piles were shutdown and following those at TMI and Chernobyl, significant upgrades were instigated and now form part of the safety cases at similar reactor designs.

The patterns of nuclides released from these incidents reflects the radionuclides in the fuel, coupled to their chemical and physical properties, notably their volatility. Thus, whilst plutonium is an important part of the spent fuel inventory (Table 1) during the fire at Chernobyl this was retained mainly as oxide particles and only locally distributed. The main isotopes released and carried away by wind were the inert gas xenon and to a lesser extent the volatile isotopes of iodine and caesium. At Windscale, xenon and iodine were the main ones released. The behaviour of these nuclides in the natural environment, at the front end of the drinking water cycle, is affected by rain wash out and dry deposition from the atmosphere and then by hold-up in soils and finally movement into surface waters or into aquifers.

The net effects of limited dispersion, hold up in soil profiles etc. and the final consequences for drinking water supplies from the incidents at Windscale and Chernobyl were probably relatively small. For example, analysis of drinking-water samples collected from reservoirs and streams revealed that concentrations of radioiodine and other radionuclides from atmospheric fallout from Windscale probably never exceeded the maximum permissible concentration permitted by the ICRP (international Committee on Radiological Protection). In the EU as a whole, concentrations of Cs-137 from the Chernobyl incident have been regularly sampled at levels of <0.1 Bq/l from 1987 to 1990, which are deemed to be of no significant health concern and activities have fallen further due to Cs fixation in soil and river sediments.[1]

2.3   Reprocessing of Reactor Fuel

After fuel is removed from civil reactors, it is allowed to cool in the reactor station water filled ponds for some months, although some stations have capacity to store spent fuel in water for longer periods. In the UK, most Magnox and AGR fuel is transferred to

Sellafield where it is reprocessed. This involves dissolution of the fuel in acid followed by chemical separation of the fission products (that are considered as waste) and of the uranium and plutonium (that are a potential energy resource). Fission products contain the bulk of the activity and are stored as high activity liquid that is then converted to a solid glass, eventually to be disposed of to a future UK geological radioactive waste repository.

To date, the only serious incident involving escape into the environment of high activity liquors at a reprocessing plant is probably that at Kyshtym, in the former Soviet Union. Here, a steam explosion expelled over 80 tonnes of salt cake, the nuclides released being residual fission products from fuel reprocessing viz: - Ce-144, Pr-144, Sr-90 and Zr-95/Nb-95.

The only major event in the UK is the well publicised leak of liquor from the THORP plant at Sellafield in 2005 when approximately 83,000 litres of dissolver product liquor, containing approximately 22,000 kg of nuclear fuel (mostly uranium incorporating around 160 kilograms of plutonium) had leaked onto the floor of a tank cell. The liquor was wholly contained in the cell and on establishing the leak, all was returned to the main process line. There are also indications of small leaks of tritium and of Tc-99 into shallow groundwater under the Sellafield site from older operations which are the subject of on-going monitoring. None threaten major aquifers or drinking water supplies.[2]

2.4 Post 9/11 Assessments of Safety at Reactor Sites and other Nuclear Facilities

Following the 9/11 terrorist attacks in New York, there was a series of assessments of the impacts of potential similar incidents on nuclear facilities in all parts of the world, including the US and UK. They included, for example, aircraft strikes on reactors or reprocessing plants, which could release activity, some of which could enter the drinking water supplies. The main assessments in the UK have been published.[3] The assessments suggest that reactor structures designed to contain the fuel etc. in the event of internal hazards such as fire or in the event of external *natural* events such as earthquakes, are also sufficient to resist this new external threat. British Nuclear Fuels (as was) has carried out similar assessments for its facilities at Sellafield, notably for the tanks which hold high activity liquors from reprocessing spent fuel. In addition, the assessments suggest the greater difficulty of steering a large aircraft at these low profile structures, especially when they are situated amongst other buildings in a complex site such as Sellafield.[4]

To guard against terrorists entering nuclear facilities, increased security measures have been put in place in the UK under the auspices of the Office of Civil Nuclear Security (OCNS). However, the sheer physical barriers already in place to contain radioactive materials (massive shielded cells and deep water ponds) makes removal of fuel or other hazardous materials a very unattractive prospect for terrorists. In addition, the most active material, the spent fuel, would pose a considerable personal risk to those attempting to misappropriate it, possibly reducing their lifespan to as little as a few hours. Finally, the fuel is composed of a very refractory uranium oxide that would be difficult to solubilise to allow it to readily be introduced into the front end of the drinking water cycle.

3 RADIOACTIVE SOURCES

Radioactive isotopes from civil power reactors are considered predominantly as a waste product. However, in industry and medicine, these isotopes have a number of useful applications and for this reason special reactors or accelerators are used to make these isotopes in a useable form (they are not usually obtained from the reprocessing of spent fuel of civil power reactors). These sources vary in size and strength from domestic smoke detectors with a few 10's of kilo-Becquerels (KBq) of activity, equivalent to micro-gram amounts of the nuclide, to industrial sources with many tera-Bequerels (TBq) of activity and containing kilogram amounts of the nuclide in solid form. They include: -

• Thickness or depth gauges, containing the widest range of nuclides such as Am-241.

• Medical sources used in imaging e.g. Tc-99m (The Environment Agency Note PPG25 gives precautions against entry of these into Controlled Waters via sewage routes etc).

• Larger medical brachytherapy sources, Gamma Knife® containing Cs-137.

• Industrial sources, used for food and other sterilisation and as heat sources in thermo-electric generators, mainly the nuclides Sr-90, Co-60 and Cs-137.

    The sources are used in a wider range of countries and locations than civil nuclear reactors, ranging from oil well logging tools in Africa and Alaska to medical sources in Brazil. In the EU alone there are over 500,000 radioactive sources in use. In the UK, users of sources over a certain size are regulated under the Radioactive Substances Act (RSA 93) by the Environment Agency (or in Scotland by SEPA). Nevertheless, there have been incidences of lost sources in the UK and internationally there have been well publicised serious accidents involving loss of control of larger sources, for example, the Goiânia incident in Brazil involving Cs-137 and resulting in several fatalities. In Europe, the HASS (High Activity Sealed Sources) Directive is meant to give increased security and control over the acquisition, use and final disposal of the larger sources by users.

3.1 Comparisons of the Sizes of Industrial Radioactive Sources with Civil Nuclear Reactor Inventories

The IAEA has formulated a method of categorising sealed nuclide sources into 5 groups, (1 being the largest and 5 the smallest) based on their potential to cause harm due to handling or inadvertent disposal.[5] The following are relative comparisons of the nuclide inventories related to ones relevant to the civil nuclear sector: -

• IAEA Category 1, 2 and 3 sources typically contain from one to as much as a thousand times more Sr-90, Pu-238, Cs-137 or Am-241 than a single AGR fuel element at discharge from a UK AGR reactor.

• Some of the largest IAEA Category 1 sources contain several thousand times more Sr-90 or Cs-137 than was predicted to have been released from the 1957 Windscale Pile 1 fire and could be of a similar magnitude to the release of these nuclides from the reactor accident a Chernobyl in 1987.

Clearly, larger sealed sources represent a very significant inventory of some radionuclides relative to releases from even the more serious historic civil reactor incidents. Whilst increased regulatory controls have been implemented, concerns remain over nuclide sources that might fall outside such control (examples are the large Sr-90 thermoelectric generators lying abandoned around Russia's northern coast.[6]). Assessments of the impacts of such sources falling into the wrong hands have been made and include risks posed to people actually handling the source, to those arising from their use in a 'dirty bomb' and also impacts on such sources entering water supplies.

Together with the inconvenience of using sealed sources (in particular, they cannot be 'turned off') these security risks are driving users to seek alternative methods of generating radiation. In radiotherapy, teletherapy units have shifted from cobalt-60 to linear accelerator (linac) sources, and companies with linac products are trying to compete with the IAEA Category 1 or 2 Gamma Knife®, radiation sources.

3.2 Potential Impacts of Sealed Sources on the Front End of the Drinking Water Cycle.

Impacts of a sealed source entering the front end of the water supply system would depend on a number of factors, including: -

- The chemical composition of the nuclide source. Earlier Cs-137 sources were based on soluble $CsCl_2$ and this was one of the reasons for the wide dispersal of this nuclide in the Goiânia incident. Newer Cs-137 sources are being based on a ceramic matrix that should be less susceptible to dispersal or solution and there are moves to replace Cs-137 with other less hazardous nuclides altogether. The larger Sr-90 sources (used in thermo-electric generators[6]) are based on a phosphate-ceramic whilst Pu-238 (also used in these devices) is based on a refractory oxide. These chemical forms are insoluble and resistant to strong acids and would be difficult to disperse or dissolve in water. Terrorists attempting to dissolve larger sources in chemicals would face serious personal radiation exposure risks.

- The point where they are introduced into the water supply chain. Sources could be placed in service or supply reservoirs. The public risk posed in these scenarios would depend on the size of the source, its solubility, the volume of water and the residence time and rate of water turnover. Sources could also be placed over a groundwater recharge zone or borehole protection zone. Here the risks posed would depend on movement of the nuclide into the soil and unsaturated zones and then into the underlying aquifer. Probably the most significant risk in the groundwater scenario is direct entry of a solubilised source into a borehole. However, supply boreholes are placed within securely fenced areas and have secure and often heavy covers.

The IAEA provides a "Plain Language" description of the effects of the Category 1-5 sources to persons handling them without proper controls and also the environmental consequences due to dispersion of the sources, including the impacts on public water supplies.[5] The "Plain Language" description for the largest sources suggests "It would be highly unlikely for a Category 1 source to contaminate a public supply to dangerous levels, even if the radioactive material was highly soluble in water".

It is not clear what quantitative assessments lie behind these plain language descriptions but clearly, in any case, the economic implications and those for public confidence in the safety of water supplies would be significantly greater than the likely health impacts due to the radioactivity itself.

4 CONCLUSIONS

A top tier assessment of the UK's civil nuclear programme suggests that the main nuclides that occur in used reactor fuel are involatile species such as Pu, fission product gases such as isotopes of Kr and Xn and volatile isotopes of Cs, Sr and I. Reactor operations centre on retaining these as far as possible within the fuel itself. Where major historic incidents have occurred (Windscale, Three Mile Island and Chernobyl) losses were dominated by the fission product inert gases and more volatile Cs and I isotopes. With dispersion in the natural environment and then hold-up in soils and the unsaturated zone, impacts of these incidents on water supplies have been generally small.

Reactor fuel is stored intact, pending a final decision for disposal or reprocessing, with some (Magnox fuel) being reprocessed as a matter of course. Reprocessing releases nuclides into high active liquors that on the UK's Sellafield site are being converted into immobile glass. This waste form, or spent fuel, would be interred in a future UK radioactive waste geological repository where migration into any groundwater would be minimised by engineered and natural barriers, adequacy of which will be demonstrated via modelling and a suitable Safety Case.

Reactor and reprocessing sites would be difficult targets for external and especially internal terrorist attack or misappropriation of used nuclear fuel or other hazardous materials. More significant risks might be posed by misappropriation of large sealed sources, a scenario that has already been widely recognised and subject to a range of mainly qualitative assessments. For attacks on water supply targets using such sources, it would be possible to set up models for specific scenarios with representative source inventories and site specific characteristics and using recognised computer models. In such analysis, it would be important to further distinguish between the absolute amounts of nuclides entering the drinking water cycle and their radiological significance in terms of public health.

Acknowledgements

This work is based on a larger survey and assessment carried out in 2005 for the Drinking Water Inspectorate within DEFRA (DEFRA Reference DWI 70/2/183). None of the opinions expressed are those of the DWI or of DEFRA.

References

1   SCOPE 50. *Radioecology after Chernobyl* - Biogeochemical Pathways of Artificial Radionuclides. Chapter 5. Radionuclide Aquatic Pathways. At: - http://www.icsu-scope.org/downloadpubs/scope50/contents.html

2   Nuclear Installations Inspectorate. *Statement of Nuclear Incidents at Nuclear Installations.* 4th Quarter of 2001. Incident 01/4/1, Sellafield. At: - http://www.hse.gov.uk/nuclear/quarterly-stat/2001-4.htm

3   Parliamentary Office of Science and Technology, *Assessing the Risks of Terrorist Attacks on Nuclear facilities.* July 2004, Report 222,. At: - http://www.parliament.uk/documents/upload/POSTpr222.pdf

4   World Nuclear Association. *The Safety of Nuclear Power Reactors.* September 2007. At: - http://www.world-nuclear.org/info/inf06.html

5   International Atomic Energy Agency, Vienna, IAEA TECDOC 1344. *Categorization of Radioactive sources. Revision of IAEA-TECDOC-1191.* July 2003.

6   L. Bolshov, R. Arutyunyan and O. Pavlosky, Radiological Terrorism. *In High Impact Terrorism.* Proceedings of an American Russian Workshop. US National Academy of Sciences, 2002, 143-144.

RAPID METHODS

K. C. Thompson

Chief Scientist, ALcontrol Laboratories, Rotherham, South Yorkshire, S60 1BZ, United Kingdom E-mail clive.thompson@alcontrol.com

ABSTRACT

The threat of malicious contamination to drinking water supplies, whether perceived or real, is of concern to water companies. At the same time, the industry's ability to deal with significant accidental contamination is also coming under scrutiny. Laboratory planning for very high impact very low probability events is extremely difficult. This is particularly applies when dealing with the analysis arising from potable water emergency pollution incidents. The main issues are: - how to rapidly detect when significant chemical, radiological or microbiological contamination has occurred; identifying the cause or convincingly proving a negative in the absence of contamination; assisting in the key initial decision whether to cut off the potentially affected water supplies and finally maintain an efficient and effective 24h/365d response system on a long-term basis for these very low frequency events. This paper considers the development of rapid methods of chemical, radiological and ecotxicological analysis. The benefits of holistic ecotoxicity screening in parallel with chemical analysis and rapid radioactivity screening are also highlighted. The numerous benefits of setting up a mutual aid water laboratory response scheme are also discussed. Microbiological emergency incidents are not covered in this paper.

Keywords: - deliberate contamination, pollution incidents, drinking water, rapid screening tests, chemical, radiological and ecotoxicological analysis.

1. INTRODUCTION AND BACKGROUND

Planning for fit for purpose analysis of samples associated with very high impact very low probability events is notoriously difficult[1-4]. This is particularly true when dealing with the laboratory analysis arising from potable water emergency pollution incidents. There are three main issues; firstly maintaining an efficient and effective 24h/365d response system on a long-term basis when there can be many months between major pollution events; secondly how to rapidly detect when potential contamination has occurred and thirdly identifying the cause or convincingly prove a negative in the absence of contamination. It is very difficult to convincingly prove a negative quickly in an emergency situation given the relatively large number of potential toxic chemical substances.

It is important to appreciate that to reduce the risk of producing an unfit for purpose emergency analytical response to zero for all emergency incident eventualities (chemical, radiological and microbiological) is prohibitively expensive. There is an asymptotic relationship between cost of running and maintaining an efficient emergency laboratory service and this risk. (See Figure 1). In the author's view, many laboratories assume their risk position is lower than it actually is in this litigious age.

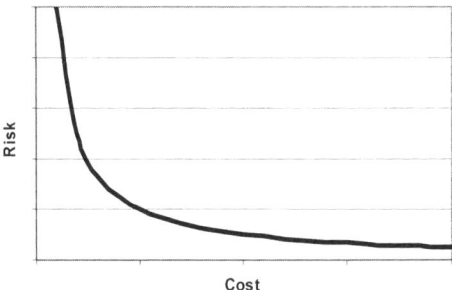

Cost

Figure 1. *Exponential risk versus cost of emergency analysis service relationship*

Ordinate. Risk of failure to provide fit for purpose analysis. **Abscissa**. Cost of maintaining the lab emergency 365d/24h system

Targeted and planned analysis of known substances in routine samples delivered at a preset time requires much less skill than trying to identify completely unknown substances in infrequent completely random samples. These can arrive without warning on a 24h/365d basis. Different skill sets are required. Equipping a changing staff base with the skill and experience to rapidly detect and identify unknown substances on a 24h/365d basis is not an easy task. For emergency incident analysis, experience and a good theoretical knowledge of water analysis is much more important than for routine pre-planned analysis. Detailed knowledge of potential false negative and positive effects of a given method is essential in order to avoid missing significant contamination or reporting false contamination. The effects of potential confounding variables need to be appreciated and addressed. These include the effect of free chlorine on ecotoxicity tests; the effect of sodium thiosulphate preservative on certain tests, the low temperature of a recently delivered sample when applying a rapid colorimetric test kit etc.

Prolonged incidents, that necessitate 24 hour working, can rapidly exhaust available expert staff and increase the chance of analytical errors. The issue of health and safety with respect to samplers and laboratory staff is another very difficult area. The chemical, microbiological and radioactivity risks need to be assessed. If staff are not fully appraised of the risks prior to potential major event there is a danger they may refuse to sample or analyse "high profile samples". A well-documented risk assessment/safe working procedure system is a pre-requisite for any robust emergency response system.

2. TARGET (PRE-PLANNED) ROUTINE ANALYSIS VERSUS SCREENING ANALYSIS

UK major water laboratories carry out high volume target analysis for all regulatory and operational parameters. However, this targeted regulatory analysis will not cover a significant number of potential terrorist threat agents. A completely different approach is needed for the handling of emergency incidents. As these incidents often occur out of working hours with limited skilled staff availability, a more focussed approach is required. It is extremely difficult to prove a negative result, especially in the context of an emergency incident situation with intense pressure to provide rapid results. The consequences of false negative or positive results must be fully appreciated by both the laboratory and the water company operational staff. The potential financial liability resulting from either scenario can be very significant. Much effort has been expended for the development of a number of robust rapid screening tests that will help to screen large numbers of samples quickly and hopefully allow prioritisation of the samples that require further investigation. These methods should have minimal false negative results (e.g. less than 0.2%) and very low false positive results. (e.g. less than 2%) It is important to estimate these key performance characteristics for any new rapid screening method.

An out of hours service cannot be as comprehensive as that available during working hours when many more staff are available. Thus it is important to tailor the normal out of hours guaranteed minimum initial response to the available limited staff resource. Rapid screening analysis allows the prioritisation of samples for more time-consuming analysis. At ALcontrol, as soon as a potential emergency pollution sample is received, a Microtox® test and a *Thamnocephalus platyurus* dormant organism screening mortality test (See 5) are immediately set up to run in parallel with the normal chemical analysis.

3. CHEMICAL SCREENING TECHNIQUES

3.1. Metals and metalloids

Almost all UK potable water laboratories have access to ICP-MS facilities. Screening for metals and metalloids using induction coupled plasma mass spectrometry (ICP-MS) is a well-proven technique and semi-quantitative scans for up to 70 elements can be rapidly carried out on potable water samples. (With a typical analysis rate of 20 samples per hour and 95% confidence limits of ± 25%.) Drinking water is a relatively easy matrix as the vast majority of samples contain more than 99.8% m/m water with the major metals present being calcium, magnesium, sodium and potassium. The potential interference effects from these four alkali and alkaline earth metals are well known and can easily be handled. Toxic elements such as arsenic, antimony, cadmium, lead, mercury, selenium, thallium, uranium etc. can all be readily detected at 1 ug/litre concentration levels. This is at least an order of magnitude below the short-term acute toxicity threshold.

3.2. Anionic substances

Ion chromatographic techniques with simple electrical conductivity detectors are capable of detecting a wide range of anions including chloride, chlorite, chlorate, perchlorate, nitrite, nitrate, sulphate, sulphite, phosphate, bromide, bromate, iodide and iodate down to ug/litre levels. More than one run would normally be required to detect all these species. This technique can also detect fluoroacetate down to about 0.1 mg/litre, although care is needed to obtain good resolution from fluoride and formate / acetate. (In the absence of fluoride and acetate/formate, a significant electrical conductivity detector IC peak at the appropriate retention time could indicate the presence of highly toxic fluoroacetate.

The use of ion chromatography linked to MS/MS is a more versatile technique for analyzing "problem" anionic species such as bromate, fluoroacetate, perchlorate and haloacetic acids. It replaces the non-specific electrical conductivity detector with a much more sensitive and very specific (lock and key) mass spectrometric detector. (See 3.2.5.2).

It is felt that there is not a requirement for novel rapid analysis techniques for the above parameters as most emergency situations should be capable of being dealt with using existing analysis routine analysis systems.

3.3. Organic substances.

Unlike toxic metals and inorganic anions, where there are a finite limited number, there are a very large number of potentially toxic organic substances and screening for these is much more problematical. (There are over 13 million known organic substances. Meinhardt[5] has pointed out that there are an estimated 64,000 chemicals in commercial use in the United States with approximately 700 new chemical agents synthesized each year. Each chemical agent may represent a "potential contaminant or parents of daughter contaminants born of reactions of these compounds with other compounds in the aquatic environment" Consequently, the use of proprietary tests for a given species such as ELISA test kits are not considered very useful in emergency screening other than as rapid tests that respond to a large number of risk agents such as cholinesterase inhibitors.

Many organisations use some form of solvent extraction and gas chromatography mass spectrometry (GC-MS) as their main screening detection technique. There are three problems with the approach, firstly many potential toxic substances are semi-polar or polar and will not be extracted and/or chromatographed efficiently unless they are initially derivatised; secondly the mass spectrometry library may not provide a correct identification and thirdly a number of natural harmless humic/fulvic acid derived substances are also detected with very poor or no identification.

Ways of overcoming these limitations are: -

i)   Use of high pressure liquid chromatography mass spectrometry (HPLC-MS/MS or HPLC-TOF) screening techniques using direct injection of the "as received" water sample. This avoids time-consuming sample pre-treatment. (e.g. derivatisation /sample pre-concentration steps). These powerful techniques are capable of rapidly detecting 1ug/litre of a wide range of polar or semi-polar organic toxic substances. In fact they are increasingly being used for routine targeted analysis to detect commonly analysed pesticides and herbicides with a regulatory detection limit of less than 0.025 ug/litre.

ii)   Development of rapid GC-MS and more recently GC-MS/MS screening techniques including a derivatisation step. It has been found possible to reduce the run time from 60 to 15 min and automate the sample pre-treatment / sample pre-concentration steps[6]. Conventional GC-MS and HPLC-MS techniques can be time consuming (up to 1 hour per sample). Thus there is a requirement to be able to screen samples quickly and prioritise the analysis so "suspect samples" are analysed first. Considerable method development work has been carried out to significantly reduce the analysis time down to 15 min for GC-MS and 5 min for HPLC/MS/MS. Another issue is that in order to run and interpret the output from GC-MS or HPLC /MS/MS screening techniques highly skilled/trained and experienced staff are required. These are becoming increasingly a "rare breed" because nowadays staff tend to be trained in carrying out high volume targeted analysis in large contract laboratories and are only infrequently involved in non-routine "one off" type analysis.

iii) Development of specialised GC-MS and HPLC/MS/MS libraries of the most likely organic substance toxic threats. There are already a number of toxic threat substance lists available in the literature and on the web [7-15]. Many of these websites have links to much useful associated background information.

In the author's view, it is important for water companies to regularly screen their water sources so that naturally occurring harmless organic substances that are normally present are detected and recorded. The chromatographic software system used can then highlight these as normal background peaks. It is important to screen over the four seasons as the pattern of natural substances often change with season.

It is also important to appreciate that almost all uncontaminated waters will give some GC-MS and HPLC-MS/MS small peaks. Before reporting any positive results for a specific toxic substance, some confirmation work should be carried out. This can be achieved by re-running the sample in mass selective mode and specifically monitoring only the main ions in the mass spectrum of the suspected substances and ensuring that the ratio of these ions are within the correct ratios. This will improve the sensitivity of the GC-MS technique by at least as an order of magnitude.

Another point to bear in mind is that quantification of polar and semi-polar substances can be grossly in error. If the system has been calibrated against a dichloromethane standard of the substance of interest, results for these types of substances can be significantly negatively biased (e.g. poor extraction and/or poor derivatisation efficiencies of the reported toxic substance). Evidence of this has been provided by the results for some polar or semi-polar substances in the LEAP emergency sample proficiency scheme where instances of under-reporting by up to three orders of magnitude have occurred. This could affect the operational response to such highly negatively biased results.

The main areas of benefit of MS/MS over single stage MS techniques are: -

- More confidence in identification
- Improved quantitation
- Better signal to noise hence lower detection limits

All these are somewhat related and stem from the two stage process which is being used.

A single ion is selected from the initial ionisation stage either from a GC or an LC. This ion selection can be achieved using a Quad or an ion trap. The next stage is to fragment this ion using some form of collision process with a gas (Argon, Nitrogen etc) either in a second Quad or within the ion trap. Once the fragmentation has been achieved the resultant product ions are detected using a third quad or the trap again.

A quad based MS/MS system often referred to as a triple Quad or QQQ can only fragment the initial ion selection once unlike a trap which can fragment the initial ion and also go on to fragment one of the product ions over many cycles so called MS to the n. However the QQQ is faster and more sensitive than the other options, Traps or time of flight (TOF).

There are other tandem configurations available to carry out MS/MS. The Q-TOF technique, where there is a quad front end but a TOF detector which detects all the product ions. The Q-Trap where there is a quad front end but with a Trap as a detector. Ions can be stored and then fragmented and MS to the n is then possible.

Because all the ions have been removed other than the one of interest, the spectra are much cleaner so the signal to noise ratio is vastly improved. This also improves identification as any ions which reach the detector must have come from that initial ion selected from the ionisation, so we can be confident that, for example an ion of mass 250 has produced a major ion of 160 and also lesser intensity ions of 175 and 58. In a single stage process, these product ions could arise from any ion produced in the original ionisation stage.

Ion traps take this principle further by being able to take mass 250 fragment it to 175 and then take 175 and produce 58 and so on.

There is an established protocol based on a points system, this is used to quantify the degree of confidence in an identification[16].

A comprehensive study carried out by the author's laboratory using an ABS 4000 QTrap LC-MS/MS system which demonstrated that it is possible to meet regulatory detection limits for the vast majority of tested pesticides and herbicides using direct injection of the water sample with the Multi Targeted Screen (MTS) approach. This is similar to a triple quadrupole quantitative multiple reaction monitoring (MRM) method. However, when exceeding a set intensity threshold level, the MRM signal triggers the acquisition of a linear ion trap enhanced product ion (EPI) scan to produce a composite MS/MS spectrum obtained at three different fragmentation voltages for library searching and compound identification. It is possible to run three methods, each with 296 compound MRMs that run sequentially, thus screening for 888 transitions in 15 minutes or less. The study showed that only six out of seventy seven potential threat agents could not be detected by LC-MS/MS by direct injection, with a detection limit better than 10 ug/l, the majority with detection limits below 1 ug/l.

A threat library would be set up using the above data. Once a threat agent library has been set up, this type of system should be capable of handling emergency incidents for the vast majority of semi-polar and polar organic chemical threat agents. In conjunction with a modern GC-MS and GC-MS/MS system almost all threat organic substances should be detectable using these two techniques.

The UK Water Laboratory Mutual Aid Group are working together to develop a UK resource for storing this information with the help of Water UK.

If a LC/MS QQQ instead of a trap is the instrument in use the most likely initial use of the QQQ would be to use a MRM method based on the threat list or more than one method depending on the number of compounds which needed to be screened. The speed of the system would allow these methods to be run sequentially and thus cover several hundred target compounds in a very short time.

An alternative less rapid approach for "unknown compounds" is shown in Figure 2 this shows a MS2 full scan chromatogram from an Agilent 6410 triple quad LC-MS system using an injection of 900 ul of a 1ug/litre standard containing 33 pesticides and herbicides. The injection is made onto an on-line SPE column and the adsorbed substances are then automatically back-flushed off the SPE on to the analytical column. This Figure clearly shows that in full scan mode the system can easily detect 1 ug/litre concentrations, which should all be well below the relevant acute toxicity levels. If an ion is detected that is not in the blank then subsequent runs can be carried to obtain fragmentation data of this unknown ion to aid possible identification.

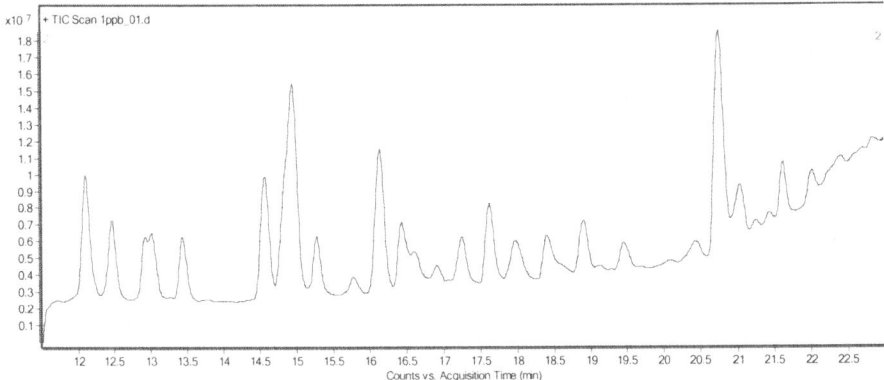

Figure 2. *Full scan direct sample injection trace trace for 33 pesticides and herbicides at 1ug/litre with an Agilent 6410 HPLC-MS/MS system*

3.4. Diode array ultraviolet spectrometry (DAUVS)

Treated potable water samples give a very simple ultraviolet spectrum from 230–400 nm and using a 50 mm path length silica cell it is possible to detect 10–50 µg/litre of most substances containing an aromatic ring in a treated drinking water. Uncontaminated waters give a simple spectrum with no obvious peaks. (See Figure 3). The sensitivity and discrimination of the technique can be improved further by using the first derivative. (See Figure 3). Each sample can be run in less than one minute.

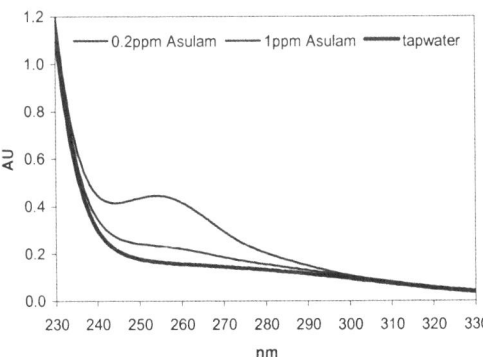

Figure 3. *UV spectra of a tapwater and a tapwater spiked with asulam (40 mm path length cell)*

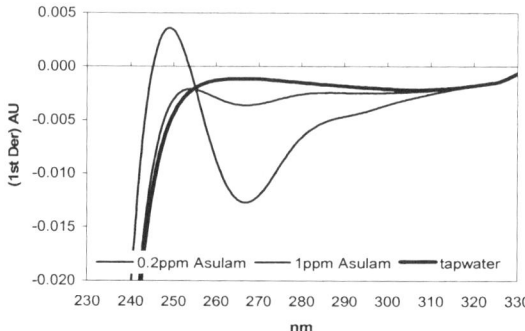

Figure 4 *1st derivative UV spectra of a tapwater and a tapwater spiked with asulam (40 mm path length cell)*

This technique has been developed further as an automatic on-line monitor for Yorkshire Water Services (YWS) and is used for screening the raw water (river) inlet at major water treatment works using a 20 mm path length cell. Some specialised software has been developed that overcomes the problem of occasional high turbidity in raw waters – allowing the instrument to operate at optimum sensitivity despite wide variations in light intensity received at the detector. Much effort was expended in trying to utilise on-line filter devices to remove all particulate matter, but these were found to be unreliable, relatively expensive and significantly increased the maintenance requirements. Without a pre-filter to remove suspended particulate and colloidal matter, the devices only require minimal maintenance (typically one visit per three or four weeks). The ultraviolet radiation appears to limit biofilm growth on the flowcell window. Many commercial on-line river intake devices have failed to be adopted because of the problems of reliably pre-treating the sample.

The device only utilises a crude wire mesh filter to remove gross particulate matter and the associated software can cope with high levels of turbidity. The system can be trained to recognise "normal signatures" which can be season dependent. It is programmed to alarm to a sudden step change or an abnormal change in the overall spectrum. It will respond to about 0.2 - 0.4 mg/litre of most substances containing an aromatic type ring (e.g. atrazine, paraquat, diquat, phenol etc.) even in turbid river waters. The detection limit is degraded with respect to treated water because of the turbidity and natural colour of raw waters relative to treated waters and also the shorter path length cell. Very few false alarms have been observed and this is essential prerequisite for a successful on-line monitor.

A fully commercialised off the shelf system is the S::CAN sealed unit diode array detector system is now available with large numbers in routine use as on-line monitors[17-19]. This unit comprises a high energy throughput optical system with a spectral range of 190 nm to 380 nm. It has no user serviceable parts. The lamp expected life is over 10 years. It can be used for measurements in treated and even in raw waters with high turbidity and/or high optical density. Turbidity is compensated by a mathematical model that reflects the particle distribution of the turbidity source, so no sample pre-treatment is necessary. It is claimed to be superior to fibre optic instruments with respect to energy and process stability. It utilises a pulsed Xenon arc lamp; double beam optics for very good baseline stability and optical path lengths up to 100 mm. It can be used either totally submersed (the system can withstand pressures up to 10 Bar) or used in a conventional bypass mode.

The impressive software "learns" the shape and features of the "normal" spectrum of the water body over the four seasons and can alarm to user selected criteria. The S::CAN system will run unattended for very long periods. It is also suitable for laboratory use.

3.5. Photoionisation Detector (PID)

A photoionisation detector suitable for volatiles in potable water typically utilises a 10.6 e.v. electrodeless discharge lamp emitting far UV radiation to ionise virtually all organic substances except for alkanes (e.g. methane, ethane, propane etc.). Ionization occurs when a molecule absorbs the high energy UV light, ejecting a negatively charged electron and forming of positively charged molecular ion. These charged particles produce a current that is easily measured at the sensor electrodes.

An RAE ppb high sensitivity PID has been evaluated in both laboratory and field situations[20]. A 500 ml sample aliquot is placed in a one litre glass bottle and sealed with aluminium foil. The bottle is shaken for 30 sec and the headspace gas simply introduced into the PID detector nozzle. (See Figure 5). With this simple system it is possible to detect down to 0.05 mg/litre diesel at an analysis rate of up to 60 samples per hour. It has been demonstrated in the author's laboratory that tap water samples containing 0.1 mg/litre of diesel could rapidly be sorted from uncontaminated tapwater using a PID device.

Diesel and gasoline contamination are thought to be the most common forms of drinking water contamination. The PID responds in a more sensitive manner to gasoline than diesel owing to the significantly higher vapour pressure of gasoline relative to diesel. This portable robust battery powered device can readily be used in the laboratory or the field by junior staff. It is a much quicker technique than extraction of the petrol or diesel into pentane and GC-MS or extraction of the petrol or diesel into tetrachloroetylene, Freon™ or carbon tetrachloride and measurement of the appropriate region of the infrared spectra. The PID does not require the use of these environmentally unfriendly organic solvents.

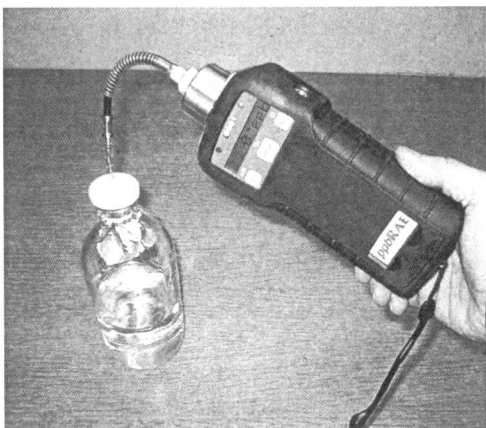

Figure 5 *Use of a PID detector in a laboratory or test room area*

A summary of the advantages and disadvantages of the PID are given below: -
Advantages.
- Simple and robust.
- Can be operated by trained field (non-laboratory) staff.
- Top of the range P1D systems are very sensitive and can rapidly detect down to ~50 µg/litre diesel or lower levels of petrol in potable water.
- Can screen up to 60 samples/ hour for volatile organics.
- For a known (single) volatile pollutant in potable water, it is possible to calibrate the system and obtain a semi-quantitative result.

Disadvantages.
- Will only detect non-polar volatile substances that will transfer to the headspace from a water sample.
- Volatile polar substances such as methanol and acetone are only weakly detected because they have low Henry's law constants and remain in the aqueous phase.
- No indication of the identity of any pollutants detected is possible.

   Diesel and gasoline contamination are thought to be the most common forms of drinking water contamination. The PID responds in a more sensitive manner to gasoline than diesel owing to the significantly higher vapour pressure of gasoline relative to diesel. Also this portable battery powered device can readily be used in the field as indicated above. It is a much quicker technique than extraction into tetrachloroethylene, Freon™ or carbon tetrachloride and measurement of the appropriate region of the infrared spectra. Also it does not require the use of these environmentally unfriendly organic solvents.

3.6. Other useful screening tests

Other useful screening tests include total organic carbon (TOC); pH; electrical conductivity; turbidity; free cyanide (CHEMetrics™); ammonia; parquet/diquat and ion chromatographic scan. For any tests carried out in the field, it is essential that method validation tests are carried out at the range of temperatures likely to be encountered. Some colorimetric tests may exhibit sensitivity changes at low temperatures. The Ticket® test, which uses a litmus-type ticket as its main component, is based on a fairly simple biochemical principal[21]. One side of the Ticket® contains a disc which is saturated with cholinesterase — an enyzme present in most living organisms, except plants — and whose main function is to control muscle performance. If the enzyme is altered or dies, so does the organism. Insecticides can inhibit an organism's ability to produce cholinesterase, and therefore, kill the organism. If enough insecticide is present in the sample, it will inactivate the cholinesterase and turn the disc white. A blue colour result indicates a negative result for the presence of cholinesterase inhibitors results are obtained within five minutes.

3.7. EPA Environmental Technology Verification Programme

The EPA has an Environmental Technology Verification program website[22] which publishes the results of standardised evaluation protocols for a wide range of instrumentation. This includes relatively mundane devices such as turbidimeters but also more recently devices specifically focussed on terrorist contamination. This includes rapid toxicity monitoring, and detection of agents and pathogens by PCR and Immunoassay.

It is important to appreciate that the vast majority of emergency incidents to date are not CBRN related, but more likely to be linked to industrial pollution or contractors working on or near to a water distribution system. Rapid screening techniques for responding to chemical and microbiological threats require an awareness of the actual potential threats or have to be broad based and non-specific (e.g. ecotoxicity testing methods). Rapid detection of groups of chemicals is generally complex even for the most sophisticated laboratory. Most of the published literature contains very little information on the development and especially the "real life" use of in-house or proprietary rapid screening techniques.

3.8. *ELISA Tests*

The excellent point-of-need systems made for large molecular weight analytes – such as the home pregnancy test show clearly what can be achieved. However, the performance of classical immunoassay is often found to be clearly lacking when applied to small molecular weight analytes such as natural and synthetic small molecular toxic substances. The problem is simply one of size. Large analytes are large enough to support the simultaneous binding of both capture and detector antibodies - allowing typical sandwich immunoassays to be formatted in which increasing analyte concentration provides an increase of observable signal over a zero background. Small molecules are simply not large enough to support simultaneous binding. Alternative 'competitive-format' systems have been used. In these analyte competes with a detectable analyte-analogue for capture antibody sites. In effect the system measures how much analyte is not present. This causes major problems not only in terms of precision and sensitivity but also read-out where, classically, reducing observable signal is indicative of increasing concentration of analyte. In practice, this can make point-of-need devices difficult to read and assess. On dipsticks, for example, zero analyte provides the maximal line against which other determinations, providing possibly less intense lines, have to be gauged. What is required is a robust generic system in which there is no observable signal in the absence of analyte (i.e. a clear strip) and even low level samples give an obvious observable line over this zero background. Devising such systems has led to the development of the Selective and Apposition Systems[22A].

A synthetic 'Apposition reagent' is first made from an antibody against the target analyte by positioning a small bindable moiety near to its binding site. It is positioned such that when analyte binds its antibody binding site the moiety can still be bound by a labelled secondary, detector, antibody. A large reagent-analogue of the analyte (or anti-idiotypic antibody) is also introduced into the system so that the remaining analyte unbound sites on the apposition agents are bound and thus blocked from binding of the secondary antibody to the moiety. Thus the more analyte present the more binding of secondary antibody occurs and the more signal is seen. This system is generic both in terms of analyte target and system format - being as applicable to high-throughput testing systems as simple point-of-need systems. Among the various areas to which these systems have now been successfully applied is that of rapid pesticide detection. Simple, self-contained, dipsticks containing all of the components necessary for final analysis have recently been produced. These allow positive pesticide detection to be conducted within a few minutes[22A].

Most existing reverse chemistry ELISA tests are not thought to be very relevant to water contamination emergencies. These "reverse chemistry" tests require considerable skill to obtain fit for purpose results. If they are only employed for infrequent emergency incidents it is highly likely that staff operating them will have great difficulty in using them in a fit to purpose manner under time-pressured conditions. Also many ELISA kits tend to detect specific parameters, thus a large number of different tests would be required.

3.9. Chlorine demand test

The chlorine demand of water is defined as the difference between the amount of the chlorine injected into a water sample and the amount of chlorine remaining at the end of a specified contact period[23], typically 20 minutes for a gross pollution incident. Because it contains fewer chemical contaminants, the chlorine demand of an uncontaminated tapwater should be significantly less than that of a tap water contaminated with chlorine reactive species. A water sample is chlorinated (typically to a free chlorine concentration of 2 – 5 mg/litre) and the concentration of free chlorine determined after given times (typically 20, and 60 min). An uncontaminated tap water is normally used as a control. Chlorine can react with sample constituents by three general pathways: -oxidation, addition and substitution. It can oxidise inorganic reduced species such as iron (II), manganese (II) and sulphide. More importantly chlorine can add to carbon double bonds or can substitute into a wide range of molecules to produce organochlorine species. The chlorine demand is a good screening test for the detection of gross organic pollution. Like diode array spectrometry, the technique does not give any information on the nature of the pollution. However, when potentially toxic substances (or the substances that these are associated with these toxic substances) that have a chlorine demand are present in suspect samples, the chlorine demand test will allow rapid differentiation of samples requiring further analysis.

3.10.   Nuclear Magnetic Resonance (NMR) Screening

Recent advances in NMR technology have resulted in a vast improvement in the sensitivity of NMR measurement. The development and implementation of the NMR cryoprobe in the past 3 years has produced sensitivity gains of 5-10 fold over comparable high specification NMR instruments. Concurrent developments in the magnetic field strength and internal architecture of the NMR spectrometer have ensured that both the sensitivity and the resolution of modern instruments are vastly superior to those produced as little as 5 to 10 years ago. Water can be difficult to analyse by other techniques (e.g. GC-MS, FT-IR) and these methods often require extensive sample preparation. However, NMR measurements can be made directly on the water samples without the need for sample preparation. As the sensitivity of the NMR measurement increases as a function of the square root of the time used to acquire the data, the detection of concentrations of analytes in mg/L range can be performed in a matter of seconds. However, to achieve detection limits in the ug/l range acquisition times of up to 36 hours are required. On a state of the art NMR system, typical detection limits[24, 25] are estimated to be: -
500 µg/litre in 15 min
200 µg/litre in 2 hours
50 µg/litre in 30 hours
     The detection limits are not mutually exclusive and therefore data acquired after 15 min of a 30 hour acquisition can be read and interpreted whilst the remainder of the data acquisition is performed. It is therefore possible to determine the presence of higher concentration contaminants during the early part of an experiment designed to detect trace levels.
     The chemical structure elucidation capability that is inherent in the NMR measurement enables unknowns to be detected and characterised. The resolution in the NMR spectrum is obtained due to changes in the local chemical environment of molecules.

As all protonated compounds give a quantitative and comparative NMR response the range of analytes that can be detected and characterised in a single NMR data set is theoretically unlimited (for organic compounds). Additionally, only a single internal standard is required for quantification of all compounds with the caveat that quantification of known compounds is likely to be more accurate using other techniques.

As NMR is a spectroscopic technique, no physical separation of a sample is required to obtain the NMR spectrum. The non-invasive NMR measurement is therefore particularly suitable for application in CBRN incidents when the nature of potential contaminants may present a risk to the analyst and the instrumentation. NMR measurements are performed in sealed glass tubes through which radio frequency waves are passed. Therefore the sample can be completely contained during analysis. Analysis is also often performed at room temperature thus further reducing any potential risk involved. As the sample remains completely intact and never comes into contact with the instrument it can be used for subsequent analysis if the amount of material available were ever to be a limiting factor.

The CSL NMR system was first applied to the May 2005 LEAP emergency proficiency sample to test the applicability of the method for water analysis (Exercise 10). Only one NMR data set was required to identify all five organic analytes which were present at a concentration range of 50-500 µg/litre. (See Table 1)

LEAP 10 Substance	Concentration (µg/litre)	No of labs	% labs detected substance present	Minimum time for NMR detection (min)
Picric Acid	504	21	10%	15
Mevinphos	80	21	100%	1800
Amitrole	50	21	10%	1800
Dicamba	50	21	71%	1800
Pyridine	200	21	52%	120

Table 1 *NMR results for LEAP emergency proficiency sample for exercise 10 (May 2005)*

Results for this and other LEAP exercises are given in this volume[25]. NMR can be used as an initial screen to indicate the presence of a contaminant. No a priori knowledge of the contaminant is required to make the measurement. With the aid of a database the identity of the contaminant can be determined (if present in the database). An understanding of structure elucidation by NMR can remove the need for a database although identification of unknowns without reference samples can be a lengthy process. This process can be expedited significantly by using complimentary chromatographic and mass spectrometry techniques.

4. SUMMARY OF CHEMICAL SCREENING TESTS

4.1. Generic   non-specific   screening   tests   relevant   to   emergency   incidents

Generic screening tests effectively use a surrogate parameter, physical property or sample toxicity to indicate that unexpected substances are present in the sample. The Table below summarises a range of non-specific tests that can be used by water laboratories. They are typically calibrated (and/or AQC checked) with a specific single relevant substance although they will summate the response from all relevant substances covered by the screening test.

Generic Non-specific Screening Tests relevant to Emergency Incidents		
Parameter	*Typical result for a non-polluted water*	*Comments*
Total organic carbon (TOC)	Up to 3 mg/l	Simple routine test
Permanganate Index	Up to 3 mg/l	No longer routinely analysed
Chlorine demand	Dependent upon water type	Not in common use for emergency incidents, but recommended
UV scan (230 – 320nm)	No distinct peaks between 240 nm and 320 nm.	First derivative spectra are easier to interpret than zero order spectra
Fluorescence screening (diesel)	<2 ug/l	Deposits in mains can cause false positives from PAHs
PID screen for VOCs and diesel/gasoline	<50 ug/l when calibrated with diesel	Very rapid test
Cholinesterase ticket screen[21]	< 1-5 mg/l	5 min reagentless test kit

Table 2   *Some generic non-specific screening tests*

4.2. Specific parameter screening tests relevant to emergency incidents

Specific parameter screening tests basically use multiparameter detection systems that separate the measured species from other species and estimate the semi-quantitative concentration by calibrating the screening test with a range of known targeted substances. Table 3 below, summarises a range of tests that can be used by water laboratories. The GC and LC screening tests may not be able to identify all detected peaks, but should be able to indicate whether the peak is a single unresolved or a multiple peak representing one or more unknown substances.

Specific Parameter Screening Tests relevant to Emergency Incidents	
Parameter	**Comments**
Hydrocarbons by IR (Diesel and gasoline)	Needs chlorinated solvents. Typical uncontaminated result <30 ug/l
ICP-MS metals/metalloids scan	Very robust technique. Should detect all relevant elements including As; Cd; Hg; Sb; Se; Tl and U etc.
IC screen	Should detect abnormal levels of all standard anions; bromate; perchlorate; chlorate; fluoride and fluoroacetate. Requires more than one run. Preferably with both conductivity and photometric detectors
GC-MS VOC screen	Very sensitive headspace or purge and trap methods. Normally based on equivalent EPA method for up to 60 targeted VOCs
Paraquat/Diquat Screen	Simple SPE pre-concentration method followed by addition of a few drops sodium dithonite solution.
GCMS Scan (using neutral acid and alkaline extraction technique)	Currently a main organics screening test for many laboratories
GCMS Scan (using neutral acid and alkaline extraction technique) plus derivitisation	Currently a main organics screening test for many laboratories
LC/MS/MS or LC/TOF scan after direct injection	Usage of this versatile technique for semi-polar and polar substances is rapidly increasing
Rapid gross alpha / beta screening test	100 ml sample volume; 12 hour count time. LOD ~ 1Bq/l for both gross alpha and beta (see 6.3)
Free cyanide and ammonia	Simple test kits with direct addition of the sample to the reagents contained in the final photometric measurement vial.

Table 3 *Some specific parameter screening tests*

4.3. Quality assurance/quality control

It is important to ensure that some simple documented robust QA/QC protocols are devised for all emergency screening methods to ensure fit for purpose results. As a minimum, each screening test should have an associated system suitability check, a blank and a QC sample containing some representative risk agent substances at typical concentrations of interest. Without these checks there is a danger of false positive or much more seriously false negative results being reported.

Also, prior to adoption, all screening methods should be validated by obtaining comparison data against a fully characterised and validated routine laboratory analysis method for the parameter in question for a range of water matrices. Any samples where there are significant discrepancies in the results should be checked using samples spiked with a known amount of the parameter by both methods to determine if one or both methods are in error.

5. ECOTOXICOLOGICAL SCREENING TESTS

5.1. Introduction

A problem with chemical testing is that there is a very large number of potentially toxic substances and it is very difficult to prove a negative particularly under pressure in an emergency situation involving large numbers of samples. A complementary approach, whilst running chemical screening, is to employ simple screening Ecotoxicity toxicity testing to detect acute toxicity. This is a holistic approach that detects toxic effects on organisms employed as biological indicators in the test used.

Aquatic biota can react to chemical stress relatively rapidly, but the effects measured do not give any information on the actual nature of the contaminants. The detection thresholds for biological reactions furthermore are "compound dependent" and may range from ppb concentrations to tens of ppm for chemicals with high mammalian toxicity. No ecotoxicity test reported to date will respond to all likely toxins, although some will respond at "relevant concentration levels" to over 80% of risk agents. It is important to appreciate that, as with all screening chemical tests, the universal ecotoxicity screening test is not available and unlikely to ever be available. It is also important to appreciate that the acute toxicity of some toxic substances can vary by more than one order of magnitude (on a mg/kg basis) among different individuals. However, some low cost, simple technology acute ecotoxicity tests respond to more than 80% of "likely" malicious risk agent toxicants at relevant concentration levels and are thought to be a useful addition to the armoury. It is possible to run large numbers of these ecotoxicity tests in an emergency situation in parallel to the chemical analysis with relatively small additional staff resources. It is crucial to be aware of confounding variables such as the need to remove free and combined chlorine and ensuring sufficient dissolved oxygen before applying any ecotoxicity test. As with chemical analyses, the potential as well as the limitations of ecotoxicity tests must be taken into account when considering water contamination issues.

5.2. Some potentially useful ecotoxicity tests

The well known Microtox® test employs a marine luminescent bacterium, (*Vibrio fischeri*) and has comprehensive sensitivity data to a range of substances[26]. Any toxicity is displayed as a decrease in light output from the bacteria within 30 min. Typically 10 min for a toxic organic substance response and up to 30 min for a toxic metal response. The overall sensitivity is not very high, consequently this very simple, reproducible and very rapid test will, in general, only detect gross pollution.

Much work has also been reported on the ECLOX® enhanced chemiluminescence light (ECL) test which is based on chemiluminescence[27]. The enhanced chemiluminescent reaction utilises the enzyme horseradish peroxidase (HRP) to catalyse the oxidation of luminol. The result is a flash of light, and in order to prolong the emission of light an enhancer is added. Light output is measured using a luminometer. Contaminants that interfere with the ECL reaction include: -

- Radical scavengers (such as antioxidants which remove oxygen from water and scavenge free radical molecules)

- Competitive enzyme inhibitors

- Non-competitive enzyme inhibitors

- Miscellaneous influences on general chemiluminescent reactions (including trace levels of manganese).

Table 4 indicates the EC50 values for a range of common toxicants determined by ALcontrol.

Substance	Eclox™ (IC$_{50}$) (mg/l)	Microtox® (30 min EC$_{50}$)	Thamnotoxkit F™ (24h LC$_{50}$)
As (III)	26	0.3	1.3
KCN	0.07	3.5	0.3
Dichlorvos	354	29	19
Thallium	113	450	0.2
Paraquat	73	20	0.6
Fluoroacetate	>1000	507	0.4

Table 4 *Comparison of acute ecotoxicity test sensitivities for three tests*

Currently ALcontrol Laboratories employs two tests for emergency incident work[28]. The well known Microtox® test and the Thamnotoxkit F® with Thamnocephalus platyurus The Microtox® test employs a marine luminescent bacterium, (*Vibrio fischeri*). Any toxicity is displayed as a decrease in light output from the bacteria.

The *Thamnocephalus platyurus* Microbiotest® employs the anostracan crustacean *T. platyurus* which is similar to *Daphnia magna,* but slightly smaller and more sensitive to most toxicants. It utilises dormant organism technology[29-32]. This technology avoids the need for culturing and effectively reduces the cost of ecotoxicity testing by an order of magnitude. The typical mortality endpoint method takes up to 24 hours before a negative result is confirmed, however if there is a significant concentration of toxic substance present, an effect is seen well before 24 hours. The unique characteristic and the major asset of Toxkit microbiotests is that these bioassays are totally independent of the (complex and costly) culturing and maintenance of stocks of test-organisms which are needed for the tests. Toxkit microbiotests make use of "dormant" or "immobilised" stages of selected test species, which can be stored for long periods of time and "activated" at the time of performance of the assays. This allows performance of these low cost and user-friendly toxicity tests "anytime and anywhere" without the need for expensive equipment or facilities.

As stated earlier these tests are seen as complementary to chemical tests and effectively increase the range of toxic substances that can be detected. Table 4 gives an indication of the sensitivity of the two tests and the range of substances that can be detected.

A further development of the *Thamnocephalus platyurus* test, to speed up the detection of toxicity, reduces the time for a negative result from 24 hours to ~1 hour[29-30]. This is based on monitoring the feeding rather than the mortality of the Thamnocephalus with highly coloured food beads after one hour. Cyst hatching must begin 30 to 45 hours prior to performing the test. The *T. platyurus* are exposed to samples for one hour, after which a suspension of red microspheres is added. The organisms ingest the microspheres, resulting in a deep red colour in their digestive tracts. Stressed (intoxicated) organisms either fail to take up particles altogether or ingest at a much lower rate. See Figure 6. The presence or the absence of coloured microspheres in the digestive tract of the larval crustaceans is observed under a stereomicroscope, and data are recorded on a sheet supplied with the test kit. The total number of *T. platyurus* in the control (standard freshwater) well(s), and the number of *T. platyurus* that have taken up the red particles are counted, and the fraction of larval crustaceans affected by the contaminant is defined as the percent inhibition. As a guideline, 30 percent inhibition of particle uptake is considered a threshold for the presence of potentially toxic compounds in the water.

A simple analogy is that a human being before succumbing to Typhoid fever (which may take 4 or 5 days from infection) will rapidly lose appetite well before death occurs. The Thamnocephalus cysts are stable for up to nine months if kept at 5°C and require ~24 hours to hatch and then should be used within 48 hours. ALcontrol Laboratories keeps some cysts hatched on a 24h/365d basis to cope with any water pollution emergencies.

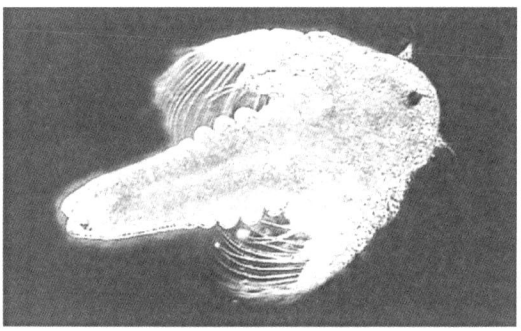

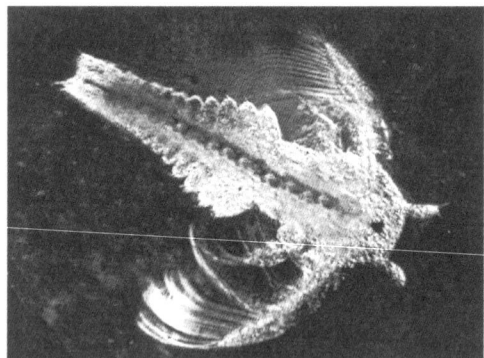

Figure 6   *RAPIDTOXKIT MICROBIOTESTS, 30-60 min test using the crustacean Thamnocephalus platyurus Microbiotest® for rapid detection of water contamination*[32].

No particle uptake (stressed organism) and particle uptake (unstressed organism) by Thamnocephalus platyurus

5.3 Some ecotox screening tests considered relevant to emergency incidents

Table 5 lists some ecotox screening tests considered relevant to emergency incidents

Substance	Eclox™ (IC$_{50}$) (mg/l)	Microtox® (30 min EC$_{50}$)	Thamnotoxkit F™ (24h LC$_{50}$)
As (III)	26	0.3	1.3
KCN	0.07	3.5	0.3
Dichlorvos	354	29	19
Thallium	113	450	0.2
Paraquat	73	20	0.6
Fluoroacetate	>1000	507	0.4

Table 5 *Ecotox screening tests considered relevant to emergency incidents*

6. RADIOLOGICAL SCREENING TESTS

6.1. Background

Gross alpha and beta determination is a standard relatively low-cost potable water screening method designed to ascertain whether further radiological examination of a water sample is necessary. Apart from the 1989 UK "Wholesomeness Definition", the 2000 European Directive requires in addition that a Total Indicative Dose be assessed (excluding Tritium, K-40, Radon and its decay products). If gross alpha is <0.1 Bq/l and gross beta <1 Bq/l no further analysis is required and the water sample can be considered to be wholesome with respect to radiological parameters (excluding Tritium, Radon and its decay products.) The Standing Committee of Analysts (SCA) has published a suitable method for carrying out this analysis[33].

In the event of nuclear accident, terrorist attack or other similar event Water Companies will need to rapidly ascertain the health and safety risk relating to radioactivity. An NRPB (HPA) study has recommended Action Levels for UK drinking water following such incidents[34]. These levels are based on Council Food Intervention Levels (CFILs) for liquid food published by the Council of European Communities. The proposed "acute" screening levels are 5 Bq/l for gross alpha and 30 Bq/l for gross beta. A subsequent Health Protection Agency (HPA) report[35] concluded that samples measuring below the screening level would not exceed these Action Levels.

If individuals were to drink water contaminated well in excess of these Action Levels for limited periods (e.g. a few weeks), this need not pose a significant radiological hazard. Thus the immediate withdrawal of drinking water supplies is in general not essential. However, every effort should be made to reduce activity concentrations in the water quickly (at least within a few weeks), in order to maximise the dose reduction achieved[35]. However, there is a problem in obtaining consistent results for samples with a gross alpha level of ~0.1 Bq/l[34].

6.2. UK Water Laboratories Mutual Aid Radioactivity Sub-group.

This sub-group has run annual intercomparison exercises for regulatory gross alpha and beta since 2001 with each participant analysing potable water samples from all the water companies represented. Since the radioactivity sub-group was established the intercomparison exercise has grown from three participating labs to the present eight, with support from DWI, LGC and other organisations.

The main exercise involves analysing drinking waters using validated methods set up to meet the UK drinking water standards for gross alpha and beta in drinking water of 0.1 Bq/l gross alpha and 1.0 Bq/l gross beta. This utilises a typical sample volume of one litre with a sample preparation time is about 12- 16 hours followed by a typical count time of 16 hours. This method is used for monitoring regulatory samples, but far too slow for screening of emergency incident samples. However, this time consuming method is far too slow for handling emergency incident situations and the sub-group developed a rapid method under the leadership of Phillipa Frewin of South West Water.

6.3. Development of a Rapid Screening Gross Alpha/Beta Method for Emergency Incidents

In order to deal with potential emergency incident situations, a rapid screening method was developed by the sub-group to meet the 5 Bq/l gross alpha and 30 Bq/l gross beta acute screening levels. It is based upon the SCA method1 using a gas proportional counter.

The method has a similar, but shortened preparation and counting stages. The method still utilises the addition of a fixed amount (0.35 g) of calcium sulphate to produce sufficient solids for thick film measurement. With the approximate 50 fold less sensitive detection limit requirement, it is practicable to use a reduced sample volume (100 ml) and a reduced ashing time (30 min) to decrease sample preparation to approximately three hours. The count time has been reduced to 30 min (from ~16 hours). Thus, a laboratory with adequate staff to carry out the relatively simple preparation stage and an eight or 10 channel gas proportional counter should be able to handle up to 150 samples in a 24 hour period with the first results being produced within 4 hours of receiving the samples. Then further results are produced at approximately half hourly intervals.

The results of the mutual aid group 2007 exercise for rapid screening of gross alpha and beta are shown in Table 6. These show, that except for the lab 4 outlier results, the agreement was remarkably good with a mean gross alpha result of 2.08 Bq/l with a standard deviation of 0.19 Bq/l. For gross beta, the corresponding figures are a mean of 2.59 Bq/l and a standard deviation of 0.60 Bq/l. Three labs also carried out a ~16 hour count which gave results that were close to the 30 min results. From this data it is concluded that this gross alpha/beta screening method is fit for purpose.

Some radionuclides will not be detected using the monitoring equipment routinely used by the water industry to measure gross alpha and beta activity. Of those listed in Table 5.3 of the radionuclides considered in the drinking water section handbook[35], those that would not be detected using this approach would be ^{75}Se, [95]Nb, [103]Ru or [169]Yb. Some of these radionuclides do not emit beta particles, while in the other cases the energy of the beta particle emission is too low to be detected by the method used. If it is suspected that these radionuclides are in the water supply it will be necessary to carry out more radionuclide specific analyses.

Laboratory	Alpha rapid (30 min)	Alpha full count time (16 hours)	Beta rapid (30 min)	Beta full count time (16 hours)
All results expressed as Bq/l				
6	2.38	1.80	2.38	2.29
7	2.32	2.29	3.26	2.97
2	1.87		3.13	
3	2.01	1.81	2.24	2.42
1	2.25		2.49	
4	8.39*		16.3*	
5	1.96		1.83	
Mean	**2.08**		**2.59**	
S.D	**0.19**		**0.60**	
* outlier result				

Table 6 *Results of the rapid alpha & beta intercomparison first exercise   (Feb 2007)*

6.4. Conclusions

The UK Water Industry Mutual Aid Radioactivity Laboratory Sub-group has achieved considerable progress since its formation in 1996: -

- A formal mutual aid system has been set up with 24h/365d contact. The industry capability of routine radioactivity regulatory analysis has been significantly improved.

- A robust rapid radioactivity screening analysis method for gross alpha and beta for emergency incident samples has been developed and validated.

- A forum to discuss methodology and related issues, and encourage development of expertise has been set up.

- Formal intercomparison exercises and associated 'wash-up' meetings to discuss results of these exercises have been set up.

- Improved awareness of training needs and the training available has been achieved.

- A radioactivity training course to meet the needs of the industry is currently being developed.

7. MUTUAL AID NETWORK

The UK Water Laboratories Mutual Aid Group was established in 1995 as an informal group discussing laboratory issues relating to emergency analysis related to water contamination incidents. This is a voluntary group with no external funding. It gained added impetus following 9/11 and became a sub-group of the National Security and Emergency Planning Working Group.

Representatives from all Water Company Laboratories, specialist laboratories, DWI and other relevant organisations have attended. A Laboratory Emergency Response Capability Statement covering almost all UK water laboratories carrying out regulatory analysis and some specialist laboratories was established and is regularly updated,. This gives all participants a 24/7 contact phone number to all laboratories that will assist, at their normal call out rates, if they are able to do so on a reasonable endeavours basis. There are no legally binding agreements in the water laboratories mutual aid scheme. This was a unanimous decision by all representatives in order to simplify the operation of the scheme. The key is 'reasonable endeavours'.

Much of the exchange of information within the industry on new rapid methods or improved screening tests occurs under the aegis of this group. It meets on an annual basis in November and rotates the location around the participating laboratories with a visit around the lab after the meeting offered. There are other meetings for sub-groups of which there are currently three: -
- Reviewing the results of each LEAP emergency proficiency scheme exercise;
- Rapid organic analysis
- Rapid gross alpha/beta radioactivity analysis

"Chatham House" rules apply to all group meetings so that nothing should be ascribed to a particular company or individual from the meeting proceedings.

8. CSL – LEAP™ EMERGENCY PROFICIENCY SCHEME

The CSL-LEAP emergency incident proficiency scheme incident has been set up by CSL-LEAP™ (the Central Science Laboratory [Ministry of Agriculture, Fisheries and Food]). Participating laboratories receive (at the same time) a simulated pollution incident sample with an associated scenario and they are judged on how quickly and accurately they can detect and semi-quantify the toxic substances present[36]. A typical exercise contains 4 – 6 different substances. The laboratory staff do not know when this will be delivered. All of these samples are treated as real pollution emergency samples except that all significant results are immediately communicated to the proficiency scheme organiser rather than the relevant water company. Sixteen circulations have been run to date and participating laboratories have gained considerable experience in handling and analysing a wide range of completely unknown samples. The early exercises indicated problems with the identification of more polar substances and the scheme has proved a very useful learning exercise. Post mortems of the exercises are held on a regular basis. (For details of the CSL-LEAP™ proficiency schemes, see http://ptg.csl.gov.uk/leap.cfm).

9. PASSIVE SAMPLING

Often, by the time samples are taken in an emergency incident situation, the peak concentration of a pollutant may be missed. It would be possible to insert simple low-cost passive sampling devices into key parts of a water distribution network that could operate for up to one month.

There would be two types, those that collect metals[37, 38] and those that collect organics[39]. Each device would be in contact the water in the relevant part of the distribution system for two to four weeks before being changed. Potable water within a distribution system typically contains very low levels of toxic metals; radioactive elements or toxic organic substances. After a significant water contamination incident occurs, the devices could be removed and quickly analysed. It should be possible to detect the captured substances from a significant transient pollution incident that would not necessarily be detected by conventional spot sampling techniques a few hours after the event. It is felt that this approach could also be very useful for detecting radioactivity incidents. If the passive devices were changed and analysed on a regular basis, it would give some indication of longer-term background levels and indication of low level diffuse type pollution. Alternatively the removed devices could be stored in a freezer and only analysed during an emergency incident. This approach would clearly show if the concentration of any sought substance was present at significant level above the typical background level.

10. CONCLUSIONS

- Analysis planning for 24h/365d very high impact, very low probability potable water emergency contamination incidents needs very careful consideration; once a response system is set up, maintaining it on a long-term basis requires much effort. Staff phone numbers, e-mail addresses, screening analysis methodology can change and careful check mechanisms are essential that the system remains robust and effective.

- Regular in-house random auditing of the emergency response system should be carried out. It is important to appreciate systems that are not in routine use, rapidly deteriorate unless they are regularly tested.

- This aspect of should also be third party audited by the appropriate regulatory body.

- The importance of a simple robust QA/QC system for this type of work must not be overlooked.

- A key consideration is how to rapidly detect when a significant contamination incident has occurred then identifying the cause of the contamination or to convincingly demonstrate absence of contamination. Background screening of uncontaminated samples should be carried out over the four seasons to establish the background organic substance spectra of all major water supplies

- All screening methods should be fully documented and validated to assess the relative amount of false negative and positive results using a wide range of risk agents.

- Mutual aid groups with 24h/365d contact numbers for specialist analysis requirements have been found to very useful in challenging emergency situations. However, these groups need managing and regular (at least annual) face to face meetings of all members are considered essential.

- Participation in a relevant emergency incident proficiency scheme is considered essential to assess laboratories on an on-going basis. In this type of scheme, emergency incident samples arrive on an unannounced basis and the receiving laboratory is judged on the speed and accuracy of its response. Thorough documented investigations should be carried out each time a substance is either missed or mis-reported.

- Occasional, completely unannounced dummy incidents need to be run to ensure that the complete emergency response system from the reporting of a potential incident; the subsequent taking of the samples; the analysis of the samples; the reporting of results and all associated communications function as intended. All shortfalls need to be fully investigated. This is expensive, but considered to be an essential requirement.

- The use of robust simple ecotoxicity screening tests running in parallel to chemical screening tests should be given serious consideration.

- The use of a rapid low technology screening method for gross alpha and beta analysis is useful for rapid screening of large numbers of samples to either show no significant radiation or to prioritise samples that need further much more complex radioactivity analysis procedures to estimate the total indicative dose

- The use of low-cost passive sampling devices should be considered

REFERENCES

1.   Eds. K. C. Thompson and J. Gray, *Proceedings of RSC/SCI/IWO International Conference Water Contamination Emergencies: Can we cope?* held on 16 – 19th March 2003, RSC Publications, 2004. ISBN 0-85404-628-3

2.   S. States, J. Newberry, J. Wichterman, J. Kuchta, M. Scheuring and L. Casson, Rapid analytical techniques for drinking water security investigations, *Jour. AWWA*, 2004, **96**:1, pp 52 – 64

3.   Eds. K. C. Thompson and J. Gray, *Proceedings of RSC/SCI/IWO International Conference Water Contamination Emergencies: Enhancing our Response* held on 12 – 15th June 2005. RSC Publications, 2006. ISBN 0-85404-658-5

4.   G. O'Neill, C. Ridsdale, K. C. Thompson and K. Wadhia, Field and laboratory analysis for detection of unknown deliberately released contaminants. *Proceedings of RSC/SCI/IWO International Conference, Water Contamination Emergencies: - Can we cope?* held on 16 – 19th March 2003. RSC Publications, 2004. ISBN 0-85404-628-3, pp 100-109

5   P. Meinhardt, http://www.waterhealthconnection.org/bt/chapter5c.asp. Last accessed 28th July 2008

6.   S. Scott, Rapid Analysis (Organics), UKWIR 2007

7.   http://www.epa.gov/emergencies/tools.htm#lol Title III Consolidated List of Lists - October 2006 Version, Last accessed 28th July 2008

8.   http://www.intox.org/databank/pages/all_pims.html Last accessed 28th July 2008

9.   http://cfpub.epa.gov/ecotox/help.cfm?sub=about Ecotox database Last accessed 28th July 2008

10.   http://toxnet.nlm.nih.gov/cgi-bin/sis/htmlgen?index.html   Last accessed 28th July 2008

11   http://www.speclab.com/compound/chemabc.htm   Last accessed 28th July 2008

11A   http://www.waterhealthconnection.org/index.asp Last accessed 28th July 2008

12   http://www.bt.cdc.gov/agent/agentlistchem.asp Last accessed 28th July 2008

13   http://whqlibdoc.who.int/publications/2004/924156638.pdf   World   Health Organization. *The WHO recommended classification of pesticides by hazard and guidelines to classification: 2004.* Last accessed 28th July 2008

14.   USEPA, *Standardized analytical methods for use during homeland security events, Revision 1.0,* September 29, 2004

15.   B. L True and R. H. Dreisbach, *Dreisbach's handbook of poisoning*, 13th Edition, the Parthenon Publishing Group, 2002. ISBN 1-85070-038-9

16   The Commission Decision 2002/657/CE   See Tables 4 - 6 in http://eur-lex.europa.eu/LexUriServ/LexUriServ.do?uri=OJ:L:2002:221:0008:0036:EN:PDF Last accessed 28th July 2008

17.   G. Langergraber, A. Weingartner and N. Fleischmann, Time-resolved delta spectrometry: a method to define alarm parameters from spectral data, *Water Science and Technology*, 2004, **50**, 13 – 20. (See also http://www.s-can.at/)

18.   J. van den Broeke, A Brandt, A. Weingartner, and F. Hofstadter. Monitoring of Organic Micro Contaminants in Drinking Water using a Submersible UV Spectrophotometer. *Proceedings of RSC/SCI/IWO International Conference Water Contamination Emergencies: Enhancing our Response* held on 12 – 15th June 2005, RSC Publications, 2006. ISBN-10: 0 85404 658 5,

19.   Joep van den Broeke, Sensitive, selective and simple - UV-spectrometry for contaminant alarm system, *Proceedings of RSC/SCI/IWO International Conference Water Contamination Emergencies: - Collective Responsibility* held on 7 – 8th April 2008. RSC Publications, 2009.

20.   Peter John Bratt, K. Clive Thompson and Peter Benke, Rapid detection of volatile substances in water using a portable photoionization detector, *Proceedings of RSC/SCI/IWO International Conference, Water Contamination Emergencies: - Enhancing Our Response,* held on 12-15 June 2005, RSC Publications, 2005. ISBN-10: 0 85404 658 5, pp 100-109

21.   Ticket® Pesticide Detection Program, Neogen, http://www.mibius.de/out/oxbaseshop/html/0/images/wysiwigpro/Agri_Screen_Pesti cide_Ticket_Lit.pdf Last accessed 28th July 2008

22.   EPA Environmental Technology Verification (ETV) Program http://www.epa.gov/etv/ Last accessed 28th July 2008

22A Colin H. Self., S. Thompson and L. A.Winger, *The Immunoassay Handbook* (Wild, D. ed.), Chapter 2. pp 41-47. Elsevier, In Press. See also http://www.selectiveantibodies.com/ Last accessed 28th July 2008

23.   Andrew D. Eaton (Editor), Lenore S. Clesceri (Editor), Eugene W. Rice (Editor), Arnold E. Greenberg (Editor), Mary Ann H. Franson (Editor) *Standard Methods for the Examination of Water and Wastewater* , 21st Edition, APHA/AWWA/WEF, 2005,   ISBN Number: 0-87553-047-8 (Page 2-41) Method 2350B Chlorine demand/requirement.

24.   Adrian J. Charlton, James A. Donarski, Stephen A. Jones, Barry D. May and K. Clive Thompson, The development of cryoprobe nuclear magnetic resonance spectroscopy for the rapid detection of organic contaminants in potable water, *Journal of Environmental Monitoring,* 2006, **8**, 1106 – 1110

25.   Adrian. J. Charlton, James A. Donarski, Barry D. May and K. Clive Thompson, Optimisation of NMR methodology for non-targeted detection of water contaminants, *Proceedings of RSC/SCI/IWO International Conference Water Contamination Emergencies: Collective Responsibility* held on 7 – 8th April 2008. RSC Publications, 2009.

26.   K. L. E. Kaiser and V. S. Palabrica, Vibrio fischeri (Photobacterium phosphoreum) toxicity data index, *Water Poll. Res. J. Canada*, 1991, **26**, 361 – 431

27.   Review of the enhanced chemiluminescence (ECL) test, *R & D Technical Report E28*, Environment Agency, 2000. (E-mail: publications@wrcplc.co.uk)

28.   G. Persoone, K. Wadhia and K. C. Thompson, Rapid toxkit microbiotests for water contamination emergencies *Proceedings of RSC/SCI/IWO International Conference, Water Contamination Emergencies*: - Can we cope? held on 16 – 19th March 2003. RSC Publications, June 2004. ISBN 0-85404-628-3, pp 122-130

(See also http://www.sdix.com/ProductSpecs.asp?nProductID=7) Last accessed 28th July 2008

29. Guido Persoone, Recent advances in rapid ecotoxicity screening, *Proceedings of RSC/SCI/IWO International Conference, Water Contamination Emergencies: - Enhancing Our Response*, held on 12-15 June 2005. RSC Publications, 2005 ISBN-10: 0 85404 658 5, pp 100-109

30. K. Wadhia and K. Clive Thompson, Low-cost ecotoxicity testing of environmental samples using microbiotests for potential implementation of the Water Framework Directive, *Trends in Analytical Chemistry*, 2007, **26**, 300 - 307

31. Eds. G. Persoone, C. Janssen and W. de Coen, *New microbiotests for routine toxicity screening and biomonitoring,* Kluwer Academic/Plenum Publishers, 2000, ISBN 0-306-46406-3 (See also http://www.microbiotests.be/product.htm),

32. http://www.microbiotests.be/toxkits/rapidtoxkit.pdf Last accessed 28th July 2008.

33. Measurement of alpha and beta activity of water and sludge samples. The determination of radon-222 and radium-226. The determination of uranium (including general x-ray fluorescent spectrometric analysis) 1985-6, *Methods for the examination of waters and associated materials*, Her Majesty's Stationery Office, 1986, ISBN 0-11-751909-X

34. Phillipa Frewin and K. Clive Thompson, Mutual aid radioactivity sub-group intercomparison exercises for gross alpha and beta analysis of potable water samples, *Poster presentation at RSC/SCI/IWO International Conference Water Contamination Emergencies: - Collective Responsibility held on 7 – 8th April 2008.* (Copies available from KCT)

35. *UK Recovery Handbook for Radiation Incidents: 2008*, HPA-RPD-042, ISBN 978-0-85951-622-8,
http://www.hpa.org.uk/webw/HPAweb&HPAwebStandard/HPAweb_C/1215416656061?p=1197637096018
 http://www.hpa.org.uk/web/HPAwebFile/HPAweb_C/1215501690470   (for the potable water section only) Both last accessed 28th July 2008

36. Barry May, Laboratory environmental analysis proficiency (LEAP) emergency scheme, Analysis methods for water pollution emergency incidents, *Proceedings of RSC/SCI/IWO International Conference, Water Contamination Emergencies: - Enhancing Our Response*, held on 12-15 June 2005. RSC Publications, June 2005. ISBN-10: 0 85404 658 5, pp 100-109.

37. W. Davison and H. Zhang, In situ speciation measurements of trace components in natural waters using thin-film gels. *Nature*, 1994, **367**, 546-548.

38 Ian J. Allan, Jesper Knutsson, Nathalie Guigues, Graham A. Mills, Anne-Marie Fouillacd and Richard Greenwood, Chemcatcher® and DGT passive sampling devices for regulatory monitoring of trace metals in surface water, *Journal of Environmental Monitoring*, 2008, **10,** 821-829

39 Branislav Vrana, Graham A. Mills, Michiel Kotterman, Pim Leonards, Kees Booij and Richard Greenwood, Modelling and field application of the Chemcatcher passive sampler calibration data for the monitoring of hydrophobic organic pollutants in water, *Environmental Pollution*, 2007, **145**, 895-904

PROCESSING AND DATABASING SPECTROSCOPIC ANALYSES AND ITS USE IN THE ELUCIDATION OF UNKNOWNS

I. Peirson

Advanced Chemistry Development Limited, Trinity Court, Wokingham Road
Bracknell, Berkshire, RG42 1PL

1 INTRODUCTION

On the analytical bench, there are a plethora of different instruments, techniques and stand-alone vendor software packages for spectroscopic analysis. The technology is now available to provide a common processing environment for all instruments with the ability to database qualitative data, notes, spectra and chromatograms along with the experimental conditions without any loss in data integrity.

This provides a number of benefits:
- It aids elucidation as the analysis software can import data from multiple spectroscopic techniques, multiple databases and multiple instruments.
- Vendor-neutral off-line processing capability allows the outsourcing of useful spectroscopic techniques such as accurate mass LC/MS and NMR whilst keeping the actual analysis and interpreted data in-house.
- The ability to create a fully searchable database of spectra, chromatograms, chemical structures and associated properties.
- Creation of systems with partly or fully automated processing, to allow rapid interpretation by non-specialists, thus significantly reducing the time required for reporting a result
- Archiving and traceability of all analytical data for legislative purposes

2 METHOD

Using ACD/IntelliXtract™, a GC/LC/MS sample from a complex matrix was rapidly componentized to individual extracted ion chromatograms, each of which was automatically quality assessed by mass spectral and chromatographic properties. During processing, the dataset was simultaneously compared against a blank and/or a spiked control to identify those components in the sample which are either unique or which differ significantly in terms of height or area.

By interrogating the results of these quality/uniqueness tests, the user can dynamically filter the results to show those that are most relevant.

Assuming that accurate mass data is present, the software can be asked to look for a list of expected contaminants. In cases where it is necessary to look for a large number of contaminants, the software can be bound to ACD/ChemFolder™, a chemical structural database, which is automatically searched during processing.

Once the relevant components have been short-listed, the pure-component spectra of these can be extracted and used to search against NIST™, Wiley™ or self-built databases. The software offers advanced reporting features to create high-quality reports and these can be formatted by a template design if the user so desires.

For a confirmation of the hit, the EI or MS/MS spectrum can be imported into the ACD/MS Processor™ window. The proposed structure is attached to the spectrum and using a secondary algorithm, a list of theoretical fragments is generated. By comparing the list of theoretical fragments to actual experimental fragments, the software then makes the spectral assignment and provides the user with an accuracy match factor. If most of the fragments are assigned then the likelihood of the hit being correct would be high.

Further understanding of the fragmentation of a compound can be obtained by passing the structure to ACD/MS Fragmenter™, which takes a source compound and renders a fragment tree, from which the relationships of parent and daughter ions can be understood.

While GC-MS (or nominal mass LC-MS) is often the first step in elucidation of unknown compounds in a mixture, there is recent interest in other techniques such as accurate mass LC-MS, IR, SERS (Surface Enhanced Raman Spectroscopy) and NMR. Large reference databases are available for IR and NMR, for spectra similarity searching.

If enough of the unknown compound can be separated, 1D and 2D NMR spectra can be meaningfully collected. In conjunction with other spectroscopic data, it is possible to use the software to verify the proposed structure, or render the likely structure of the unknown.

It is envisaged that the spectra from each analytical test will be stored in an investigational database as they are acquired. This will provide a centralised repository of live spectra that can be further interrogated at a later late as more information is obtained, as well as growing into an in-house knowledge store which can be used in the future for the elucidation of unknown structures.

ACD/Labs also provides specialised method development software, that can be used to fully screen column, buffer and solvent choices as well as to optimise gradient and temperature programs, to provide the most efficient chromatographic separation in the shortest run time.

3 CONCLUSION

For the full identification of unknown compounds in a mixture, multiple analytical techniques may need to be employed. Using the vendor-neutral software processing tools, all data types can be imported into a uniform processing interface. Advanced processing algorithms assist the user in interpreting the data and making assignments to chemical structure.

Creation of a multi-technique database helps organise the investigation and allows the scientist to revisit earlier tests, either as part of the current study or to assist in future investigations.

HANDBOOKS TO ASSIST IN THE MANAGEMENT OF A RADIOLOGICAL INCIDENT INVOLVING THE CONTAMINATION OF DRINKING WATER SUPPLIES

J. Brown, B.T. Wilkins and D. Hammond

Health Protection Agency - Radiation Protection Division, CRCE, Chilton, Didcot, Oxon. OX11 0RQ, UK.

1   INTRODUCTION

If a radiological event occurred near an open source of a water supply, then the water would probably pass through an established treatment works prior to being supplied to the consumer. Consequently, any such incident could lead to exposure to radiation for both consumers and the operatives that work in the affected water treatment works. It is important therefore that the drinking water industry has information and guidance to assess the radiological impact on both consumers and operatives. Two complementary Handbooks have been produced that provide guidance to the water industry and those decision-makers that would be responsible for managing the response to a radiological incident.

The first of these is the UK Recovery Handbook for Radiation Incidents (HPA 2005) (referred to hereafter as the UK Recovery Handbook), which provides guidance on recovery options to aid prompt decision-making in the first few months after a radiological incident. The Handbook covers radiological aspects, descriptions of the options and decision trees to assist in the choice of options. To help people become familiar with the content of the Handbook and how to use it effectively, some training aids have been devised. The Handbook considers agricultural and domestic food production, inhabited areas and drinking water. • The remainder of this paper discussion is focused on the drinking water section only.

A further Handbook has been produced to help the water industry to make decisions on how water treatment works can be operated in the event of a radiological incident and to manage any radiation exposures to the operatives [Brown et al 2008a]. To aid clarity, this Handbook is referred to in the rest of this report as the 'Treatment Works Handbook'. The Handbook covers the effectiveness of drinking water treatment processes in removing radionuclides, a methodology to assess radiation exposures to operatives working within drinking water treatment works and guidance on where radionuclides may concentrate within treatment works and the impact this may have. Worked examples are given to assist users both in planning for a radiological incident and in the management of a radiological incident.

This paper provides a summary of what is included in the two Handbooks. It sets out how they can be used together as part of contingency planning for a radiological incident and in dealing with the aftermath of an incident. As illustration, the paper also contains limited extracts from both Handbooks.

2   SCOPE AND AUDIENCE OF HANDBOOKS

The water industry has a responsibility to provide a potable source of drinking water. The Handbooks are intended to help the water industry in two ways. These are as follows.

- To assess the impact that any radiological incident may have on the drinking water that it supplies to the public.
- To assess the impact that any radiological incident may have on the people carrying out operations at a water treatment works.

In terms of drinking water, the aim of the UK Recovery Handbook is to provide guidance on recovery options for reducing doses from ingestion by members of the public. Guidance is given to help users to identify important problems and to evaluate options for the management of contaminated drinking water. Emphasis is placed on the management of the radionuclide content in drinking water as supplied 'at the tap' and not that in drinking water sources, e.g. reservoirs. The main focus of the UK Recovery Handbook is to give guidance that is relevant for an accidental release for a nuclear site or from the transport of nuclear weapons. However, many of the recovery options will also be relevant to other radiological emergencies and the Handbook considers a wide range of radionuclides.

The aim of the Treatment Works Handbook is to provide a tool for the water industry to manage the potential risks to operatives working at a treatment works and it complements the UK Recovery Handbook. It also provides guidance on the likely effectiveness of drinking water treatment in removing radionuclides from water and as such provides important additional and up-dated information to support the UK Recovery Handbook[1]. Radionuclides have been chosen to reflect the hazards that operatives of drinking water treatment works could be exposed to and to exemplify a range of chemical and physical behaviours in drinking water treatment works. These radionuclides include all of those considered in the UK Recovery Handbook.

The audience for the UK Recovery Handbook is expected to be primarily those organisations likely to be represented on the Recovery Working Group (RWG) [NEPLG, 2007]. The RWG is a multiagency group that consists of representatives of both local and central Government bodies; it would normally be chaired by the local authority but may be delegated to the relevant Environment Agency. The main audience for the Treatment Works Handbook is expected to be the Water Companies. However, it is also envisaged that it will also be used by those that regulate the drinking water industry.

3   THE 'TREATMENT WORKS' HANDBOOK

This Handbook contains information to help the user to answer the following questions.

- What happens to radioactivity within a drinking water treatment works?
- How do I estimate activity concentrations in treated drinking water and the waste products sludge and filter bed media?

---

[1] An up-dated version of the Drinking Water Section within the UK Recovery Handbook is due for publication in autumn 2008. Some of the material presented in this paper is not included in the 2005 version.

- What are the potential radiation exposure pathways for operatives?
- What tasks and exposure pathways are likely to be most important in terms of radiation doses to operatives at treatment works?

The Handbook is supported by a scientific report that provides additional technical information [Brown et al, 2008b]. In particular, the supporting report includes a full review of the effectiveness of drinking water treatment in removing radionuclides from water and the methodology developed to estimate radiation doses to operatives from radioactive contamination within a treatment works.

3.1   Estimating potential doses to operatives in drinking water treatment works

A step-wise approach is presented to assist the water industry estimate what the potential doses could be for a specific water treatment works or for a typical works in advance of a radiological incident. The main steps are:

- compile essential information about the treatment works;
- consider tasks carried out within the treatment works;
- estimate doses for each identified task assuming a unit activity concentration in the medium giving rise to the exposure;
- estimate doses for each identified task assuming a unit activity concentration in the water entering the works; this takes into account the partitioning of radioactivity in the works and enables doses to be compared across tasks;
- identify tasks and radionuclides that give rise to the highest doses within normal operation of the works and identify the exposure pathways that are likely to be the most significant for these doses.

The Handbook contains a worked example that takes the user through these steps for a notional drinking water treatment works. An extract from this worked example covering the first 3 bullet points above is given in Figure 1.

3.2   Identifying risks to operatives in the event of an incident

A generic tool has been provided to help the water industry to assess and manage the potential radiation exposures to operatives working in a drinking water treatment works. In addition, three scenarios have been developed so that users can see how the tool is intended to operate. For each scenario the user is taken through the steps involved and guidance is given on the factors that need to be considered. These include the need and urgency for specialist radiation protection advice and the ways in which the tool can be adjusted to deal with a particular works and a specific incident. Illustrative worked examples are provided for each scenario to support training in the use of the Handbook. The three scenarios are as follows.

1   It is suspected or known that contaminated water has passed through the treatment works. No measurements in input or output water are available.

2   Contaminated water has passed through the treatment works and measurements indicate that the input water is still contaminated.

3   It is suspected that contaminated water is going to pass through the treatment works in the near future, for example via an airborne release from a nuclear reactor that has deposited radioactive material on to the water catchment area.

Compile essential information about the treatment works	Required information	Answer / result
	Daily water throughput	80 megalitres (Ml) per day
	Treatment processes	Flocculation / clarification (dissolved air flotation) (DAF) and rapid gravity filtration (RGF).
	Number of rapid gravity filters = 6 with area = 5m x 10m and depth 0.8 m	total volume of sand in RGF is therefore 6 x 40 m^3 = 240 m^3 for a daily 80 Ml throughput
		mass of sand = 4.8 10^5 kg (assuming a density of sand of 1600 kg m^{-3})
	Amount of sludge produced	5000 kg sludge produced per 80 Ml
Work involving handling and management of waste sludge	Filling and emptying sludge press (2.5 h every day)	17.5 h per week
	Shovelling sludge to bunkers and trailers (0.5 h every day)	3.5 h per week
	Transporting sludge to landfill (once a week)	1.5 h per week
Work involving handling and management of used filter media	Emptying and replacing filter sand (5 days every 5 years)	Infrequent task 35 h
Other tasks carried out	Daily checking of DAF (1 h every day)	7 h per week
	Daily checking of treatment works (1 h every day)	7 h per week
	Inspection of back-washing (0.5 h every day)	3.5 h per week

Doses mSv received per Bq kg^{-1} in the contaminated media giving rise to the exposure

Day - to-day tasks, doses calculated for 1 weeks operation of treatment works

	Doses, mSv in a week / Bq kg^{-1} in contaminated medium	Calculation
Filling and emptying sludge press	^{60}Co = 7.0 10^{-6}	4.0 10^{-7} mSv/h x 17.5 (hrs/wk)
	^{239}Pu = 1.1 10^{-5}	6.0 10^{-7} mSv/h x 17.5 (hrs/wk)
Shovelling sludge to bunkers	^{60}Co = 1.9 10^{-6}	5.5 10^{-7} mSv/h x 3.5 (hrs/wk)
	^{239}Pu = 2.1 10^{-6}	6.0 10^{-7} mSv/h x 3.5 (hrs/wk)
Transporting sludge	^{60}Co = 2.0 10^{-7}	1.3 10^{-7} mSv/h x 1.5 (hrs/wk)
	^{239}Pu = 2.1 10^{-12}	1.4 10^{-12} mSv/h x 1.5 (hrs/wk)

Figure 1 *Extract from worked example for estimating potential doses to operatives at a treatment works*

Figure 2 shows an extract from Scenario 1. Figure 3 gives the corresponding extract from the worked example for Scenario 1, dealing with the steps shown in Figure 2.

Step 1: Prioritise sampling within the treatment works	Doses may have been received by the operatives as the contaminated water passed through the plant.
If waste sludge handled on the site the highest priority is to measure activity concentrations in the sludge at the points that it is handled.	Contamination may remain in the works, primarily in sludge and filter media.
If treatment works contains rapid gravity filters and their back-washing is checked on a day-to-day basis, measure activity concentrations in the filter media. Consider reducing the time spent on inspections / stopping them until further information is available.	The highest priority is, therefore, to assess and control any further doses that could be received by operatives due to routine day-to-day tasks.

If routine maintenance is imminent and it is difficult to delay this, the assessment of doses to operatives undertaking these tasks is the next priority.

Any future doses from routine inspection and water testing are likely to be very low as the contaminated water and any associated floc has already passed through the plant. Therefore, these tasks can continue as normal as the risks of any exposure, if any, are very low. |
| If any routine maintenance is imminent (within a few days) that involves working with sludge or filter bed media and it cannot be delayed, the next priority is to measure activity concentrations in the filter bed media or sludge that the operatives will be working with or will be close to. | However, doses from inspection of the back-washing of rapid gravity filters should be considered separately as this task can potentially give rise to inhalation doses over the longer term due to accumulation of contamination in the filter media over the duration of the incident. Exposure will continue until the filter media are replenished / replaced or the activity concentrations reduce due to radioactive decay. This pathway is only of potential concern if contamination contains alpha-emitting radionuclides. |
| Step 2: Using measured activity concentrations, estimate potential future doses from tasks identified in Step 1 for 1 week of exposure for a *'critical individual'*.[2]

If alpha emitting radionuclides are measured, include estimates of potential future doses from the inspection of back-washing of rapid gravity filters.

If urgent routine maintenance tasks need to be carried out, estimate doses from these tasks. | This step is based on cautious, default assumptions. The doses calculated as part of emergency planning assume that the activity concentration in each contaminated material is at a constant level for 1 week and that day-to-day tasks are carried out for 1 week with operatives being exposed to this contamination. Doses from routine maintenance tasks are estimated by assuming that the activity concentration remains constant for the length of time that the maintenance takes place. For routine maintenance tasks involving sludge or filter bed media, the contamination is static within the environment where the maintenance is taking place. It is therefore reasonable to assume that operatives will be exposed to this contamination throughout the length of the maintenance task.

Future doses are only likely to be received from most day-to-day tasks over a short period and, indeed, may not be received at all if the contamination passed through the works a while ago. |

---

[2] A "critical individual" is the term used in the Treatment Works Handbook for someone who is assumed to carry out all of the day-to-day tasks within the works. This may not apply to any single individual within the works, but provides a useful first stage in assessing radiological impact (see Brown et al 2008a for further details).

Figure 2   *Extract from Scenario 1 showing the steps involved in identifying risks to operatives in the event of an incident*

Step 1: Priorities for sampling within treatment works	Sludge is handled on site. Measure activity concentrations in sludge.	Measurements indicate that the contamination is from ^{60}Co. This is a long-lived gamma-ray emitting radionuclide.		
	Inspection of back-washing of rapid gravity filters occurs. Measure activity concentrations in sand in filters.	Measured activity concentration is 10,000 Bq kg^{-1} in the sludge from the pressing done at the time of the incident.		
	No routine maintenance is imminent.	Measured activity concentration in filter media is 20 Bq kg^{-1}.		
Step 2: Using measured activity concentrations, estimate potential future doses for *'critical individual'* carrying out all day-to-day tasks	Day-to-day tasks	mSv in a week / Bq kg^{-1} in sludge (^{60}Co) [calculated as part of planning]	Activity concentration in sludge or filter media, Bq kg^{-1}	Dose from ^{60}Co, mSv in a week
	Filling and emptying sludge press	7.0 10^{-6}	1.0 10^4	7.0 10^{-2}
	Shovelling sludge to bunkers	1.9 10^{-6}	1.0 10^4	1.9 10^{-2}
	Transporting sludge	2.0 10^{-7}	1.0 10^4	2.0 10^{-3}
	'Critical individual'			9.1 10^{-2}

Figure 3   *Illustrative worked example for Scenario 1, steps 1 and 2.*

The early stages of managing the actual or potential incident are shown in the scheme in Figure 4. This illustrates how the Treatment Works Handbook and the UK Recovery Handbook link together.

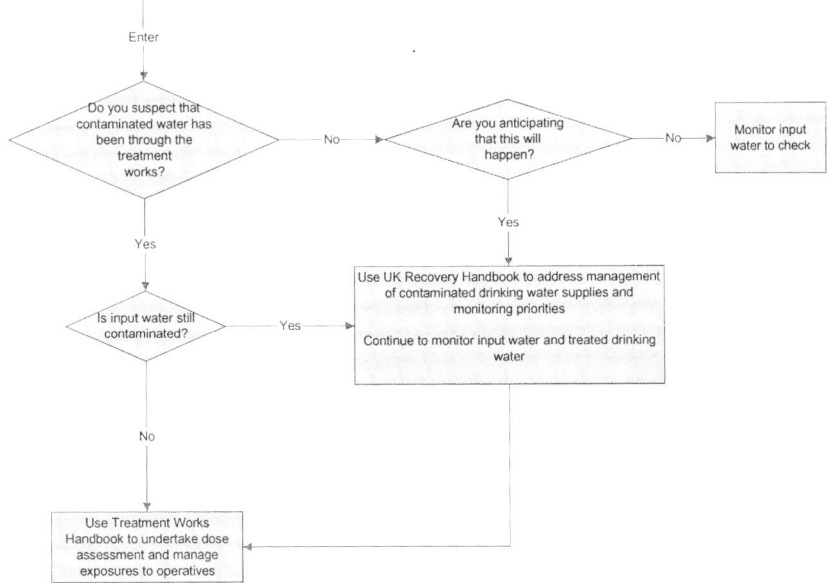

Figure 4   *Early stages of managing an incident showing links between the Handbooks*

4   UK RECOVERY HANDBOOK: DRINKING WATER

The Drinking Water Section of the UK Recovery Handbook contains the following information to guide decision-makers through the available recovery options.

- UK criteria and legislation relevant to radiation incidents and drinking water supplies.
- Data sheets for a number of options for managing drinking water. The data sheets contain fairly general information for broad categories of options to allow flexibility in their implementation by different Water Companies. Figure 5 shows the recovery options for drinking water that are included in the Handbook and an extract of a data sheet, illustrating their format. Information such as objectives, constraints, feasibility, waste, costs and side-effects are included.
- Top-level guidance on monitoring of water supplies.
- A decision framework. This is built around UK advice and includes consideration of different types of contaminating event, including a nuclear accident and deliberate contamination of water both before and after treatment. Decision trees lead to a sub-set of options being selected for further consideration.
- The use of environmental measurements to estimate activity concentrations in drinking water in the absence of direct measurements.
- Guidance on the use of the Handbook and, in the 2008 edition, training material using worked examples. These examples have been chosen to take the user through all parts of the Handbook.

4.1   The use of generic scenarios in the UK Recovery Handbook

Scenarios and worked examples have been developed to help users become familiar with the content of the UK Recovery Handbook and its structure. They can also be used for training purposes. They take the user, in a very general way, through the main decision steps and types of problems that they would need to address in the development of a recovery strategy. The worked examples provided are only illustrative and should not be used as proposed solutions to the contamination scenarios selected. The scenarios and worked examples included are:

- contamination of water due to deposition from a contaminated plume;
- direct contamination of water before treatment;
- direct contamination of water after treatment.

Figure 6 gives an extract of the worked example for the scenario 'direct contamination of water before treatment' to illustrate the type of information provided in the Handbook for familiarisation and training. It has been assumed that the radionuclide of interest is ^{90}Sr.

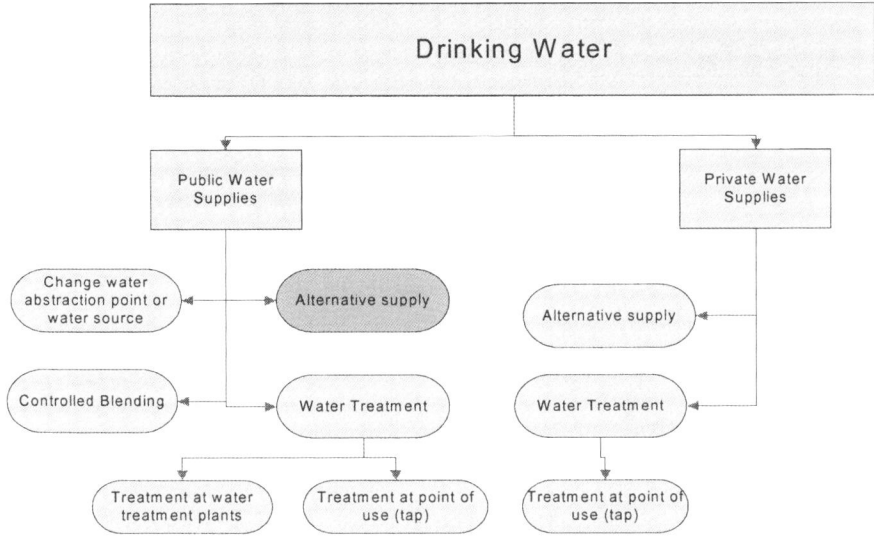

I Alternative drinking water supply	
Objective	To reduce ingestion doses to consumers by providing an alternative supply of suitable drinking water in the event of activity concentrations exceeding intervention levels.
Other Benefits	None
Countermeasure description	If restrictions were placed on the use of drinking water supplies due to activity concentrations exceeding intervention levels, alternative sources of water would need to be provided for drinking water and water used for culinary purposes. This data sheet considers the use of:
	1. Bottled water
	2. Water from other sources of drinking water provided by water companies via tankers and bowsers at distribution points
	Advice is likely to be given that continued use of the water supply for sanitation is expected and this will not give rise to any significant hazard.
	If the level of contamination was sufficiently high, then, in extreme cases, the water supplies could be turned off completely. This has not been considered in detail in this data sheet.
Target	Drinking water
Targeted radionuclides	This countermeasure would target all radionuclides associated with the contamination.
Scale of application	Small - medium
	Sufficient drinking water would need to be provided to sustain the population affected by any restrictions to their normal drinking water supply. Also sufficient drinking water would need to be provided to meet any legal obligations placed on the supplier. In general, the supply of alternative water could only be maintained for a short period (days) and then only to relatively small numbers of people in local or regional communities. Distribution of bottled water or water via tankers and bowsers is likely to take at least 8 hours to plan and arrange. It is important, therefore to encourage use of existing water supplies for sanitation purposes to avoid other public health issues...........

Figure 5   *Options for managing contaminated drinking water in the UK Recovery Handbook and extract from a data sheet*

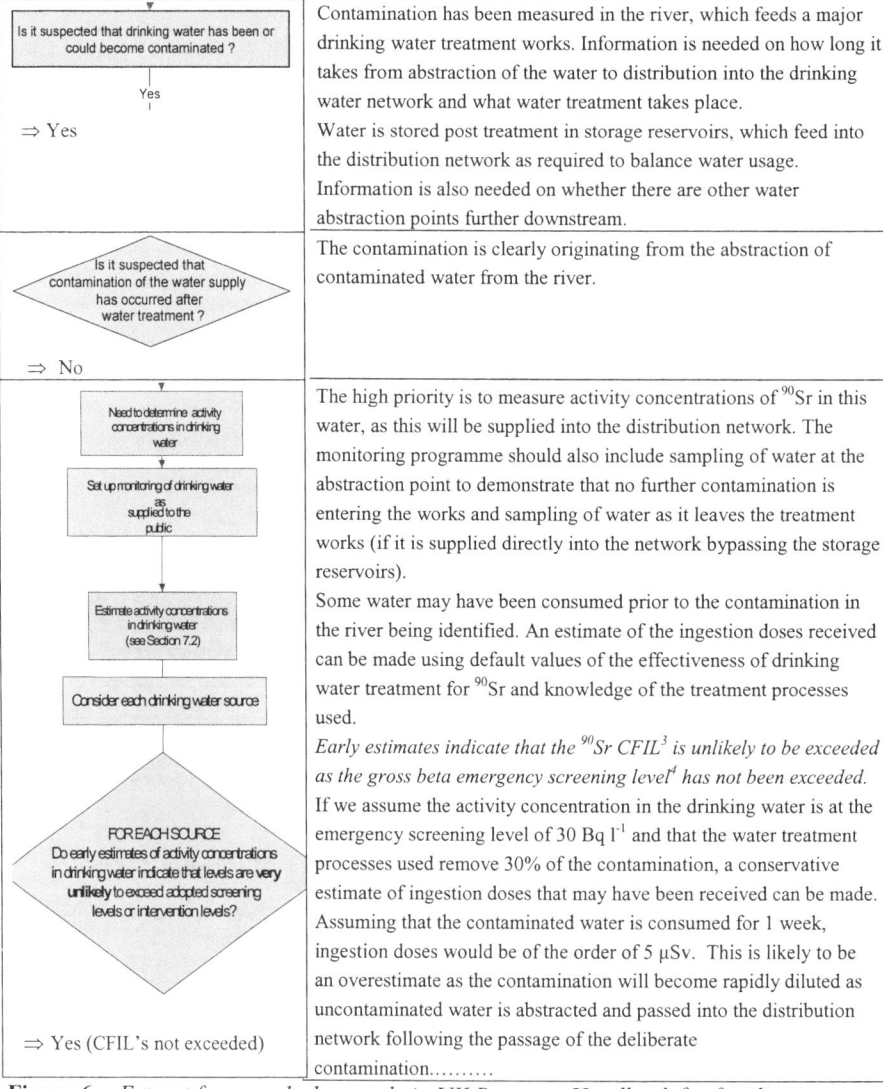

Is it suspected that drinking water has been or could become contaminated ?

Yes

⇒ Yes

Contamination has been measured in the river, which feeds a major drinking water treatment works. Information is needed on how long it takes from abstraction of the water to distribution into the drinking water network and what water treatment takes place.

Water is stored post treatment in storage reservoirs, which feed into the distribution network as required to balance water usage. Information is also needed on whether there are other water abstraction points further downstream.

Is it suspected that contamination of the water supply has occurred after water treatment ?

⇒ No

The contamination is clearly originating from the abstraction of contaminated water from the river.

Need to determine activity concentrations in drinking water

Set up monitoring of drinking water as supplied to the public

Estimate activity concentrations in drinking water (see Section 7.2)

Consider each drinking water source

FOR EACH SOURCE
Do early estimates of activity concentrations in drinking water indicate that levels are **very unlikely** to exceed adopted screening levels or intervention levels?

⇒ Yes (CFIL's not exceeded)

The high priority is to measure activity concentrations of ^{90}Sr in this water, as this will be supplied into the distribution network. The monitoring programme should also include sampling of water at the abstraction point to demonstrate that no further contamination is entering the works and sampling of water as it leaves the treatment works (if it is supplied directly into the network bypassing the storage reservoirs).

Some water may have been consumed prior to the contamination in the river being identified. An estimate of the ingestion doses received can be made using default values of the effectiveness of drinking water treatment for ^{90}Sr and knowledge of the treatment processes used.

Early estimates indicate that the ^{90}Sr CFIL[3] is unlikely to be exceeded as the gross beta emergency screening level[4] has not been exceeded. If we assume the activity concentration in the drinking water is at the emergency screening level of 30 Bq l^{-1} and that the water treatment processes used remove 30% of the contamination, a conservative estimate of ingestion doses that may have been received can be made. Assuming that the contaminated water is consumed for 1 week, ingestion doses would be of the order of 5 µSv. This is likely to be an overestimate as the contamination will become rapidly diluted as uncontaminated water is abstracted and passed into the distribution network following the passage of the deliberate contamination..........

Figure 6   *Extract from worked example in UK Recovery Handbook for familiarisation and training purposes*

---

[3] The CFIL (Council Food Intervention Level) is an Action Level for foodstuffs specified by the European Commission [CEC, 1989a; CEC, 1989b]. In the UK, the values for liquid foods are considered to be appropriate for drinking water [NRPB, 1994].

[4] The Environment Agency has issued advice on the use of gross measurements of activity in the aftermath of a radiological incident [EA, 2002].

5 CONCLUSION

This paper provides a summary of the information contained in two Handbooks that have been produced to assist decision-makers in the management of contaminated drinking water supplies following a radiological incident. Both Handbooks have been designed to assist in the planning for such incidents and include material for familiarisation and training. Only brief extracts can be given here. Both Handbooks contain further guidance, more worked examples and supporting technical information.

References

1 HPA, 2005. UK Recovery Handbook: UK Recovery Handbook for Radiation Incidents: 2005, *HPA-RPD-002*, Chilton, UK.
2 J Brown, D Hammond and B T Wilkins, 2008a. Handbook for assessing the impact of a radiological incident on levels of radioactivity in drinking water and risks to operatives at water treatment works, *HPA-RPD-XXX*, Chilton, UK.
3 J Brown, D Hammond and B T Wilkins, 2008b. Handbook for assessing the impact of a radiological incident on levels of radioactivity in drinking water and risks to operatives at water treatment works: Supporting Scientific Report, *HPA-RPD-XXX*, Chilton, UK.
4 NEPLG, 2007. Nuclear emergency planning liaison group: *consolidated guidance*, http://www.berr.gov.uk/energy/sources/nuclear/safety-security/emergency/neplg /guidance / page18841.html.
5 CEC, 1989a. Council Regulation (Euratom) No 3954/87 laying down the maximum permitted levels of radioactive contamination of foodstuffs and feeding stuffs following a nuclear accident or any other case of radiological emergency. *Off J Eur Commun L211/1.*
6 CEC, 1989b. Council Regulation (Euratom) No 770/90 laying down the maximum permitted levels in minor foodstuffs following a nuclear accident or any other case of radiological emergency. *Off J Eur Commun L101/17.*
7 NRPB, 1994. Guidance on restrictions on food and water following a radiological accident. Documents of the National Radiological Protection Board, **5**(1), Chilton, UK.
8 Environment Agency, 2002. Review of alpha and beta blue book methods: Drinking water screening levels. National Compliance Assessment Service *Technical Report, NCAS/TR/2002/003*, UK.

ROBUST ON-LINE TOTAL ORGANIC CARBON (TOC) ANALYSER FOR SECURITY MONITORING

E. Milks

Municipal Applications Specialist, GE Analytical Instruments, Boulder, CO

1 INTRODUCTION

In 2006, the USEPA singled out the distribution system as the most vulnerable component of a utility's drinking water system.[1] In response, many waterworks are taking a closer look at how they may protect the public's drinking water. One option is to place on-line water quality panels throughout the distribution system to detect a change in the baseline water quality.

The cost of implementing a water quality panel is not only a capital investment, but also includes labour to support the instrumentation on that panel. More robust analytical equipment that necessitate limited operator interface, infrequent replacement of reagents or carrier gases, and extended calibration periods can save a utility manpower, and therefore money, over time.

From testing performed by the USEPA, total organic carbon (TOC) was identified as one of the most important parameters for detecting a threat compound in drinking water. Therefore, a robust TOC analyser is essential for efficiently monitoring public drinking water.

2 CONTAMINATION WARNING SYSTEM AND TOC IMPORTANCE

To fully protect the distribution system and public drinking water, the USEPA has proposed a Contamination Warning System[2] (CWS) that would include on-line water quality monitoring to measure components such as TOC, chlorine, conductivity, and pH, where a contaminant would be indicated by a significant, unexplained deviation from the baseline water quality. An example of this panel is shown in Figure 1.

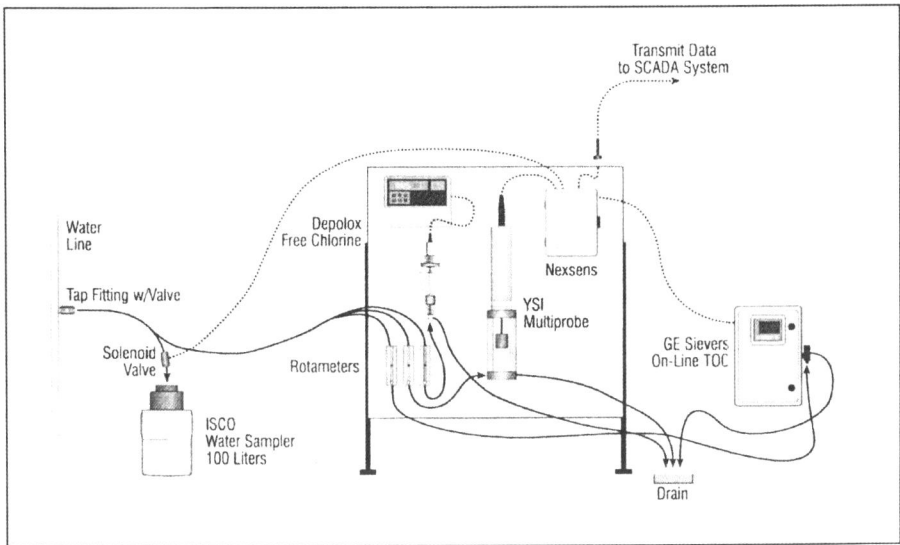

Figure 1 *Potential Water Security Panel*[2]

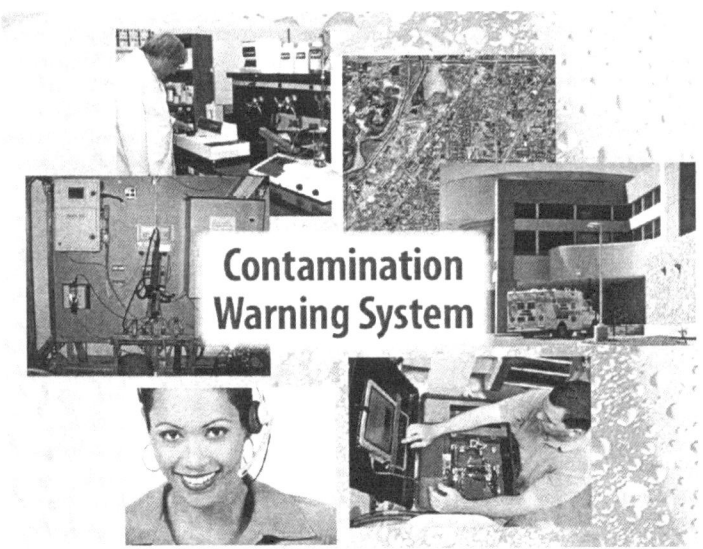

Figure 2 *USEPA webpage*[3] .

In addition to the on-line water quality panel, the CWS program would also include off-site sampling and analysis of the distribution system samples analyzed for various contaminant classes and specific compounds which typically cannot be monitored easily on-line. In addition to water analysis, enhanced security monitoring (such as cameras or motion activated lighting), consumer compliant surveillance via customer service hotlines, public health surveillance of local hospitals, and over-the-counter medications can be utilized to determine a security threat to the public water system. Figure 2 depicts the different components of the CWS program.

In order to determine the most useful parameters for on-line water quality monitoring, the USEPA ran a number of pipe loop and in-field tests. According to the USEPA's studies[4] from the group of parameters tested including TOC, free and total chlorine, specific conductance, oxidation reduction potential (ORP), pH, and turbidity, TOC and free chlorine detect the widest array of contaminants. TOC and free chlorine also produce the largest, and most easily detectable, water quality changes. In the USEPA Interim Guidance on Planning for CWS Development, published in May 2007, on-line TOC analysis is shown to be useful for detecting the presence of many organic compounds (see Table 1). Most organic contaminants should trigger a change in TOC at concentrations well below the lethal concentration.

Class	Description	Example Contaminant	TOC	Cl2[1]	COND
1	Petroleum products	Diesel	X		
2	Pesticides (reactive)	Aldicarb	X	X	
3	Inorganic compounds	Arsenite salts		X	X
4	Metals	Mercuric salts			X
5	Pesticides (non-reactive)	Fluoroacetate	X		
6	Chemical warfare agents	VX	X	X	
7	Radionuclides	Cesium-137			X
8	Bacterial toxins[2]	Botulinum toxin	X	X	
9	Plant toxins	Ricin	X		
10,11	Pathogens[2]	Vibrio cholerae	X	X	
12	Persistent chlorinated organic compounds	PCBs	X		

Acronyms: TOC - total organic carbon, Cl₂ - chlorine residual, COND - conductivity
1) Indicated contaminant classes have been shown to consume free chlorine residual. Results are not applicable to chloramine residual.
2) These contaminants are chlorine sensitive, thus it would be necessary to neutralize the chlorine residual in order to maintain potency. Many neutralizing agents would also increase TOC.

Table 1 *Impact of Contaminant Detection Classes on Water Quality Parameters*

Data in Table 2 again supports TOC as being one of the most important sensors for response to a contaminant. The table was formulated from data in an article[5] written by the John Hall, EPA, concerning a pipe loop test done at the EPA's T&E Facility in Cincinnati, Ohio. It specifically shows TOC responding to wastewater, potassium ferricyanide, glyphosphate, malathion, aldicarb, E coli, terrific broth, and nicotine.

In addition, many plants are switching disinfectant strategies from chlorine to chloramines, or other alternative disinfectant alternatives. If a utility makes this switch away from chlorine, TOC becomes even more important for detecting threat compounds in the water, and the chlorine monitor loses its effectiveness

SENSOR RESPONSE FOLLOWING INTRODUCTION OF CONTAMINANT									
CONTAMINANT	Chloride	Free Chlorine	Total Chlorine	Ammonia-nitrogen	Nitrate-nitrogen	ORP	Specific Conductance	TOC	Turbidity
Wastewater	⇧	⇩				⇩	⇧	⇧	⇧
Potassium ferricyanide	⇧	⇧		⇧	⇧	⇧		⇧	
Glyphosphate formulation	⇧	⇩				⇩		⇧	
Malathion formualtion		⇩	⇩			⇩		⇧	⇧
Aldicarb		⇩	⇩			⇩		⇧	⇧
Escherichia coli in Terrific Broth		⇩	⇩	⇧				⇧	⇧
Terrific Broth		⇩	⇩			⇩		⇧	⇧
Arsenic trioxide		⇩	⇩	⇧	⇩	⇩			⇧
Nicotine	⇧	⇩		⇧	⇩	⇩		⇧	

Table 2 *Data, on-line water quality parameters as indicators of distribution system contamination*

3 BEYOND SECURITY THREATS

It is important to remember that water quality panels, including TOC, can be used not only for intentional threats but also for unintentional threats that may cause cross connection or accidental contamination of the public's drinking water.

For example, an American city with a water population of over 250,000 experienced an accidental contamination in March 2008. While fighting a fire, the local fire department hooked up their fire foam hose incorrectly and introduced foam into the distribution system via the hydrant. The city's water utility ran analyses with GC/MS, pH, alkalinity and TOC to determine the extent of the contamination. A utility spokesman said, "By far, the TOC results were the most useful in determining the contaminant concentration and scope of the contamination throughout the distribution system. And having your TOC instrument on hand was invaluable in determining the 'Do not drink' alert and when the water was safe for consumption."

4 OVERVIEW OF TOC AND HOW A TOC ANALYSER WORKS

Total Organic Carbon (TOC) is the measure of all covalently bound, carbon-hydrogen containing compounds amenable to oxidation. TOC in drinking water is formed from the decay of naturally occurring vegetation, and can include algae, sediment, and particles in water. The TOC content in a water source varies from region to region, by type of water body, and even seasonally within a single water source.

Inorganic carbon (IC) is also a significant factor when measuring TOC. Inorganics are typically high in comparison to TOC for many municipal drinking water applications. To minimize the background noise of IC when measuring TOC, it should be fully removed to obtain accurate TOC readings.

The majority of TOC analysers function in the same manner. They convert inorganic carbon to carbon dioxide (CO_2) by reducing the pH of the sample, and this CO_2 is measured. The organics in the water sample are then oxidized completely to CO_2 and the CO_2 produced is measured. To differentiate TOC from IC, there are two basic options. The first is to purge the IC and measure TOC directly. In this manner, we assume that all of the organics are non-purgeable, and therefore NPOC equals TOC. Secondly, we can measure both the IC and TC and obtain TOC by difference, which captures both purgeable and non-purgeable organic carbon (NPOC + POC).

Several different TOC methodologies are available in the marketplace. The options for oxidation technology are UV/Persulfate, Heated Persulfate, Combustion, and Supercritical Water Oxidation. Detection technology choices are Conductometric, Membrane Conductometric, and Non-Dispersed Infrared (NDIR). The two options for handling inorganic carbon are sparging or "measure and subtract." All methodologies are similar in that they oxidize organic compounds to CO_2, where CO_2 is measured and reported as TOC. However, external reagents and carrier gases are typically required for an NDIR or combustion technology.

GE Analytical Instruments' Sievers TOC Analysers are entirely self-contained (see Figures 3 and 4). The Analysers use an internal acid cartridge that typically last three to six months (depending on the application), instead of a large external carboy of acid, which needs frequent replacement. Removal of inorganic carbon from drinking water samples is performed with an ICR module that uses only ambient air—not carrier gases which are required for purging procedures. Oxidation of organics is performed using a UV lamp and the addition of oxidizer from an additional internal cartridge. The aqueous $CO2$ is then diffused across a gas permeable membrane into deionized water. The increase in conductivity is measured using a temperature compensated conductivity cell

Sievers TOC technologies utilize UV/Persulfate Oxidation, a patented Membrane Conductometric Detection method, and the measure and subtract inorganic carbon handling. The entire process is so stable that calibration only needs to be performed yearly. The Sievers TOC Analysers comply with Standard Methods 5310 C and the USEPA Regulation for TOC, 415.3. These instruments are currently used in municipal drinking water applications, as well as in the semiconductor, pharmaceutical, and power industries.

The Sievers TOC technology was originally developed for the International Space Station for continuous on-line monitoring for water reuse. This instrument's core requirements excluded external chemicals or carrier gas. The environment required the instrument to be robust and self-contained, similar to what the security monitoring application requires.

GE's TOC Analyser is robust for security monitoring because it does not require purge gases, carrier gases, or external reagents. The instrument's calibration is stable for a year, and the calibration can be performed in less than two hours of unoperated sampling. The instrument easily runs quality control or verification standards to confirm calibration stability without the need to take the instrument offline from the original on-line water source.

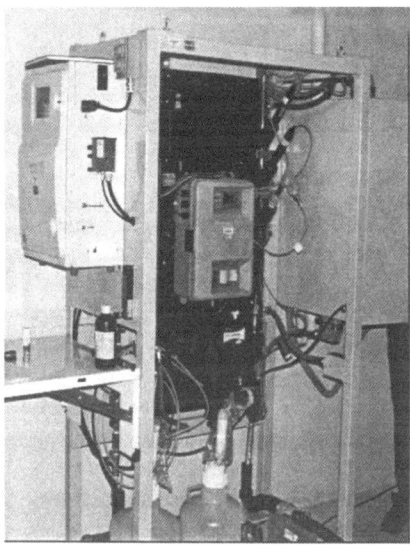

Figure 3: *Sievers Online TOC Analyser installed on a Water Security Panel*

Figure 4: *Sievers On-line TOC Analyser installed on a Water Security Panel*

The Sievers TOC Analyzer also includes extended unattended operation and easy start-up. The reagents typically last three months, so maintenance is only required quarterly. No warm up time is necessary; the instrument generates results twenty minutes from when it is turned "On." In on-line mode, the Analyzer gives results every four minutes while collecting fresh samples continuously. It thus allows a great response to any incident that impacts TOC.

The instrument easily interfaces with SCADA systems through 4-20 mA output and alarms. No special training is required to operate the Analyzer; anyone can handle the straightforward menu screens and wizard interface prompts.

5   CONCLUSIONS

Public drinking water distribution systems have been identified as vulnerable points within a waterworks system. On-line TOC analysis has been shown to be an effective indicator of many intentional or accidental threat compounds in drinking water. In order for a utility to support an on-line water quality panel, a more hands-off, robust TOC analyser is required to reduce operator time and costs. GE Analytical Instruments' TOC Analyser uses no external reagents, no carrier gases, and offers 12 months' calibration stability, making it one of the most robust TOC analysers found in the drinking water market today.

References
1   D. Schmelling, EPA Drinking WaterSentinel Program, Presentation to the AWWA Water Security Congress. Washington, 2006
2   John Hall, Jeff Szabo and Greg Meiners, Contaminant Warning System, Design, Implementation and Evaluation Presentation.
3   USEPA Website, 2007. http://www.epa.gov/safewater/watersecurity/pubs/guide_watersecurity_securityinitiati ve_interimplanningpdf.pdf.
4   USEPA Research Highlights News. 2006, On-line Water Quality Monitoring. Natl. Homeland Security Res. Ctr., www.epa.gov/NHSRC/news/news112706.html.
5   J. Hall, A. D. Zaffiro, R. B. Marx, P. C. Kefauver, E. R. Krishnan, R. C. Haught, J. G. Herrmann, On-line Water Quality Parameters as Indicators of Distribution System Contamination. *Journal of the American Water Works Association, 2007,* **99** *(1) 66-77.*

WATER UK EMERGENCY PLANNING

P. Fenton

Southern Water Services, Southern House, Yeoman Road, Worthing,BN13 3NX

1 INTRODUCTION

Emergency Planning in the UK Water Industry is driven by the provisions set out in Section 208 of the Water Industry Act, 1991. The Security and Emergencies Direction (SEMD), given under the powers of section 208, requires the industry to 'Make, keep under review and revise such plans as it considers necessary to ensure the provision of essential water supply or, as the case may be, sewerage services at all times, including a civil emergency or any event threatening national security.'

This paper supports the presentation given to the Water Contamination Emergencies Conference at The Royal Society of Medicine, London on 7th April 2008.

2 WATER COMPANIES RESPONSIBILITIES IN EMERGENCIES

When emergencies arise, water companies may have to:
- Take all reasonable steps to restore normal supplies;
- Protect consumers by advising them quickly and appropriately, for example to avoid using water for cooking and drinking;
- Avoid supplying water which is unfit for human consumption to ensure that an offence under section 70 of the Water Industries Act, 1991, is not committed. *Section 70 allows that it shall be a defence if the water undertaker can show that all reasonable steps were taken and due diligence exercised;*
- Supply water for drinking by tanker or other means, paying particular attention to the need of special groups (e.g. bottle-fed babies);
- Monitor and keep records; and
- Consider, under the guidance set out for SEMD, applying to the Secretary of State for an emergency relaxation of regulation. The SEMD guidance states that water companies should provide a minimum of 10 litres per person per day within 24 hours of the emergency arising. In large events this may not be possible, in which case the water company would inform the Secretary of State what could be

provided within the 24 hour timeframe and how long before the 10 litres per person can be provided.

3 OTHER KEY AGENCIES

In addition to the water companies, the following agencies would be key interactions during a water contamination emergency:

- The Drinking Water Inspectorate (DWI) will liaise nationally with water companies;
- Environment Agency (EA) would provide environmental advice to water companies;
- DEFRA would act as lead government department for co-ordinate the Government National Response Plan to major incidents;
- MOD would respond to defence related incidents;
- Government Regional Offices; and
- Local Resilience Forums would prove local liaison and links with Social Services, NHS and local businesses.

4 THE WATER UK EMERGENCY & SECURITY FOCUS GROUP

This group, formerly the Water Industry Mutual Aid Group, was formed initially to set up a formal mechanism for industry-wide mutual aid in emergencies. Over the years it has expanded its remit to include security (in the form of a standards sub-group) and to ensure best practice in the field of emergency planning.

Membership of the group is designed to give adequate representation across both water and sewerage companies and water only companies and includes the Scottish and Welsh water companies.

The group owns, updates and publishes a joint document, The United Kingdom Water Industry Emergency Planning & Security Manual. The manual covers the following:

- Conditions of use;
- Contact lists;
- Equipment schedule;
- Standard procedures;
- Commercial companies;
- Laboratory capability; and
- Military aid.

5 COMMUNICATING AN EMERGENCY

The water industry has developed a means of rapid communications for confidential emergency information. The 'CASCADE' procedure covers all emergency communications utilising personal contact and facsimile machines to ensure the information only goes to the intended recipient.

The use of PC access under very strict controls is under development.

In a potential national emergency, when an incident could impact more than one company, the response to media enquiries would be directed to DEFRA as the lead government agency. Water companies would in turn supply frequent situation reports at timings set by DEFRA or at a time of significant change to ensure the Secretary of State has total visibility of the situation.

6 MUTUAL AID

The Water Industry Mutual Aid Scheme is the means by which the resources held across the industry can be made available to a water company in an emergency at any time. It also highlights the limits to which the industry can be expected to help itself.

The scheme requires all water companies to play their part by making available their resources. It is also essential that the register of equipment (found in the Water Industry Emergency Planning & Security Manual) is maintained up to date.

In addition to the scheme, the Water UK Emergency & Security Focus Group have published an agreed protocol which limits the amount of equipment or resources each company would supply. This recognises the current threat situation leading to larger scale or multiple incidents. It also puts into place a dispute mediation mechanism.

THE SCOTTISH WATERBORNE HAZARD PLAN

M. McGuinness

Scottish Water, Castle House, 6 Castle Drive, Carnegie Campus, Dunfermline, KY11 8GG

1. INTRODUCTION

The Scottish Waterborne Hazard Plan was developed as a multi-agency approach to the management of waterborne hazard incidents within Scotland.

2. DEVELOPMENT

The plan was developed via a working group with representatives from Scottish Water, NHS Health Boards, Local Authority Environmental Health and Emergency Planning, Health Protection Scotland, Food Standards Agency and the Association of Chief Police Officers.

3. SCOPE OF THE PLAN

The plan addresses water quality incidents ranging in size from a single property to a large population but it does not include outbreaks of illness, radiological releases from controlled sites, private water supplies or deliberate contamination incidents.

4. STRUCTURE OF THE PLAN

There are seven main sections to the plan
- Initiation
- Investigation
- Risk Assessment
- Risk Management

- Risk Communication
- Annexes
- Appendices

The Annexes provide information and guidance, standard templates for press releases, notices to the public and agendas etc. along with scientific data. Whereas, the Appendices provide incident specific guidance on 23 previously identified common hazards.

5. RISK IDENTIFICATION

The identification of the risk occurs through a number of processes, the initial investigation looks at all relevant past and current information on the systems involved, the kevels and trends of contamination, the sampling and analytical methods being used, the effectiveness of the water treatment process and the storage and distribution characteristics of the supply, this will include travel time through the system and the identification of the number and type of properties affected. This information is gathered by the Recovery Command.

The likely outcomes of any Risk Assessment carried out are, there is a significant risk, there is insufficient evidence to indicate whether or not there is a risk, or there is no risk. At this stage decisions may need to be made on the basis of best available information.

6. RISK MANAGEMENT

The purpose of risk management is to minimise the risk of harm occurring in a population served by the affected water supply. It is not possible to stipulate in advance the nature of mitigation as different options will be appropriate in different situations. Control measures can be put in place if there is an identified risk, or where there is insufficient evidence to decide with confidence and a precautionary approach is favoured. Certain control measures may be exercised in the face of risk such as re-iterating existing advice to vulnerable customers only regarding the safe use of drinking water or issuing general advice to the public on taking action to ensure the safe use of existing water supplies and to major consumers such as food and drinks manufacturers. The advice to customers may be "Do not use for drinking, cooking or washing", "Do not use for drinking or cooking" or "Boil tap water for consumption".

7. INCIDENT MANAGEMENT TEAMS

When Scottish Water becomes aware of a suspected or actual waterborne hazard the Public Health Team will initially contact the local Consultant for Public Health Medicine and the Environmental Health Department. This communication is normally via telephone and the discussions will focus on making an early risk assessment with a view to deciding whether or not there is a risk to Public Health.

In this initial assessment the "virtual team" will consider the initial information on the nature and severity of the problem, the potential population affected, the results of any analyses carried out, both current and historical, the details of any investigations and the possible mitigation measures that may be carried out to minimise any risk to the public. Another consideration at this point in time is whether there is a need to call in additional expertise. Accurate early information is vital.

The decision is then made as to whether there is a need to form an Incident Management Team. This will be formed if there is a clear threat to public health or if it is not possible to determine a conclusive assessment of risk. If the Incident Management Team is to be formed the Scottish Water is responsible for advising and calling in the core members of the group from all organisations. In order for the decision to be made appropriately Asset Operations must clearly identify the area affected, with clearly defined boundaries, determine the Health Board, Local Authority and Police areas affected. If however there is any doubt as to whether an organisation administers within that affected area then they should be included in the first instance.

The core membership of the Waterborne Hazard Incident Management Team is local NHS Boards, the Chair of the group being the Consultant for Public Health Medicine with a press officer from the NHS; a Local Authority Environmental Health Officer; the Public Health Manager, Asset Operations Manager and external communications from Scottish Water; and, if required, a senior police officer will be invited to attend. It is also possible to call on additional members such as Health Protection Scotland, Food Standards Agency, SEPA, divisional veterinary officers and others as thought appropriate.

The Scottish Water representatives on an Incident Management Team will come under the direct command of the Scottish Water Incident Commander. They report directly to him and he is responsible for ensuring that the information required by the Incident Management Team representatives is provided. In large scale incidents the Incident Commander may appoint a liaison link between himself and the staff on the Incident Management Team.

The role of the Recovery Information Co-ordinator is to collect and collate information from the recovery command required by the Incident Management Team to ensure that the information is available in a coherent form. Any information passed to the Incident Management Team has to be approved by the Incident Commander.

The plan requires that a communications sub group be set up. This is normally led by the NHS Board Head of Public Relations. This is because the Incident Management Team is primarily concerned with measures to protect public health. However, the strategy adopted will have a significant impact on Scottish Water's reputation. A Scottish Water Communications sub group member will work closely throughout the incident with the Health Board public relations staff to ensure a consistent message is going out.

Scottish Water is not precluded from issuing press statements relating to operational and legal responsibilities but these must not detract from the public health message.

A view of the internal management team structure can be seen in Figure 1.

8. IN SUMMARY

Scottish Water has found the plan invaluable when incidents do occur. The advantages are obvious, as everyone knows the role they play and have exercised this regularly the plan just swings into place as soon as it is required. Having pre agreed statements and messages that only require the addition of pertinent information saves lengthy discussion at the

Incident Management Team as does having a set agenda. Incident meetings should be brief and to the point and anything that facilitates this is of great value to all parties involved.

Figure 1 *Internal management team structure*

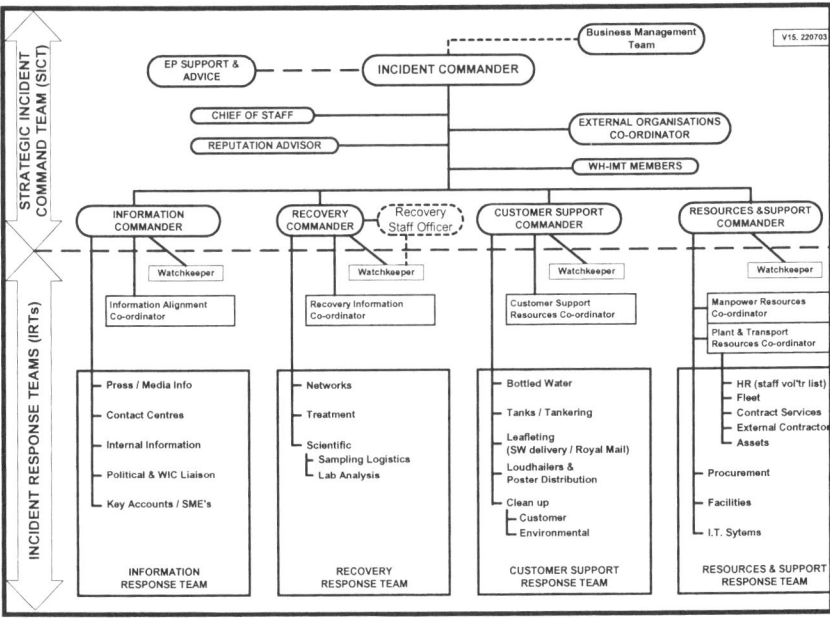

RESEARCH RELATED TO WATER SECURITY

K. R. Fox

National Homeland Security Research Center, U. S. Environmental Protection Agency, 26 W. MLK Drive, Cincinnati, OH 45230, U.S.A.

1 INTRODUCTION

The Public Health Security and Bioterrorism Preparedness and Response Act (Bioterrorism Act) of 2002 is the legislative mandate for the U.S. Environmental Protection Agency's (EPA) work in water security. This law, coupled with executive directives and the Agency's own strategic plan for homeland security, guide the Agency's research and technical support activities to protect the water infrastructure. The Homeland Security Presidential Directive (HSPD) 7, Critical Infrastructure Identification, Prioritization, and Protection, reinforces EPA's role as the sector-specific lead for water infrastructure. It also assigns the responsibility of coordinating the overall national effort to protect critical infrastructure and key resources of the United States to the Department of Homeland Security.

As the sector-specific federal lead for protecting the nation's drinking water and wastewater infrastructures, EPA plays a critical role in the homeland security arena. To meet these responsibilities, the Agency's Office of Water established the Water Protection Task Force. The Task Force was formally organized as the Water Security Division (WSD) in August 2003. Additionally, the Agency's Office of Research and Development (ORD) officially established the National Homeland Security Research Center (NHSRC) in February 2003. These organizations work together in providing research and technical support to the drinking water and wastewater sectors.

NHSRC's Water Infrastructure Protection Division (WIPD) contributes by conducting applied research and then reporting on ways to better secure the nation's water systems from threats and attacks. The Division is producing analytical tools and procedures, technology evaluations, models and methodologies, decontamination techniques, technical resource guides and protocols, and risk assessment methods. All of these products are for use by EPA's key water infrastructure customers — water utility operators, public health officials, and emergency and follow-up responders. Other research programs in NHSRC deal with the protection of buildings and rapid risk assessment.

To better understand the security problems of the water industry in the United States, EPA has engaged with numerous water experts and stakeholders from government, industry, and academia. Other key participants are representatives from public health organizations, emergency responders and follow-up responders, law enforcement officials, environmental groups, and related professional associations.

As a result of these meetings, EPA has gained valuable insights on the vulnerabilities and technical challenges facing the water industry for which research and technical support are crucial. With assistance from other federal agencies and contractors, both WSD and NHSRC are addressing these challenges.

Research being conducted has resulted, and will continue to result, in products that support the water sector in several ways. These products are: 1) strategies to prevent damage from attacks, 2) tools to rapidly detect contamination, 3) techniques to rapidly contain contamination and mitigate contamination impacts, and 4) methods to support rapid recovery following an attack. Each of these areas will be discussed in further detail in this paper.

2 PROTECTING WATER SYSTEMS FROM PHYSICAL THREATS

Physical attacks range from minor acts of vandalism to major incidents that can result in damage to structures such as pumps, pipelines, and storage facilities. These attacks can disrupt service, resulting in loss of potable water to residences and industries, and loss of water for emergency services such as fire control. Disruption in service by the water sector can quickly result in a public health emergency due not only to shortages in drinking water but failure to supply water to critical and interdependent facilities, such as medical facilities and energy plants. A need was identified to prepare updated identification and prioritization of physical threats to, and vulnerabilities of, drinking water infrastructure. This would take into account information gained from vulnerability assessments and from other assessments water systems and their infrastructure.

With support from EPA and the water industry, the American Society of Civil Engineers (ASCE) has developed voluntary design standards for new construction, reconstruction, and retrofitting water facilities with a focus on security. The standards include minimum security standards for SCADA and other computer systems used by the water industry. The design standards are available on the ASCE Web site (www.asce.org/static/1/wise). In addition, EPA is developing recommendations for countermeasures, such as redundant systems to mitigate the impacts of attacks. The Agency has also issued information on establishing collaborations with other water utilities and sectors to share information and resources in preparation for and response to service disruptions.

3 DETECTION OF CONTAMINATION

Unlike a physical attack, the intentional introduction of a biological, chemical, or radiological contaminant into a drinking water system is not always evident. Correctly identifying contaminants after they have entered a drinking water system is extremely important, particularly when they are not apparent through general observation or conventional testing.

When a contamination is suspected, it may not be evident what that contaminant is and the analysis of unknown samples is a problem. Thus a major need identified here was to develop, test, and validate a protocol for the analysis of "unknowns" that is specific to drinking water supplies and systems.

Systematic updating of the analytical modules based on results from improved analytical projects is a continuing research effort. This may result in new analytical hardware and associated laboratory analysis methodologies for biological, chemical and radiological contaminants in water. Sampling and analysis must be standardized and validated within a given set of parameters. A survey and summary of all available sampling, concentration, and analysis techniques (both presumptive and confirmatory) will help meet this need. An inventory of environmental monitoring methods for contaminants is currently being developed by EPA and is being extended to encompass methods applicable to specific water threats. This inventory will: (1) identify information gaps and help to focus the development of reliable methods and hardware, (2) support the identification of needs for improved analytical methods, and (3) produce documented acceptable analytical techniques for water samples.

One example of improved hardware is the development of an improved concentration technique and device for biologicals in water that would be both portable and safe to use by water utility personnel.

In addition to laboratory techniques, EPA is evaluating field detection of contamination. Common water quality sensors are being tested because many water utilities already use them to monitor general water quality and are knowledgeable about their operation and maintenance. It is anticipated that these same sensors will be successful in alerting a utility to intentional contamination events. Experiments conducted to date involved challenging 20 water quality sensors to detect over 25 contaminants injected into the distribution system simulators.

4 TECHNIQUES TO CONTAIN AND MITIGATE CONTAMINATION

Improved distribution system models that can be used to more effectively protect drinking water in the event of deliberate contamination is a critical need identified by the water industry. Distribution systems models need to be better understood and improved for use in preventing, containing, and mitigating contamination of drinking water systems. This includes an analysis of how the models can help when contaminants are injected into a water system. Commercially available and non-proprietary hydraulic models can be used to evaluate water movement in distribution systems. These models can also be used to model how non-hazardous tracer chemicals move through distribution systems. There are several large water utilities that are currently using hydraulic models for a basic understanding of their distribution system, but the use of these models in medium and small water utilities is limited.

The ability to contain a contaminant depends on:
1) how quickly after the event is detected;
2) how fast the contaminant is dispersed in a distribution system;
3) how far the contaminant spreads;
4) how active the chemical contaminant is with disinfection materials; and
5) how effective residual disinfection materials are for inactivating biological

contaminants.

All these questions were asked of the modeling area. Current water distribution system models may help to answer some of these questions. Another area of support is assistance to the regulatory authorities, public health officials, and first responders for tracking the contaminant flow in drinking water attacks.

Additional effort is needed to determine how models can be used to establish optimum locations for in-line monitoring devices or sampling strategies for grab samples. The water industry identified a need for a better understanding of pumping devices, storage facilities, and various connections. The capabilities of hydraulic models to effectively respond to a contamination event needs to be better understood. If in-line monitoring or detection can be networked with distribution system operations, containment or direction of a contaminant within a system may be possible. Hydraulic models would show where to shut down a system, isolate the contamination to prevent it from spreading, or direct the contaminated water to an off-line storage system. Such action during a contamination would minimize exposure and facilitate cleanup.

During an event, when sections of a water distribution system are shut down, the hydraulic models could be used to determine how to get clean drinking water to those pen sections of the distribution system. That same information would be useful in determining where in the distribution system alternative water supplies may be needed during an emergency. During a contamination event, exposure to contaminants will depend on their concentration and the length of exposure.

5 RECOVERY OF WATER SYSTEMS AFTER AN EVENT

Historically when unintentional contamination of water systems have occurred (backflow incidences or cross connection contamination) flushing and/or disinfection practices have been followed. In incidences where the distribution systems have had intrusions of microbial contaminated water, water utility personnel have increased the disinfection residual in the water and then implemented uni-directional flushing to move the microbial contaminants out of the system. Customers of the water utility were advised to boil their water prior to consumption, but other uses were permitted during the cleanup period.

There have been a few documented (and undocumented) chemical contamination events of water systems where some have been intentional and some unintentional. In cases where fire suppression systems or refrigeration systems contaminated drinking water distribution systems (back flow or cross connection) flushing was the technique used to removed the contaminated water and to strip the contaminant from the interior of the pipe walls. There have been a couple of other contaminating events where certain pesticides were intentionally introduced. The bulk of the contaminant was flushed from the system, but residual contaminant remained and in some cases, the material that had been in contact with the contaminant had to be replaced. In another case, mercury found its way into a system and all of the materials that had been in contact with the mercury had to be replaced.

In the event of an intentional contamination of a drinking water system, decontamination of the system would be necessary prior to giving the all clear signal. If the decontamination can be simple flushing, then the expense and time can be minimized. If simple flushing and increase disinfection is not enough to remove the contaminants, then other practices must be developed and implemented. If no practical decontamination technique is available, then replacement becomes the final option. The expense and time parameters increase drastically with increasing difficulty of decontamination.

The US EPA is trying to determine the types of contaminants and treatment/decontamination techniques that may be required during an intentional attack on a drinking water system. There are several research projects that are addressing this need and they include 1) an improved understanding and documentation of the environmental fate of contaminants in source waters, within drinking water plants, and in distribution systems, 2) new and more effective treatment and decontamination technologies and processes for waters that have been contaminated, and 3) an improved understanding and documentation of decontamination and disposal of pipes, equipment, and other materials and of when a decontaminated system can be returned to safe use.

6 SUMMARY

The U.S. Environmental Protection Agency (as the sector-specific federal lead for protecting the nation's drinking water and wastewater infrastructures) is continuing to work with the water industry to protect against and respond to terrorist attacks. The NHSRC will continue to develop tools and information that can be used by Federal, State, Local authorities and water systems to accomplish this difficult task.

EARLY WARNING AND REPORTS

V. Murray

Chemical Hazards and Poisons Division, Health Protection Agency, 7th Floor, Holborn Gate, 330 High Holborn, London, WC1V 7PP, UK

1 INTRODUCTION

This paper presents a summary of the work of the Health Protection Agency and how it relates to other organisations, agencies and stakeholders. Information about chemical incident surveillance, its process and the data relating to chemical incidents involving water contamination is presented. The advantages of the London Early alerting system and how it operates to provide enhanced surveillance is included. Recommendations of how to develop improved early warning and reports are offered.

2 THE HEALTH PROTECTION AGENCY

The Health Protection Agency (HPA) is a non-departmental Governmental Body and its role is to provide an integrated approach to protecting UK public health through the provision of support and advice to the NHS, local authorities, emergency services, other Arms Length Bodies, the Department of Health and the Devolved Administrations. It provides authoritative scientific and medical advice to the NHS and other bodies about the known health effects of communicable diseases, chemicals, poisons and other environmental hazards.

The HPA Centre for Radiation, Chemical and Environmental Hazards comprises the Radiation Protection Division (formerly the National Radiological Protection Board) and the Chemical Hazards and Poisons Division. The headquarters for the Centre is based at Chilton in Oxfordshire.

Every day in Britain, serious chemical incidents occur which threaten people's health. Such potential health threats might involve chemical fires, chemical contamination of the environment, or the deliberate release of chemicals and poisons. Old tyres catch fire releasing clouds of toxic smoke, acid leaks out of a tanker creating noxious gas or an explosion rips through an industrial plant. Exposure to hazardous substances can also occur during accidents at home and work, and as a result of deliberate and malicious releases. In particular it provides clinical information and guidance for health professionals dealing with cases of chemical poisoning arising as a result of a deliberate release.

The Chemical Hazards and Poisons Division (CHaPD) of the HPA provides chemical advice and support. The CHaPD Strategic Goal is *to anticipate and prevent the adverse effects of acute and chronic exposure to hazardous chemicals and other poisons.* This advice covers clinical issues such as:

- personal protective equipment

- decontamination and evacuation

- toxicological and epidemiological advice on impact on public health

- clinical advice on antidotes and medical treatment

- the public health impact of industrial sites

- health effects from chemicals in the environment (including water, soil, waste)

The Chemical five supra-regional teams (Birmingham, Cardiff, Chilton, London, and Nottingham) are located nationwide.

Guidance is available round-the-clock from medical toxicologists, clinical pharmacologists, environmental scientists, epidemiologists and other specialists. The division also advises doctors and nurses on the best way to manage patients who have been poisoned, through a contract with the National Poisons Information Service (NPIS).

The National Poisons Information Service (NPIS) is a clinical toxicology service for health care professionals working in the NHS and is a service commissioned by the HPA. The service consists of a network of units across the UK, providing information and advice on the diagnosis, treatment and management of patients who may have been accidentally or deliberately poisoned. Information on management of poisoning is available on its Internet database TOXBASE (available to registered professionals) or via a 24 hour telephone service for more complex cases requiring specialist advice. These services are not for members of the public; NPIS supports NHS Direct and NHS 24 in providing advice to members of the public.

The Division has identified related research themes for its initial research programme in the area of toxicology (the study of the harmful or toxic effects of chemicals on health). These aim to identify the key gaps in knowledge about the assessment and management of health risks from exposure to hazardous chemicals and poisons. They include:

- Biomarkers of exposure or toxic effect, with the aim of identification of robust markers, such as the presence of the chemical in human urine or blood, following exposure to hazardous chemicals, will enable much better health risk assessments following chemical incidents. This work is a priority for the HPA and its partner organisations and is a long-term undertaking in view of the current extent of the knowledge gap.

- the long-term consequences of low level, chronic exposure to chemicals and poisons are currently not well understood and there is increasing public concern about the possible impact, especially in relation to reproductive health, asthma and cancers.

Areas of specific interest for those interested in water contamination incidents include:

- Compendium of Chemical Hazards: CHaPD is producing a series entitled a *Compendium of Chemical Hazards.* The style and content have been developed following extensive joint working across the HPA and the chemicals chosen have been selected following an audit of queries.

- Children's Environment and Health Action Plan for Europe Children's Environment and Health Action Plan for Europe (CEHAPE) is an initiative led by the World Health Organization Regional Office for Europe (WHO Euro). It was launched in June 2004 and signed by all the member states (currently 53 in number), including the UK. The aim of the CEHAPE is to protect the health of children and young people from environmental hazards.
- Chemical Incident Surveillance: Chemical incidents need to be detected early so that effective action can be taken to protect public health. Response and surveillance systems are therefore an essential component of the Agency's role, along with research to evaluate the immediate and longer-term health effects of exposure.

3 CHEMICAL INCIDENT SURVEILLANCE

The principles of public health surveillance of chemical incidents are summarised in Figure 1.

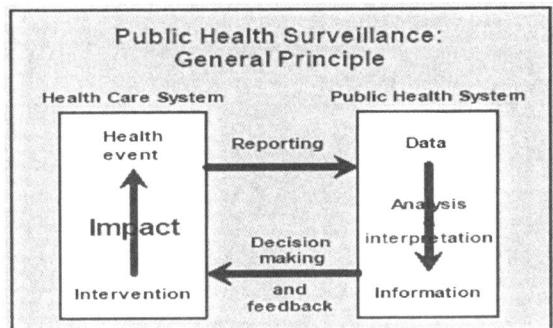

Figure 1 *The "surveillance wheel" from*
http://www.whoint/om/slideshows/nationalsystem

The aims of the CHaPD public health system for monitoring chemical incidents are:
- To describe the frequency and characteristics of chemical incidents in the UK, including identifying the chemicals most commonly involved.
- To describe the effects on the population's health resulting from chemical incidents in the UK.
- To compare rates of incidents and identify trends in the different regions to better prepare and respond, or to evaluate changes in policy.
- To create networks for collaboration and sharing of expertise.
- To provide the mechanism for disseminating lessons learnt in enhancing preparedness, timeliness and appropriateness of response

The Division manages a chemical surveillance system (CISS) for England and Wales. This report provides a summary of the characteristics and distribution of chemical incidents recorded in England and Wales during 2006 and 2007. The definition of a chemical incident is summarised.[1] During 2006 and 2007 a total of 2063 incident were reported with 1571 (74%) as classed as actual, with 377 (18%) as potential and 165 (8%) for information. All the incidents are analysed below with the subset of water related chemical incidents being specifically analysed separately.

Table 1 shows the regional distribution of chemical incidents and rate per 1,000,000 populations. The regional distribution of chemical incidents for 2006 and 2007 are shown in Figure 2.a with figure 2.b showing the regional distribution of water related chemical contamination incidents for 2006 and 2007. It is apparent in both these sets of data that London region dominates for the number of incidents reported. Of note, no water related chemical incidents were reported in Yorkshire and Humber and North West regions.

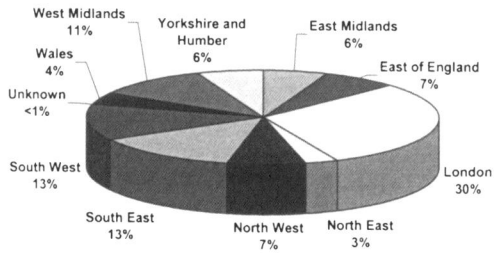

Figure 2a *Regional distribution of chemical incidents for 2006 and 2007*

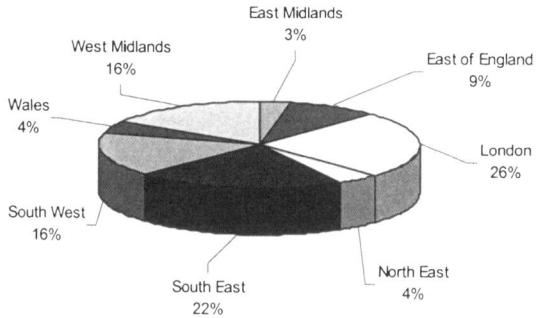

Figure 2b *Regional distribution of water related chemical contamination incidents for 2006 and 2007*

---

[1] Definition of a chemical incident: All incidents representing *"an acute event in which there is, or could be, exposure of the public to chemical substances which cause, or have the potential to cause ill health"* should be included in the National Database. All incidents with an off-site impact are to be included, as well as on-site incidents where members of the public are affected. (For the purposes of the definition, hospital staff and emergency services personnel should be regarded as members of the public).

Table 1 *Regional distribution of chemical incidents and rate per 1,000,000 population*

Geographical Region	Total number of incidents	Rate per 1,000,000 population
North East	64	24.6
North West	149	21.6
Yorkshire & The Humber	126	24.7
East Midlands	121	27.5
West Midlands	219	40.6
East of England	143	27.0
London	625	83.3
South East	272	33.2
South West	260	51.0
Wales	82	27.3

The monthly distribution of chemical incidents for 1st January 2006 to 31st December 2007 is shown in Figure 3. No significant variation is shown except that the last six months of each year appears to be busier that the first six months.

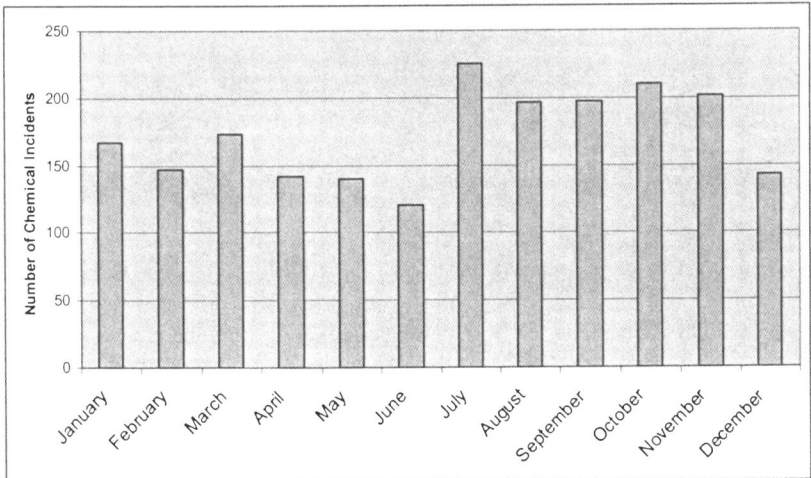

Figure 3 *Monthly distribution of chemical incidents for 1st January 2006 to 31st December 2007*

The location of chemical incidents in England and Wales are shown in Figure 4.a with the water related incidents in Figure 4.b

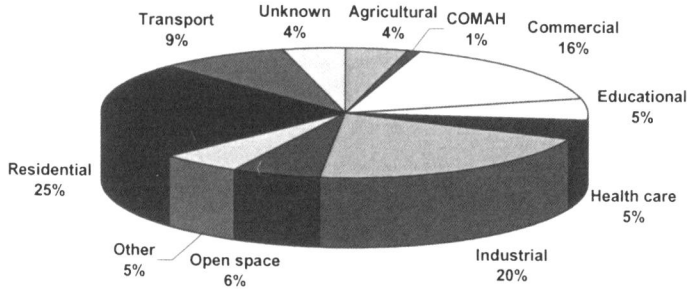

Figure 4.a *Location of chemical incident in England and Wales*

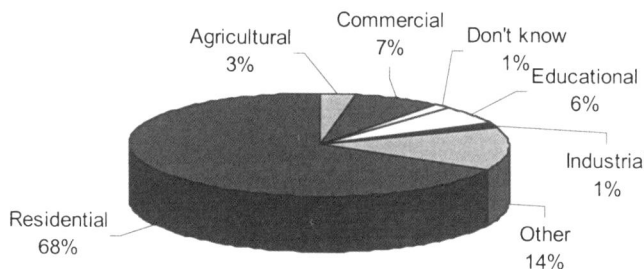

Figure 4.b *Location of water related chemical incident in England and Wales*

Chemicals associated with individual chemical incidents in 2006 and 2007 are shown in Figure 5.a with those associated with water related incidents shown in Figure 5.b. It is apparent that metals are more frequently reported in water related incidents at 23% of incidents than in all incidents at 6%. Equally pesticides are reported at 6% in water related incidents with total incidents at 1%, whereas petroleum oils are reported in 6% of water incidents but occur in all incidents at 4%. This data reflects the types of incidents that CHaPD are requested to provide advice and support for water related events.

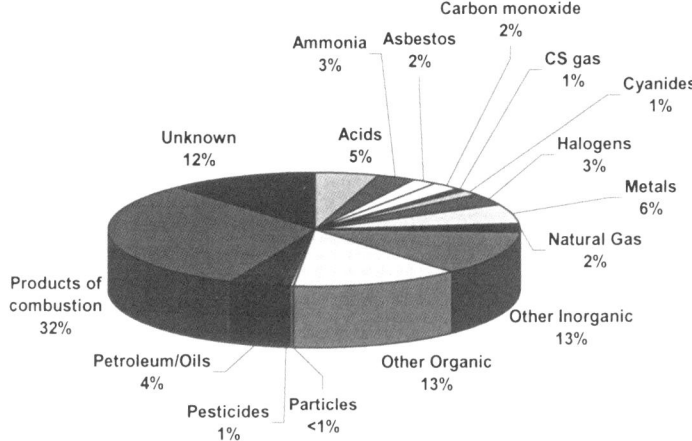

Figure 5.a *Chemicals associated with chemical incidents 2006-2007*

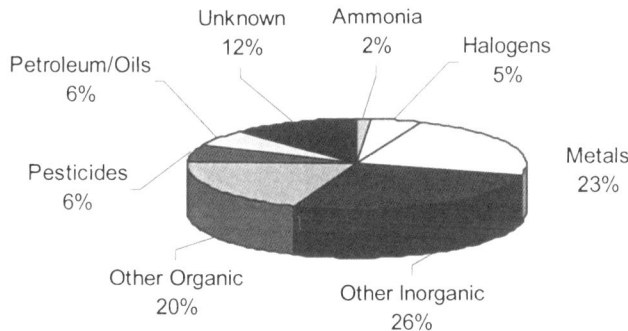

Figure 5.b *Chemicals associated with water related chemical incidents*

In order to determine the source of notification the data was analysed by group of notifying organisations (figure 6). Of significance is the wide range of organisations who report chemical incidents to HPA. The main notifiers to CHaPD are our Local and Regional Health Protection Units who are working at the front line. Water companies report only 1% of our incidents but most of these reports are channelled via the Drinking Water Inspectorate. Water companies may well contact local Health Protection Units first as the HPA protocol for notification recommends and then the enquiries may require expert advice and support form CHaPD as required.

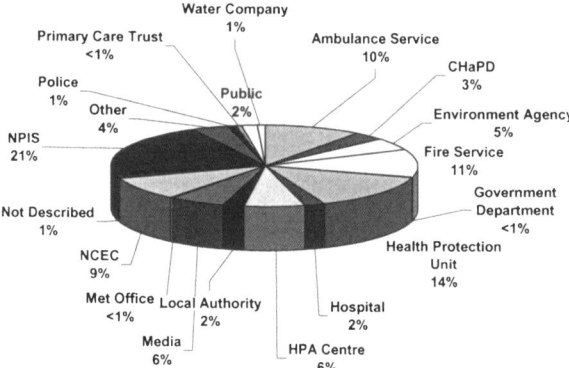

Figure 6 *Notifying organisation of chemical incidents for the reporting period 01/01/2006 – 31/12/2007*

4 LONDON EARLY ALERTING

In 2004 a project was undertaken to review and compare, in retrospect, chemical incidents that occurred in London and those actually reported to CHaPD, London. Visits were made to London Ambulance Service (LAS) and London Fire Brigade (LFB) and information obtained from visits used to review chemical incident reporting in 2003. Concern was generated at the findings are shown in figure 7.

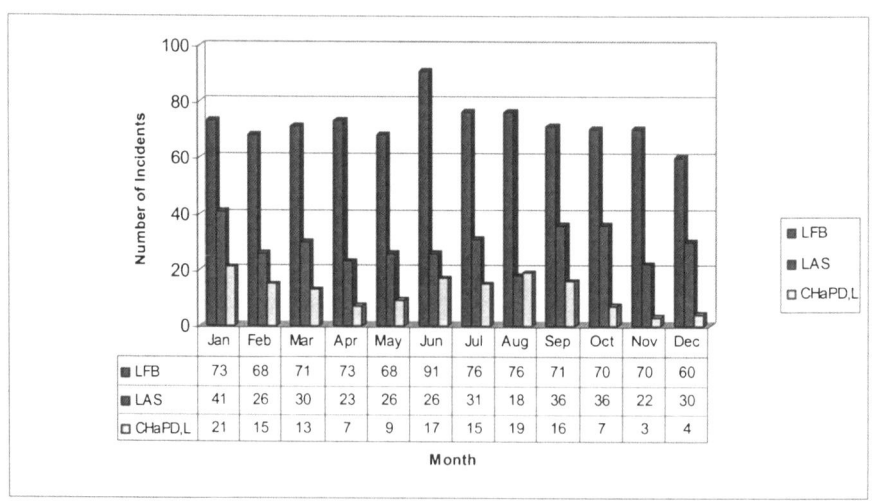

Figure 7 *Results of 2003 retrospective review (Comparison of chemical incidents recorded by London Ambulance Service, London Fire Brigade and CHaPD, London in 2003 Chemical Hazards and Poisons Report January 2005)*

The 2003 data showed that number of incidents reported to HPA is limited in comparison to numbers reported to LFB and LAS and analysis provided strong evidence that a chemical incident Early Alerting system should be implemented in London. The aims of London Early Alerting System are to:

- result in more timely public health response;
- improve the ability to act more quickly to minimise adverse health effects and to assist in saving life;
- improve the ability to help to protect NHS resources; and
- have the ability to share London Early Alerting data with stakeholders.

The chemical incident early alerting system for London was established in consultation with many stakeholders:

- London Ambulance Service (LAS);
- London Fire Brigade (LFB);
- HPA Health Emergency Planning Advisors (HEPA);
- National Security Advice Centre (NSAC);
- Environment Agency (EA);
- Drinking Water Inspectorate (DWI);
- HPA London Regional Epidemiology;
- Guy's and St Thomas' Poisons Unit (GTPU);
- Chemical Hazards and Poisons Division, HPA (CHaPD); and
- Department of Health (DH).

This system has been set up to cascade early alert in the sequence shown in Figure 8. However, all may contact CHaPD London. The management of the calls is closely supported by the Health Emergency Planning Advisers who also provide 24 hour cover for London.

From the audit undertaken in November 2006, notification of chemical incidents to CHaPD London improved (at least 86% of chemical incidents attended by LFB and 100% of those attended by LAS). A wide range of chemical incident types occur in London on a daily basis. CHaPD is able to notify other agencies and professionals of these incidents to provide real time incident management advice and also to provide written fact-sheets, checklists and other forms of guidance to health care and emergency service personnel.

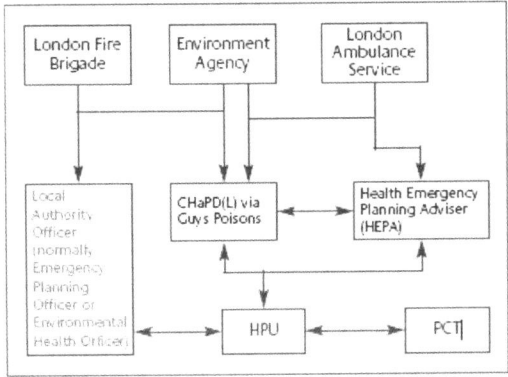

Figure 8 *The chemical incident early alerting system for London*

A stakeholder interview survey was also undertaken. The conclusion from this survey included a series of recommendations which included:

1. The London chemical incident early alerting system should be recognised as a core function for CHaPD, London.
2. Aims and objectives for the system should be updated, against which the system can be better evaluated in future.
3. Terms of reference for the early alerting stakeholder group should be updated.
4. Incident capture should be improved by broadening membership of the early alerting stakeholder group.
5. Suggested projects for CHaPD to pursue, to improve reporting include:
   - Pilot Emergency Department reporting
   - NHS Direct chemical incident recognition project
   - Continue work with local authorities to improve reporting
6. Communication systems should progress to allow for the dissemination of information updates from all sources, as incidents progress.
7. More detailed feedback is recommended regarding incident outcomes in real-time and for long term surveillance.
8. A fit-for-purpose, managed, secure, early alerting system database should be developed to fulfil criteria set out by participating organisations.
9. The delivery of chemical incident training in London should be reviewed, in line with emerging threats and the need to reach key operational staff.

5 CONCLUSION

The HPA provides an integrated approach to protecting UK public health through the provision of support and advice to the NHS, local authorities, emergency services, other Arms Length Bodies, the Department of Health and the Devolved Administrations. It provides authoritative scientific and medical advice to the NHS and other bodies about the known health effects of communicable diseases, chemicals, poisons and other environmental hazards.

The CHaPD manages a chemical surveillance system (CISS) which provides a summary of the characteristics and distribution of chemical incidents recorded in England and Wales. In addition the enhanced the London Early Alerting System clearly shows the benefit of early warning and reports with improved links with emergency services and other organisations involved in chemical incidents, the ability to ascertain which incidents are/are not being reported to CHaPD, London, links with emergency services to update on developments with Detection, Identification and Monitoring equipment and the ability to share information and share best practice for incident response.

The HPA is therefore committed to early warning and reporting with a multi-agency and multi-disciplinary approach which is part of the Civil Contingency Act requirements.

6 ACKNOWLEDGEMENTS

The author would like to thank colleagues within CHaPD and in particular Dr Patrick Saunders, Lorraine Stewart, Dr Ruth Ruggles, Dr Richard Mohan and Dr Giovanni Leonardi for their assistance and support.

References

1 Health Protection Agency
http://www.hpa.org.uk/webw/HPAweb&Page&HPAwebContentAreaLanding/Page/11538
22623851?p=1153822623851 (accessed 17.06.08).

2 Health Protection Agency. Centre for Radiation, Chemical and Environmental Hazards.
http://www.hpa.org.uk/webw/HPAweb&Page&HPAwebAutoListName/Page/1199451944
035?p=1199451944035 (accessed 17.06.08).

3 Health Protection Agency. Chemical Hazards and Poisons Division.
http://www.hpa.org.uk/web/HPAweb&Page&HPAwebContentAreaLanding/Page/1153386
734384 (accessed 17.06.08).

4 Health Protection Agency. Chemical Hazards and Poisons Division. Deliberate Release
- Information for Health Professionals - Chemical Agents.
http://www.hpa.org.uk/web/HPAweb&HPAwebStandard/HPAweb_C/1195733813305
(accessed 17.06.08).

5 Health Protection Agency. Chemical Hazards and Poisons Division.
http://www.hpa.org.uk/webw/HPAweb&Page&HPAwebAutoListName/Page/1207293977
308?p=1207293977308 (accessed 17.06.08).

6 Health Protection Agency. Chemical Hazards and Poisons Division. National Poisons
Information Service.
http://www.hpa.org.uk/web/HPAweb&Page&HPAwebAutoListName/Page/115893460758
3
(accessed 17.06.08).

7 Health Protection Agency. Chemical Hazards and Poisons Division. National Poisons
Information Service TOXBASE.
http://www.hpa.org.uk/web/HPAweb&Page&HPAwebAutoListName/Page/115893460758
8 (accessed 17.06.08)

8 Health Protection Agency. Chemical Hazards and Poisons Division. Research and
Development Priorities.
http://www.hpa.org.uk/web/HPAweb&HPAwebStandard/HPAweb_C/1195733752652
(accessed 17.06.08).

9 Health Protection Agency. Chemical Hazards and Poisons Division. Compendium of
Chemical Hazards.
http://www.hpa.org.uk/webw/HPAweb&Page&HPAwebAutoListDate/Page/11538466734
55?p=1153846673455 (accessed 17.06.08).

10 Health Protection Agency. Chemical Hazards and Poisons Division. Children's
Environment and Health Action Plan for Europe.
http://www.hpa.org.uk/web/HPAwebFile/HPAweb_C/1194947328059 (accessed 17.06.08)

11 Health Protection Agency. Chemical Hazards and Poisons Division. Chemical Incident
Surveillance.
http://www.hpa.org.uk/webw/HPAweb&Page&HPAwebAutoListName/Page/1158934607
635?p=1158934607635 (accessed 17.06.08).

12 G Leonardi, SURVEILLANCE, Chemical Incident Report, July 2002, 24-30.
http://www.hpa.org.uk/web/HPAwebFile/HPAweb_C/1194947350265 (accessed
17.06.08).

13 Health Protection Agency. Chemical Hazards and Poisons Division. Chemical
Incidents Surveillance Review: January 2006 - December 2007.
http://www.hpa.org.uk/web/HPAwebFile/HPAweb_C/1211184033548 (accessed 17.06.08)

14 L. Stewart, Personal communication, March 2008.

15 Health Protection Agency. Local and Regional Services.
http://www.hpa.org.uk/webw/HPAweb&Page&HPAwebContentAreaLanding/Page/11538
22623816?p=1153822623816 (accessed 17.06.08).
16 R. Paddock and V. Murray, Early Alerting: The Future for chemical incident reporting.
Chemical Hazards and Poisons, 2005, Report 3, 22-24.
http://www.hpa.org.uk/chemicals/reports/chapr3_jan2005.pdf (accessed 17.06.08).
17 R. Cordery, R. Mohan and R. Ruggles, Evaluation of the London Chemical Incident
Early Alerting System: (1) an audit of chemical incident reporting in London, two years on.
Chemical Hazards and Poisons Report, May 2007, 11-14.
http://www.hpa.org.uk/web/HPAwebFile/HPAweb_C/1197382231164 (accessed 17.06.08).
18 R. Cordery, R. Mohan and R. Ruggles, Evaluation of the London Chemical Incident
Early Alerting System: (2) stakeholder interview study Chemical Hazards and Poisons
Report, May 2007, 15-18.
http://www.hpa.org.uk/web/HPAwebFile/HPAweb_C/1197382231164 (accessed 17.06.08).

OK, WE'VE GOT A PROBLEM, SO WHO DO WE TELL? INTER-AGENCY COMMUNICATIONS – A WATER COMPANY VIEW

D. A. Woolloff

Manager of Water Quality, Yorkshire Water Services Ltd. Western House, Halifax Road, Bradford, BD6 2LZ, England

1 INTRODUCTION

Whether it be due to deliberate act, a natural catastrophe or equipment failure, any event which results in the contamination of public water supplies and threatens to harm public health has the potential to severely damage public confidence in the supplying water company, its reputation, share value and ultimately staff morale. Managers will have to deal with a huge variety of issues and at the same time demonstrate a 'duty of care' to both staff and customers. Fortunately such contamination events are uncommon, but their very rarity can, in itself, be one of the most important factors hindering an individual manager's preparedness for the unthinkable.

During a serious incident, managers will be required to transition from their normal comfort decision making processes, dealing with the day to day events in the confines of their own organisation, into an environment where other organisations have a significant input or even overall control. Such a transition, at least for the unprepared, can lead to a "Crisis of Control" whereby the decisions that quite rightly should be taken by the company may appear to be moving out of the company's hands.

Furthermore, communicating risk to and the subsequent reassurance of public audiences in the event of a water contamination incident is a challenge fraught with difficulties, where any battle between facts and emotions almost invariably results in the emotional aspects winning the day at the expense of truth and logic. Engineers and scientists tend to base communications on facts, yet the emotional aspects of a risk situation, especially when reported by the media, can be much more dominant and more difficult for the scientifically-minded to accept.

2 CRISIS OF CONTROL?

Water supply networks are by their very nature dynamic whereby changes are required on a regular basis to ensure varying consumer demands are met. Operational managers within the Water Industry are well versed in event management using systems and procedures to accommodate planned and unplanned asset outage whilst maintaining the supply-demand balance.

The regulatory obligations of the Water Industry Act 1999[1] and the various Regulations[2,3,4,5,6,7] made under its auspices are well established and well understood by practitioners within the industry, whereas, although familiar to the Industry's Emergency Planners, the Civil Contingencies Act 2004[9] and the Contingency Planning Regulations 2005[10] are not necessarily well understood by Operational and Scientific staff. Nor is the plethora of Guidance associated with them.

Furthermore, water is a unique foodstuff in that it is delivered direct to the customer and in many cases may be consumed within hours of its leaving the water treatment works. Any event involving contamination of the public water supply has to be treated as an immediate threat to public health and will involve the Health Protection Agency and the Department of Health.

The challenge facing the water company is that an escalating incident will rapidly move from one which is managed and controlled from within the confines of the company's own organisation into a multi agency response incident managed under the "Gold", "Silver" and "Bronze" structures under the overall control of the emergency services. At this point the water company becomes a Category 2 (or "co-operating") responder and is no longer in direct control.

Despite no longer being in direct control, as a Category 2 responder a water company participates in a local resilience forum on a "right to attend, right to invite" basis and should be engaged when it can add value. During an emergency affecting its operations it retains control of its operations and still has to meet its own legal obligations. Water companies should not expect to be the lead body in any incident but should concentrate on their core competencies providing engineering solutions and customer information within a Category 1 lead command structure.

3 CRISIS OF CONFIDENCE?

The potential impact of a contamination incident upon the reputation of the water company involved and the water industry as a whole cannot be overstated. The organisation's reputation is often its most valuable asset and when that reputation comes under attack, protecting and defending it becomes the highest priority yet there is also a moral dimension to incident management. Irrespective of any potential risks to health, water companies are expected to be risk averse and to act in ways that reinforce this public expectation.

In any incident, no matter how large or how small, managing and directing information flow is essential if water companies are to discharge their moral obligations yet maintain credibility with all their various stakeholders who want, need, and deserve to know the risks involved and what can be expected.

Key questions for the water company are exactly what information to release, to whom it should be released and when it should be released. In many water systems the water will have been used before the results of formal laboratory analysis are available and hence unless there is clear physical evidence that contamination has taken place it will always be a judgement call when to "go public". Questions to be considered in determining whether or not early notification is appropriate include balancing the risk of false alarms and unnecessary public concern against the very real risk of illness if warnings are delayed, including the extreme case of warnings no longer being heeded should a company "cry wolf" once too often. Here the water company will benefit from having a close working relationship with their Health Protection Units. Discussions on a very regular basis will provide a robust platform to manage a developing incident

4 BUSINESS CONTINUITY AND EMERGENCY RESPONSE PLANS

Whenever an incident has the capacity to cause a significant disruption to an organisation by either interrupting its normal mode of operation or by impacting on its reputation, a response plan can help minimize the disruption. A well defined and well understood set of Business Continuity or Emergency Response Plans will help scientific and technical managers overcome the Crises of Control and Crises of Confidence issues that can arise when managing within a complex control environment. The best Plans will recognise that the priorities of the front line emergency responders may well differ from those of the water company yet will still provide a sound framework to enable a Company's own front line staff to respond appropriately.

Preparing contingency plans in advance, as part of an overall crisis management or business continuity plan, is the first step to ensuring an organisation is appropriately prepared to deal with the unexpected.

Communication is critical during an emergency and needs to be addressed thoroughly within the plan to ensure that the organisation rapidly mobilises response teams, provides guidance and instructions to employees, and communicates with appropriate authorities and external stakeholders. Companies also need a way for external stakeholders to call in to provide information, as well as receive it. In order to maintain credibility, the company must "speak with one voice" even if different individuals are involved. A plan that does not address the communications issues is a plan destined to fail at an early stage of the incident and, therefore, in addition to providing detailed guidance to front line staff in the field it should:-

a) identify who should represent the organisation at the various levels of multi agency control team; and
b) identify who are authorised to speak publicly about the situation, whether this be a designated spokesperson or specified members of the incident response or crisis control teams.

The first hours after a crisis breaks are the most crucial, so working with speed and efficiency is important, and the plan should indicate how quickly each function should be performed. Delays in communication can be particularly damaging as rumour will inevitably fill the gap left by an absence of information and once started, rumours can be almost impossible to control. One of the most challenging aspects of communication management is reacting - with the right response – and reacting quickly. A good communications plan should include template style scripts for the most common events and if possible include anticipated questions (and answers). Templates and the "Question and Answer" scripts should be regularly vetted by appropriately qualified technical staff to ensure that they are up to date and factually correct and also by the organisation's Legal Advisor to ensure that any statements made do not provide ammunition for possible future litigation.

Finally, simply developing a series business continuity and emergency response plans and placing them on the bookshelf is not the answer. It is crucial that the plan is regularly tested and exercised in conjunction with other key organisations likely to be involved in a major incident such that gaps, oversights and weaknesses are exposed in the relatively safe exercise environment. Gaps or errors in plans will lead to communication failures and exercising a plan can reveal these before the organisation has to rely on it in a real emergency. Even the best produced plans, refined by thorough testing, still need to evolve as the business evolves and regular ongoing exercising will help ensure that the plan is ready when needed. Involving employees at all levels in the exercising process is crucial in ensuring they are familiar with the plan as no matter how good the plan, it will fail if key personnel are not familiar with its operation.

5 KEYS TO EFFECTIVE COMMUNICATION

When preparing to offer a statement externally as well as internally, information must be accurate. Providing incorrect or manipulated information has a strong tendency to backfire and will only make matters worse whereas timely, accurate and sympathetically delivered messages can help reduce damage to an organisation's credibility or even enhance it.

Sadly, many technical managers can appear cold, arrogant, unfeeling, and corporately driven and run the risk of ruining the best prepared statement through poor delivery. The value of good media training cannot be overstated. When it goes wrong, say so and apologise rather than attempt to defend the indefensible. Furthermore, many people find the concept of percentages difficult to grasp and hence a risk of '1 in 1,000' is more often easily understood than '0.1%'. It is crucial that scientific and technical statements avoid the use of percentages especially when discussing risk associated with an incident. It is far less emotive to state that "risk rises from one case in 1,000,000 to two cases in 1,000,000" than the equally correct "there is a 100% increase in risk."

Finally, the growth of the internet has resulted in an increase in the overall understanding of scientific research and people are often no longer satisfied with reassuring statements by officialdom. Today customers tend to want to know how the research has been funded, and who stands to benefit which leads to one of the most fundamental barriers to effective communication of risk – the 'well, they would say that' factor. This, however, can be overcome by getting third party endorsement from a credible source. The very best plans should include strategies on engaging the media to accurately communicate messages and context as well as identifying and recruiting credible third-party allies who can support the communication process.

6 CONCLUSION

Events such as droughts, floods and power outages and the like have shown that interruptions to our businesses, not to mention our daily lives, are never far away. Even the best thought out response plans are inadequate if they are not updated and exercised regularly and do not include effective means of communicating to the emergency responders, customers, stakeholders and the public at large.

REFERENCES

1 Water Industry Act 1991 (c.56).
2 The Water Supply (Water Quality) Regulations 2000, SI 2000 No. 3184.
3 The Water Supply (Water Quality) (Amendment) Regulations 2001, SI 2001 No. 2885.
4 The Water Supply (Water Quality) Regulations 2001 (Wales), SI 2001 No. 3911.
5 The Water Supply (Water Quality) 2000 (Amendment) Regulations 2007, SI 2007 No. 2734.
6 The Water Supply (Water Quality) 2000 (Amendment) Regulations 2007 (Wales), SI 2007 No. 3374.
7 The Water Undertakers (Information) Direction 2004.
8 The Security and Emergency Measures (Water and Sewerage Undertakers) Direction 1998.
9 The Civil Contingencies Act (c.36).
10 The Civil Contingencies Act 2004 (Contingency Planning) Regulations 2005, SI 2005 No.2042

REVIEW AND EVALUATION OF WATER CONCENTRATION TECHNOLOGIES FOR ANALYSIS BY REAL-TIME PCR

S.L. Cunningham[1], B.M. Dowling[2]

[1] Dstl Porton Down, Salisbury, Wiltshire SP4 0JQ, UK
[2] Health Protection Agency, Newcastle laboratory, Newcastle General Hospital, Westgate Road, Newcastle upon Tyne NE4 6BE, UK

1 INTRODUCTION

The water industry routinely screens drinking water supplies for common chemical and microbiological contaminants according to the Water Supply (Water Quality) regulations 2000[1] and amendments. The methods currently employed rely on bacterial culture and can take up to several days to produce a result. Following the invention of the polymerase chain reaction (PCR) in the mid-1980s[2], assays can be designed to detect specific genomic information of target micro-organisms. PCR assays can be used in addition to abbreviated culture methods[3] or on water samples directly[4], negating the requirements for culture altogether.

The Defence Science and Technology Laboratory (Dstl) Porton Down and the Health Protection Agency (HPA) Newcastle laboratory have collaborated in research to develop rapid methods for the concentration of biological agents from large volume water samples for analysis by real-time PCR. Having previously developed a method for analysing water directly to obtain a result within the working day (8 hours), further research was desirable to investigate improvements to the method. An extensive review of the literature was undertaken to identify candidate water concentration technologies for evaluation. The review considered a broad range of technologies which were divided into primary and secondary water concentration techniques, commercially available water filtration and DNA extraction kits and emerging technologies.

Primary concentration techniques were considered to be those that enabled samples of one litre or more to be reduced in volume to between 1 and 10 mL. The majority of commercially available concentration techniques available were filtration based, i.e. traditional flat membrane filters, depth filters, hollow fibre filters, gas pressure filters and tangential flow filtration. High speed centrifugation has also been utilised[5]. Secondary concentration techniques further concentrated water samples to volumes of less than 1 mL which are more suitable for DNA extraction and PCR. The secondary concentration techniques considered were centrifugal filtration[6], solid phase extraction[7] and immunomagnetic separation[8]. Emerging technologies still undergoing development included continuous channel centrifugation[5], ultrasonic[9] and dielectrophoretic separation[10] and DNA extraction by sonication[11] and electroporation. Commercially available filtration kits for PCR were also investigated.

PCR uses temperature cycling to produce copies of specific target nucleic acids using a pair of oligonucleotides, known as primers, which are complementary to specific regions of a unique gene in the target organism's genome. Real-time PCR makes use of fluorescent dyes to follow the exponential amplification of the target DNA as it happens. In the TaqMan® PCR chemistry, a third oligonucleotide, called a probe, is used. The probe has a fluorescent dye (known as the reporter dye) conjugated at one end and a second dye that can "quench" the fluorescence of the reporter dye is conjugated to the other end. The probe hybridises to a unique DNA sequence between the 3' ends of the primers, which increases the specificity of the assay as three hybridisation events have to occur. Copies of DNA are made by the DNA polymerase enzyme adding nucleotides to the 3' end of the primers and the subsequent nascent strand. When the enzyme encounters the hybridised probe, its 5'-3' exonuclease activity hydrolyses the probe, liberating the reporter dye from the proximity of the quencher with a resultant increase in fluorescent signal. Assays are optimised with an excess concentration of the probe to ensure that probe hydrolysis occurs with each new copy of DNA. The level of fluorescence therefore increases in real-time as the reaction progresses.

2 METHOD AND RESULTS

2.1 Technologies Chosen from the Review for Experimental Evaluation

2.1.1 Primary Concentration Techniques. Standard vacuum assisted membrane filtration as used by routine water testing laboratories[12], with replacement of 47 mm membranes by 25 mm membranes. The filter substrates identified for testing were mixed cellulose esters (MCE) membranes (Millipore, France), polyethersulphone (PES) membranes (low protein binding; Millipore, France) and NanoCeram alumina fibre depth filters (Argonide Corp., USA). Vivaflow 50 tangential flow filtration cassettes were also evaluated (Sartorius, Germany).

2.1.2 Secondary Concentration Techniques. Microcon centrifugal spin columns (500 µL capacity; Millipore, France), for DNA concentration and VivaSpin 20 centrifugal filtration devices (15 mL – 20 mL capacity; Sartorius, Germany)) for use following sample concentration with tangential flow filtration cassettes.

2.1.3 Commercially Available Water Filtration and DNA Extraction Kits. UltraClean™ Water DNA Isolation kit (MoBio, available from Cambio, Cambridge, UK), AquaScreen® *FastExtract* kit for conventional filtration systems (Minerva Biolabs available from Cambio, Cambridge, UK) and AquaScreen® filtration and extraction kit (Minerva Biolabs, available from Cambio, Cambridge, UK).

2.2 Dstl Methods for the Evaluation and Results

2.2.1 Primary Concentration Techniques. The evaluation of the chosen technologies was carried out using a surrogate for the waterborne pathogens of interest. The chosen organism, *Burkholderia cepacia* ATCC 17759 (type strain), is an ACDP hazard group 2 organism which has been demonstrated to have good survivability in water[13]. A TaqMan® PCR assay for *B. cepacia* had previously been designed at Dstl.

The assay demonstrated good correlation with plate counts as illustrated in figure 1. The bacterial cell count correlated well with cycle threshold (ct) value, the value used to determine the amount of DNA present in a positive PCR result.

The sample concentration technologies were treated according to one of two protocols: 1) initial primary concentration followed by elution of bacterial cells captured on the membrane into 1 mL of a proprietary Chelex resin for DNA extraction; 2) as at 1 but with a secondary concentration of the extracted DNA. This scheme for sample concentration and DNA extraction was adapted from a well characterised method used by Dstl and HPA to analyse environmental samples. The vacuum filtration method was similar to that used in routine water testing laboratories[12], allowing for the simultaneous filtration of three samples and using 50 mL syringe bodies as the sample reservoir. PCR assays were performed on the LightCycler PCR instrument (Roche) with assays that use TaqMan® hydrolysis probes.

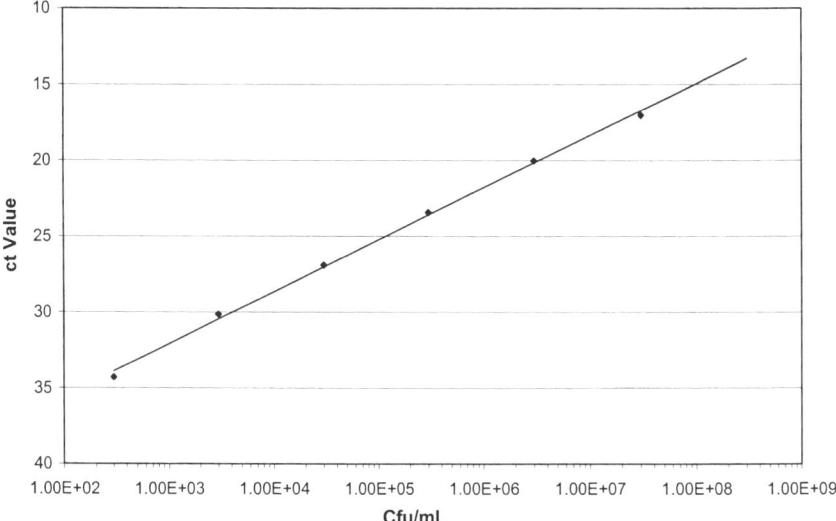

Figure 1   *The relationship between cell numbers of B. cepacia in a serial dilution of an overnight culture and the ct value assigned by the LightCycler when the serial dilutions were assayed using the RecA real-time PCR assay [threshold set to 0.5 units].*

Evaluation of MCE, PES and NanoCeram© Alumina Fibre Membranes was achieved using 100 mL water samples inoculated with 100 μL aliquots from 10-fold serial dilutions of an overnight culture of *B. cepacia*. The samples were enumerated using spread plates of 100 μL aliquots on L-agar prior to filtration and 10 μL aliquots were assayed by PCR, both before and after filtration.

MCE membranes are used routinely in the water industry and by Dstl and HPA for environmental sample testing[12]. 100 mL water samples inoculated with *B. cepacia* were filtered through these membranes. The results, obtained from duplicate filtration runs, following DNA extraction and PCR for pre- and post-filtration aliquots, are shown in figure 2. The lower limit of detection (LLOD) of the assay system, pre-filtration, was 2×10^3 colony forming units (cfu) mL^{-1}. Following filtration the sample containing 20 cfu mL^{-1} yielded a positive PCR result. This equates to a 100-fold improvement in the LLOD following filtration. Filtration resulted in the eluate from a 100 mL water sample being recovered in 1 mL of DNA extraction reagent, therefore the theoretical maximum concentration factor was 100-fold. Comparison of the detection limits pre-and post-filtration indicates 100% efficient capture and recovery of bacterial cells.

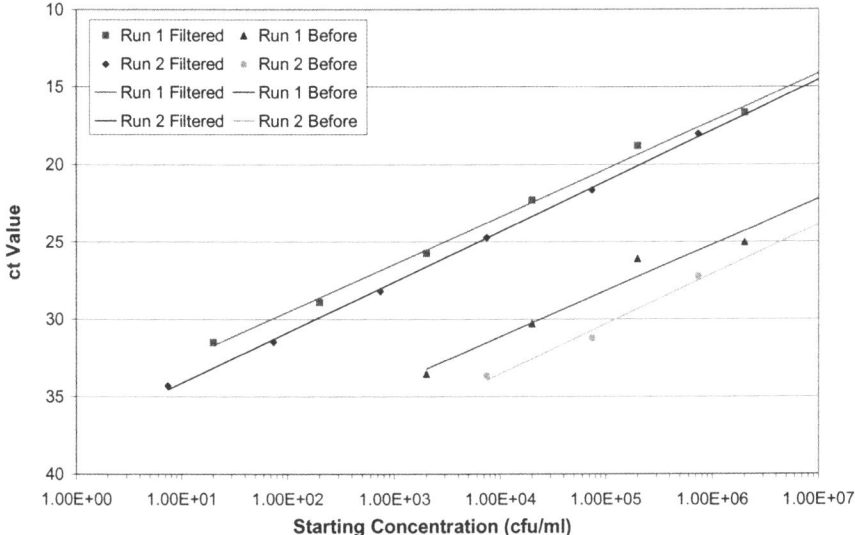

Figure 2 *PCR results from 100 mL water samples seeded with B. cepacia before and after filtration and recovery from mixed cellulose esters membranes [threshold set to 0.5 units].*

Polyethersulphone membranes were tested in the same way. The results from duplicate filtration runs are shown in figure 3. The lower limit of detection before filtration was approximately 1.3×10^3 cfu mL^{-1}. After filtration the sample containing approximately 13 cfu mL^{-1} yielded a positive PCR result. Therefore PES membranes also attained approximately 100-fold concentration and 100% efficiency of cell capture and recovery of DNA.

NanoCeram® depth filters have a much larger nominal pore size than MCE or PES membranes[13]. The alumina fibres are highly charged therefore biological particles are retained on the filter by electrostatic forces rather than by size exclusion. From performance data released by the manufacturer, it was anticipated that this characteristic would result in increased flow rate and sample throughput whilst achieving 99% capture and recovery of bacterial cells. Evaluation of the membranes was performed as described above, processing of 100 mL water samples seeded with 10-fold serial dilutions of *B. cepacia*. Bacteria captured by the membrane were eluted through a change in buffer conditions when the membrane was agitated in 1.5 mL elution buffer. Eluted bacteria were centrifuged at 6,000 x g for 5 minutes and re-suspended in 1 mL of the DNA extraction reagent used previously. The LLOD before filtration was found to be approximately 1.7×10^3 cfu mL^{-1} and had a mean ct of 39.7 (figure 4). No log increase in the detection limit of the assay system was seen following filtration however the mean ct was reduced to 36.9, indicating that a positive PCR result had been detected earlier during the reaction due to an increase in DNA concentration post-filtration; this suggested an approximate 8 fold increase in the LLOD.

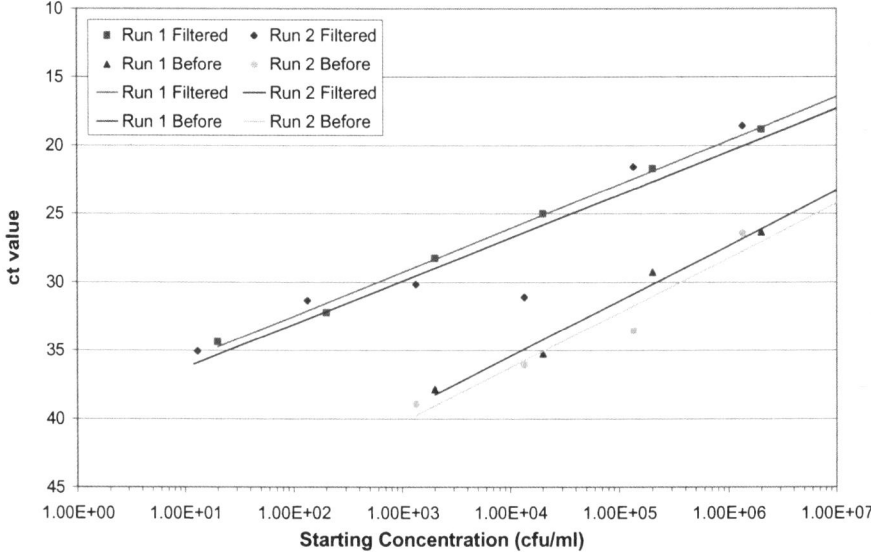

Figure 3   *PCR results from 100 mL water samples seeded with B. cepacia before and after filtration and recovery from PES membranes [threshold set to 0.5 units].*

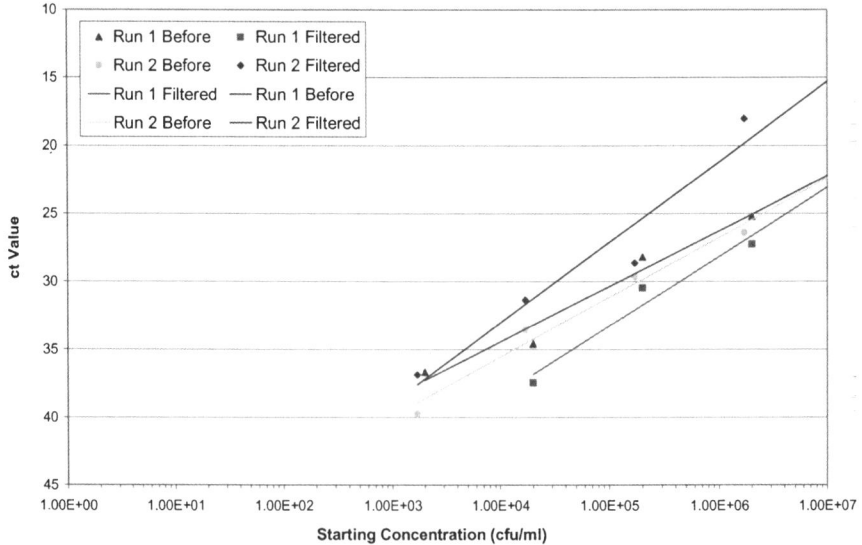

Figure 4   *PCR results from 100 mL water samples seeded with B. cepacia before and after filtration and recovery from NanoCeram® membranes [threshold set to 0.5 units].*

Tangential flow filtration cassettes are traditionally used for applications such as protein purification[14]. The sample is pumped across the surface of the membrane under pressure and the bulk liquid is forced through the membrane whilst particulates are retained and re-circulated. The elution volume from tangential flow filtration cassettes is determined by the inherent dead volume of the filtration cassette and associated tubing. As the elution volume was predicted to be in excess of 10 mL, VivaSpin 20 centrifugal spin columns with 20 mL sample capacity, were chosen to provide secondary sample concentration for this method. Vivaflow 50 Tangential flow filtration cassettes were evaluated using 1 L water samples seeded with 100 μL aliquots from 10-fold serial dilutions of an overnight culture of *B. cepacia*. Preliminary trials indicated that the filtration time for 1 L of sterile distilled water was in excess of four hours and the elution volume from the cassettes was found to be approximately 13 mL (table 1). The filtration time was deemed to be prohibitively long therefore further testing with seeded water samples using the VivaSpin 20 centrifugal spin columns was not undertaken.

Trial	Volume	Filtration time (mins)	Filtration rate (mL min-1)	1L filtration time (mins)	Elution volume (mL)
1	1000 mL	255	3.9	255	11.5
2	500 mL	150	3.3	300	12.0
3	500 mL	135	3.7	270	15.0

Table 1 *Filtration rates attained by tangential flow filtration cassettes.*

2.2.2 Secondary Concentration Techniques. Microcon® spin columns were evaluated for concentration of the DNA extracted after vacuum filtration of water samples using MCE and PES membranes. Each 500 μL DNA, in extraction reagent, was concentrated approximately 10-fold using the Microcon® spin columns to yield approximately 50 μL of DNA. PCR was performed in duplicate for each concentration of DNA and the LLODs are shown in table 2.

Before DNA Concentration		Following DNA Concentration		Concentration Factor Achieved
Detection Limit (cfu mL^{-1})	Mean ct value	Detection Limit (cfu mL^{-1})	ct value	
7.4	37.0	0.74	40.0	≤ 10
20	37.0	2.0	39.1	≤ 10

Table 2 *Results of secondary concentration using Microcon® centrifugal filtration units to concentrate samples following vacuum filtration with mixed cellulose esters and PES membranes.*

In each case a 10-fold increase in the limit of detection was seen, as measured in cfu mL^{-1}. The ct values increased following DNA concentration indicating a decrease in DNA concentration relative to the pre-concentration end-point. This suggested that the recovery of DNA from the spin columns was not 100% efficient.

The preferred method of primary concentration was vacuum filtration using MCE or PES membranes. Both resulted in almost 100% efficient cell capture and DNA recovery. A direct comparison of the two membrane types using water samples seeded with aliquots from a 3-fold dilution series was also performed. The LLOD of the complete assay system using MCE membranes was approximately 30 cfu mL^{-1} and had a flow rate of 50 mL min^{-1} or 20 minutes to filter a 1 L sample. The LLOD using PES membranes was approximately 90 cfu mL^{-1} with a flow rate of 67 mL min^{-1} or 15 minutes to filter 1 L. Small differences in the overall sensitivity of the assay systems and flow rates were seen.

2.3 HPA Methods for the Evaluation and Results

2.3.1 Commercially Available Water Filtration and DNA Extraction Kits. Three kits were chosen for evaluation: UltraClean^{TM} Water DNA Isolation kit (MoBio), Aqua Screen® *FastExtract* kit for conventional filtration systems (Minerva Biolabs) and AquaScreen® filtration and extraction kit (Minerva Biolabs). The resulting DNA extracts were analysed by PCR on the LightCycler instrument (Roche, Switzerland) with an assay that used TaqMan® hydrolysis probes.

A three phase study of the kits was conducted at the HPA, Newcastle. In phase one, for each filtration and DNA extraction technology, two replicate 100 mL sterile distilled water samples were seeded with approximately 10^3 cfu of the target organism. A third 100 mL water sample was left unseeded (negative process control). Standard 500 mL water collection bottles containing sodium thiosulphate pentahydrate, at a final concentration of 18 mg L^{-1}, were used for all samples. The viable count of the inoculum was determined by enumeration of colonies from aliquots of each water sample on Columbia Blood Agar (CBA; Oxoid, UK) spread plates. The plates were incubated for 18-24 hours at 37 °C. The manufacturer's method for filtration and DNA extraction was followed for each technology.

In phase two each technology was compared with the standard method, using 25 mm MCE membranes, for water sample filtration and DNA extraction. 1 mL aliquots from 10-fold serial dilutions of a 16-24 h, 10 mL single strength nutrient broth (SSNB) culture of the target organism were used to seed two replicate sets of 100 mL of sterile distilled water samples in standard bottles as above, to produce two dilution series from 10^{-1} to 10^{-8}. Viable counts of the SSNB culture dilutions and the seeded water samples were determined by enumeration of colonies on CBA spread plates. These two sets of water samples were processed simultaneously, each with a negative process control, one using the standard method and the other according to the relevant manufacturer's method to produce PCR ready DNA. The LLOD of the PCR assay with the resulting DNA extracts was correlated with the viable count. Phase two was performed twice, for each of the technologies that were thought to be potentially viable, in order to check repeatability.

In phase 3, verification of the optimum method was performed using the standard 1 litre sample volume. The inoculum was prepared as in phase two. One litre seeded water samples were prepared by initially inoculating one set of 500 mL water samples (maximum bottle capacity available) with 1 mL volumes of the dilutions of interest, namely 10^{-4} to 10^{-7}. Viable counts of the SSNB culture and the seeded water samples were determined by plating aliquots onto CBA. Following filtration of the four seeded samples and a negative process control, 500 mL water was added to an additional five standard water bottles. The water bottles were rinsed to achieve a uniform concentration of sodium thiosulphate and then one 500 mL volume was added to each of the bottles which had contained the seeded samples and the negative control. These bottles were then rinsed and the samples filtered through the same set of filters. Further processing to produce PCR ready DNA was then undertaken using the selected method.

The method for the UltraClean^{TM} Water DNA Isolation kit[15] stipulated that the water sample was filtered through a 47 mm diameter cellulose acetate membrane in a disposable cup (supplied) and then inserted into a 15 mL bead tube with lysis buffer and vortexed for

10 minutes. A vortex adaptor (MoBio) which holds four 15 mL tubes was used. The DNA was recovered from the supernatant following centrifugation (2,500 x g for 1 minute) and purified using the reagents provided. It was subsequently eluted from the 50 mL spin filter into 3 mL, 10 mM Tris, provided as part of the kit.

The method for the AquaScreen® *FastExtract* kit[16] for conventional filtration systems used 47 mm polysulphone membranes (supplied) for sample filtration. The membrane was placed upside down in a 52 mm Petri dish, 2 mL lysis buffer was added and rinsed over the membrane which was incubated for 30 minutes at 37°C. The DNA was extracted and purified from 500 µL (25%) of the lysate, over a number of steps, using reagents provided and eluted into 60 µL of elution buffer.

The AquaScreen® filtration and extraction kit was a novel, closed system from sample filtration to DNA extraction[17]. Filtration was achieved using a specialised manifold with 1 litre glass filtration vessels connected to 25 mm filter holders containing modified polyethersulphone membranes (all supplied). Following filtration, DNA extraction was performed through the addition of extraction reagents to the filter holder, which was sealed above and below with plastic stoppers, vortexing and two periods of incubation for 30 minutes at 37°C. Subsequently, ethanol was added to the filter holder and vortexed vigorously for two minutes. DNA was removed from the filter using the vacuum manifold to wash the DNA onto a spin column attached beneath the filter. DNA was eluted from the column with 60 µL of elution buffer by centrifugation (1 minute at 6,000 x g then 20 seconds at 16,100 x g).

The results of these evaluations were recorded as either a positive or negative PCR result or where serial dilutions were analysed, the LLOD by PCR with respect to the viable count per original water sample (table 3). The AquaScreen™ filtration and extraction kit showed potential for high recovery of DNA in phase 1 and had the advantages of being a closed system, thus reducing the possibility of loss of sensitivity or cross contamination. However, sealing of the filter holder with small stoppers was awkward and vortexing of these resulted in some leakage, which could cause false positive results. Because of these factors, this method was not evaluated in phases 2 and 3.

	UltraClean™	AquaScreen® *FastExtract* kit	AquaScreen® filtration and extraction kit
Phase 1 Two spiked samples	1 x negative PCR 1 x weak positive PCR	2 x positive PCR	1 x positive PCR Repeat: 2 x positive PCR
Phase 2 100 mL dilution series compared to standard method	LLOD[1]: 10^4 cfu 100 mL^{-1} water sample	LLOD: 10^2 cfu 100 mL^{-1} water sample	Not done
Phase 3 1L dilution series	Not done	LLOD: 10^2 cfu 1 L^{-1} water sample	Not done

[1]Lower limit of detection

Table 3 *Results from Phases 1, 2 and 3.*

The UltraClean™ kit had the advantage of using 47 mm diameter membranes which enabled faster filtration with less potential for blockage. Good quality tubes were supplied by the manufacturer but numerous centrifugation steps were required and only four tubes could be vortexed at one time using the vortex adaptor for this ten minute step. In phase 1, only one of the spiked samples produced a weak positive PCR result. The limit of detection by PCR in phase 2 was 10^4 cfu 100 mL^{-1} water sample. This was inferior to the AquaScreen® *FastExtract* kit which achieved a limit of detection of 10^2 cfu 100 mL^{-1} water sample in phase 2; the sensitivity was not compromised by the increase in sample volume to one litre in phase 3 being 10^2 cfu L^{-1}.

Comparison of the sensitivity data for each method (table 4) indicated that the range in overall sensitivity seen was strongly influenced by features of the methods. For example, the final volume into which the DNA was eluted ranged from 30 µL to 3 mL however the same volume of DNA extract, 4 µL, was added to each PCR assay. Therefore a different percentage of the original sample for each kit, namely, 0.13% for the UltraClean™ method to 6.7% for the Aqua Screen™ filtration and extraction kit and the standard method was analysed. Each of the kits was able to detect a low level of target bacteria as can be seen from the comparison of the data for cfu per assay, 1 – 13 cfu, so the efficiency of sample concentration and DNA extraction was considered to be similar. The AquaScreen™ filtration and extraction kit detection level was recorded as ≤67 cfu as this concentration was tested in phase 1 and no further evaluation was undertaken.

Feature	UltraClean™	AquaScreen® *FastExtract* kit	AquaScreen® filtration and extraction kit	Standard Method (MCE)
Sensitivity per water sample	10^4 cfu 100 mL^{-1}	10^2 cfu 100 mL^{-1}	≤10^3 cfu 100 mL^{-1}	10 cfu 100 mL^{-1}
Final volume DNA	3 mL	60 µL	60 µL	30 - 60 µL
% original sample per PCR assay	0.13 %	1.7 %	6.7 %	3.3 – 6.7 %
Cfu per PCR assay	13 cfu	2 cfu	≤67 cfu	~ 1 cfu
% sample fully processed	100 %	25 %	100 %	50 %
Membrane diameter	47 mm	47 mm	25 mm	25 mm
Membrane material	cellulose acetate	polysulphone	modified polyethersulphone	mixed cellulose esters
Shelf life	< 1 year	2 months from first use		> 1 year
Reagent cost	£21.50 / test	£11.80 / test	£12.50 / test	£5.00 / test

Table 4 *Comparison of Commercially Available Water Filtration and DNA Extraction Kits.*

3 CONCLUSIONS

The standard method using MCE membranes remained the preferred technology with almost 100% efficient cell capture and recovery of DNA. This method had the advantages of being well established and cost effective but the use of 25 mm diameter membranes restricted the flow rate and sample handling was compromised by the capacity of the sample filtration vessel. Use of a larger sample reservoir would improve handling safety. Based on their performance, PES membranes could be substituted for MCE membranes.

The AquaScreen® *FastExtract* kit for conventional filtration systems was considered to be a feasible alternative, having comparable efficiency with the standard method. It had the advantages of incorporating a 47 mm diameter membrane and a larger sample reservoir. However, the reagent cost was double that of standard vacuum filtration; the use of the Petri dish as a vessel to house the membrane for cell lysis was a cause for concern with regard to health and safety; and the shelf life of the kit was short once opened. In addition, only 25% of lysate was fully processed restricting the performance of the kit. Further research may be able to resolve these issues.

This work was funded by the Drinking Water Inspectorate. Thanks to Dave Cleary from Genetic Detection, Dstl Porton Down for providing the *B. cepacia* RecA real-time PCR assay.

© Crown copyright 2007. Published with the permission of the Defence Science and Technology Laboratory on behalf of the Controller of HMSO.

References

1. DWI (2000) The Water Supply (Water Quality) Regulations 2000 No.3184 London: Drinking Water Inspectorate.
2. Gibbs, R.A. (1990) *DNA amplification by the polymerase chain reaction.* Analytical Chemistry, 62 pp1202-1214.
3. Varughese, E. A., Wymer, L. J. & Haugland, R. A. (2007) *An integrated culture and real-time PCR method to assess viability of disinfectant treated Bacillus spores using robotics and the MPN quantification method.* J. Microbiol. Methods. 71, pp66-70.
4. Polaczykb, Amy L. *et al.* (2008) *Ultrafiltration-based techniques for rapid and simultaneous concentration of multiple microbe classes from 100-L tap water samples* J. Microbiol. Methods, 73, May 2008, pp 92-99.
5. Zuckerman, U. and Tzipori, S. Portable continuous flow centrifugation and method 1623 for monitoring of waterborne protozoa from large volumes of various water matrices. (2006) Journal of Applied Microbiology, 100, pp 1220-1227.
6. Data sheet: *Vivaspin 6 & 20 Technical data and operating instructions.* Available at http://www.bioexpress.com/specs/Vivaspin6And20DataSheet.pdf [accessed May 2008].
7. Product information: *Solid Phase Extraction (SPE).* Available at: http://www.whatman.com/SolidPhaseExtractionSPE.aspx [accessed May 2008].
8. Product Information: GeneCatcher™ Magnetic Beads. Available at http://www.invitrogen.com/site/us/en/home/Products-and-Services/Applications/ Nucleic-Acid-Purification-and-Analysis/DNA-Purification/DNAP-Misc/ GeneCatcher-Magnetic-Beads.html [accessed May 2008].

9. Martin, S. P. *et al. Spore and micro-particle capture on an immunosensor surface in an ultrasound standing wave* (February 2005) Biosensors and Bioelectronics, 21, pp 758-767.

10. Aldaeus, F *et al. Multi-step dielectrophoresis for separation of particles* (June 2006) J. Chromatogr. A, 1131, pp 261-266.

11. Taylor, Michael T. et al. Lysing Bacterial Spores by Sonication through a Flexible Interface in a Microfluidic System (February 2001) Anal. Chem. 2001, 73, pp 492-496.

12. The Microbiology of Drinking Water (2002) – Part 1 – Water quality and public health, Methods for the examination of waters and associated materials, The Environment Agency, 2002.

13. Leff, L. G. *et al. Identification of aquatic Burkholderia (Pseudomonas) cepacia by hybridization with species-specific rRNA Gene Probes,* Applied and Environmental Microbiology, Apr. 1995, p. 1634–1636.Tepper, F. Kaledin, L., Nano Fiber Biological Filter. Argonide Corp., Sanford, Florida, USA. Available at: http://www.argonide.com/publications.html [accessed May 2008].

14. Pall   technical   support: *Tangential   flow   filtration.*   Available   at http://www.pall.com/laboratory_7063.asp [accessed May 2008].

15. Mo Bio Laboratories Inc. UltraClean™ Water DNA Isolation kit (0.45µm) catalog #14800-10   10   preps.   Instruction   Manual.   Version   06262006 http://www.mobio.com/files/protocol/14800-10.pdf [accessed March 2007].

16. AquaScreen®   *FastExtract*   Version   10-2007   http://www.minerva-biolabs.com/download/AS_FastEx_1007.pdf [accessed May 2008].

17. AquaScreen®   extraction   kit.   Version   10-2007   http://www.minerva-biolabs.com/download/ASE_1007.pdf [accessed May 2008].

SCIENTIFIC AND TECHNICAL ADVISORY CELL (STAC) – GETTING TIMELY PUBLIC HEALTH ADVICE TO MULTI-AGENCY FRONTLINE RESPONDERS

R. Carr (1), S. Ibbotson (2), VSG Murray (3)

(1) HPA West Midlands HPU, Health Protection Agency; (2) HPA West Midlands, Health Protection Agency, (3) Chemical Hazards and Poisons Division, London, Health Protection Agency.

1 INTRODUCTION

This presentation outlines the arrangements for convening a Scientific and Technical Advisory Cell (STAC) and the central co-ordinating role that the Health Protection Agency[1] can take in this process. STAC is part of the command and control structure for major incident response in the UK (figure 1). The presentation also reflects on recent experiences of the authors as part of the first national STAC, which was convened to support the frontline response to the July 2007 floods, lessons learned from this experience, and progress with their implementation.

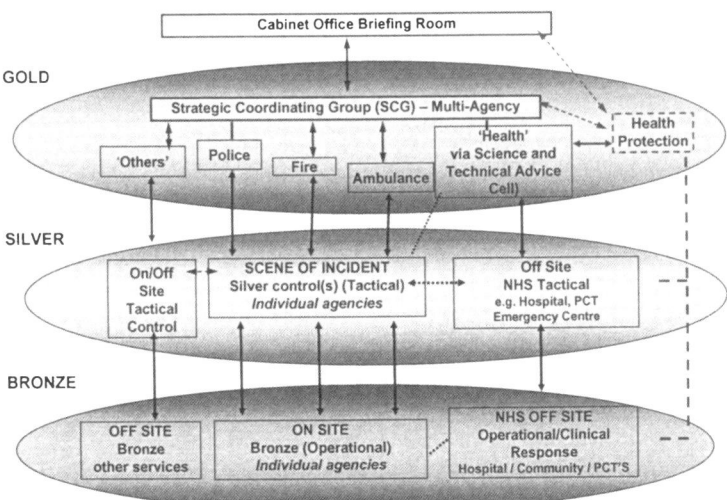

Figure 1 *Command and Control Structure for Major Incident Response (updated from Rowe, 2005[2])*

2 POLICY BACKGROUND

Following the Buncefield oil depot fire in 2005[3,4,5], the Cabinet Office recognized the need to ensure co-ordination of health and environmental scientific advice to multiple agencies responding Strategic Co-ordination Groups (SCG).[6] Previously, health advice was given separately from environmental advice during emergencies, through a Health Advisory Team (HAT).[7] From April 2007 the national JHAC arrangements and legislation have been replaced by those for a STAC.[8] However, the current guidance does not specifically cover arrangements which might apply in situations where national coordination of scientific and technical advice might be helpful.

A STAC, when requested by the Incident Commander (usually police), will provide the SCG with advice during both the acute and recovery phases of an emergency.

Regional Directors of Public Health are responsible for ensuring that there are robust STAC arrangements in their areas. However, the Health Protection Agency, through its Local and Regional Services, is often responsible on their behalf for ensuring the operational arrangements to convene a STAC are in place.

Other environmental and health experts, for example the Environment Agency, may be called upon as appropriately to support the deliberations of the STAC in their particular fields of scientific expertise.

3 STAC FUNCTIONS

A critical role within the STAC is the STAC Chair. STAC Chairs are most likely to be either Directors of Public Health or senior staff from the Health Protection Agency, but could be from another agency if the primary issues being considered were not health related. A national training programme has been established and is now being rolled out for both these groups of professionals by the Health Protection Agency on behalf of the Department of Health.

The Chair of the STAC is responsible for:
- Identifying and corralling the right scientific and technical experts,
- Liaising with national specialist advisors from agencies represented in the STAC,
- Fielding and predicting issues and questions from local responders (via the SCG),
- Arriving, as far as is possible, at a common view on the scientific and technical merits of different courses of action,
- Providing a common brief to inform local decision makers (the SCG),
- Maintaining a written record of STAC advice, decisions made and the rationale behind both of these.

4 PUBLIC HEALTH EFFECTS OF FLOODING

Forty percent of natural disasters worldwide are due to flooding or its sequelae. There is a growing weight of opinion to suggest that flooding disasters are becoming more frequent and more severe in impact.

The health impact of flooding is variable and depends on the severity, time of day or year, timeliness of warnings, emergency preparedness of the victims and the existing social and economic structures to deal with its effects.

The primary early public health effects of flooding in developed countries are mainly attributed to:

- Drowning and injuries (e.g. fractures, burns and electrocution) [9],[10]
- Skin rashes [11]
- Gastroenteritis [11]
- Respiratory infections [11]
- Physical and mental stress [10]
- Chemical exposures including carbon monoxide [12], [13]

Positive health outcomes are believed to be promoted through the early provision of advice and practical assistance, emotional and social support, and through coordinated efforts to rebuild communities and social relationships. However the evidence base for specific interventions is not particularly robust.

5 THE JULY 2007 FLOODS AND THE NATIONAL STAC

July 2007 was the wettest July on record in England. Severe flooding was recorded in the rivers Severn, Avon and Thames. Flash flooding also affected some communities which were not subject to river floods and were not used to dealing with the effects of such a crisis. This led to the establishment of three separate Strategic Co-ordination Groups (SCGs) in the West Midlands and South West Regions.

There was an early recognition by the Health Protection Agency, and other partners, that it would be helpful to ensure a nationally coordinated approach to delivering expert national advice to local and regional responders. However, this was not intended to circumvent local arrangements and a local STAC was operating at the same time in the South West Region. On this basis, the Health Protection Agency convened a national Scientific and Technical Advisory **Group** (STAG) on Sunday 22 July.

The core STAG group was established, in an emergency operations centre in central Birmingham, which met via twice daily teleconferences with national health and environmental experts from a number of national agencies. Besides expert infectious disease and chemical hazards support from the Health Protection Agency, additional advice was requested by the HPA and provided by:

- The Drinking Water Inspectorate [14]
- The Environment Agency [15]
- The Food Standards Agency [16]
- The Health and Safety Executive [17]
- The Government Decontamination Service [18]
- A representative of the Chartered Institute of Environmental Health [19] to provide Local Authority input
- Severn Trent Water,[20] the water utility most involved in response in Gloucestershire

The core functions of this group were both, to offer scientific and technical advice to the three SCGs, and to produce real-time surveillance outputs for infectious diseases and chemical exposures in the affected areas.

A broad range of issues were considered by the STAG, regarding the safety of drinking and other water supplies, risks of infection and environmental contamination, safety during of procedures for cleaning up and the restoration of normal functions in communities.

6 OUTPUTS FROM THE STAC

The national STAG was operational for 6 days, and the main outputs included:

- Near real time surveillance reports
- Risk assessments [21]
- Ad hoc advice and briefings
- Rolling brief for the SCGs
- Rolling public facing brief (via agency websites[22] and leaflets[23, 24, 25, 26, 27])
- Advice about carbon monoxide hazards and risks[28]
- Chemical exposure risk assessment tool [29]

7 LESSONS AND ISSUES FOR FURTHER CONSIDERATION

The July floods provided an opportunity to test the new STAC arrangements during a national emergency.

The national group was able to provide timely advice on many, but not all, issues to local and regional multi-agency responders through a process of achieving consensus amongst national scientific and technical experts. Some of the more challenging issues dealt with by the group required further work to produce 'off the shelf' tools and advice which could be made widely available in advance of future similar emergencies.

The group also provided consistent public facing information through co-ordination of the various national agencies' websites.

A considerable amount of information and advice was generated through this incident that would be invaluable in the light of similar future flooding incidents, and this is being collated nationally by the HPA for that purpose.

Since the STAC arrangements and procedures were new there were inevitably issues which arose regarding the need for further clarification of these arrangements, particularly regarding when a national STAC or equivalent would be beneficial and how this should relate to a number of separate SCGs and local STACs (since the guidance is really designed to cover situations where a single STAC gives advice to a single SCG.).

The HPA gave evidence to the subsequent Pitt Review on flooding[30] and a number of these issues have been picked up through recommendations in the Review's reports, and are now being implemented nationally.

A national training programme has now been developed for STAC Chairs by the Department of Health and is now being successfully rolled out through regions by the health protection Agency on behalf of the Department of Health.

8 ACKNOWLEDGEMENTS

R Hall, HPA South West, Health Protection Agency; J Simpson, Centre for Emergency Preparedness and Response, Health Protection Agency; B McCloskey, CEO Office, Health Protection Agency.

9 REFERENCES

[1] Health Protection Agency http://www.hpa.org.uk/ (accessed 20.06.08)

[2] Rowe A. Emergency Planning: Management and Co-ordination of Major Incidents Chemical Hazards and Poisons Report Health Protection Agency June 2006 23- 25

[3] Aus Charlotte, Mohan Richard, Murray Virginia Incident Response: Buncefield Fire Summary Chemical Hazards and Poisons Report Health Protection Agency June 2006 4-5

[4] Webster H. Modelling the plume from the Buncefield Oil Depot fire. Chemical Hazards and Poisons Report Health Protection Agency June 2006 6-7

[5] Troop Pat and the Health Protection Agency in collaboration with the Dacorum and Watford and Three Rivers Primary Care Trusts. The Public Health Impact of the Buncefield Oil Depot Fire. July 2006ISBN: 0 901144 82 7 http://www.hpa.org.uk/web/HPAwebFile/HPAweb_C/1194947375551 (accessed 20.06.08)

[6] Civil Contingencies Secretariat, Cabinet Office Preparing for Emergencies. Draft Guidance on Part 1 of the Civil Contingencies Act 2004,: ts associated Regulations and non-statutory arrangements http://www.ukresilience.gov.uk/~/media/assets/www.ukresilience.info/guidance1%20pdf.ashx (accessed 20.06.08)

[7] UK Resilience Provision of scientific and technical advice to Strategic Co-ordinating Groups during a major incident. Last updated: 13/04/2007 http://www.ukresilience.gov.uk/news/stac_guidance.aspx (accessed 20.06.08)

[8] Provision of scientific and technical advice in the strategic co-ordination centre: STAC Guidance to Local Responders (April 2007) http://www.ukresilience.gov.uk/news/~/media/assets/www.ukresilience.info/stac_guidance%20pdf.ashx (accessed 20.06.08)

[9] Ahern MJ, Kovats RS, Wilkinson P, Kew R, Metthies F (2005) Global health impacts of floods: epidemiological evidence. Epidemiological reviews, Vol. 27: 36 - 46

[10] Hajat S, Ebi KL, Kovats SR, Menne B, Edwards S, Haines A (2004) The human health consequences of flooding in Europe: a review. Chap. 4 in Extreme weather events and public health responses,: 185 – 196

[11] Reacher M, McKenzie M, Lane C, Nichols T, Kedge I, Iverson A, Hepple P, Walter T, Laxton C, Simpson J (2004) health impacts of flooding in Lewes: a comparison of reported gastrointestinal and other illnesses and mental health in flooded and non-flooded households. Communicable Disease and Public health, Vol. 7 (No. 1): 1 – 8

[12] Euripidou E and Murray V (2004) Public health impacts of floods and chemical contamination. Journal of Public Health, Vol. 26(No. 4): 376 – 383

[13] Centres for Disease Control and Prevention (2005) carbon monoxide poisoning after hurricane Katrina - Alabama, Louisiana, and Mississippi, August - September 2005. Mortality and Morbidity Weekly Report, Vol. 54 (No. 39): 996 – 998

[14] The Drinking Water Inspectorate http://www.dwi.gov.uk/ (accessed 20.06.08)

[15] Environment Agency http://www.environment-agency.gov.uk/ (accessed 20.06.08)

[16] Food Standards Agency http://www.foodstandards.gov.uk/ (accessed 20.06.08)

[17] Health and Safety Executive http://www.hse.gov.uk/ (accessed 20.06.08)

[18] Government Decontamination Service http://www.defra.gov.uk/gds/ (accessed 20.06.08)

[19] Chartered Institute of Environmental Health http://www.cieh.org/ (accessed 20.06.08)

[20] Severn Trent Water
http://www.stwater.co.uk/server.php?search_word=flooding+2007&submit=Go&change=SearchResults&show=nav.5627 (accessed 20.06.08)

[21] Health Protection Agency Flooding Chemical Event Checklist
http://www.hpa.org.uk/web/HPAwebFile/HPAweb_C/1194947351380 (accessed 20.06.08)

[22] Health Protection Agency. Information for people who have been affected by the flood 28 August 2007
http://www.hpa.org.uk/web/HPAweb&HPAwebStandard/HPAweb_C/1195733854183 (accessed 20.06.08)

[23] Health Protection Agency Health advice following flooding
http://www.hpa.org.uk/web/HPAwebFile/HPAweb_C/1194947339369 (accessed 20.06.08)

[24] Health Protection Agency Cleaning up after a flood – health advice
http://www.hpa.org.uk/web/HPAwebFile/HPAweb_C/1194947420817 (accessed 20.06.08)

[25] Health Protection Agency Top tips for coping with water shortages in the floods
http://www.hpa.org.uk/web/HPAwebFile/HPAweb_C/1194947340948 (accessed 20.06.08)

[26] Health Protection Agency Public health advice – water consumption
http://www.hpa.org.uk/web/HPAwebFile/HPAweb_C/1194947341382 (accessed 20.06.08)

[27] Health Protection Agency Public Health Implications of Flooding of Sports Playing Fields http://www.hpa.org.uk/web/HPAwebFile/HPAweb_C/1194947416548 (accessed 20.06.08)

[28] Health Protection Agency Health advice following floods –
chemical and environmental hazards
http://www.hpa.org.uk/web/HPAwebFile/HPAweb_C/1194947410746 (accessed 20.06.08)

[29] The South Yorkshire Flooding Working Group Development of a risk assessment framework for potential chemical contamination during flood events Health protection Agency Chemical Hazards and Poisons Report May 2008: 33 – 36 http://www.hpa.org.uk/web/HPAwebFile/HPAweb_C/1211266315123 (accessed 20.06.08)

[30]Pitt Michael. Learning lessons from the 2007 floods: An independent review. Cabinet Office December 2007 http://www.cabinetoffice.gov.uk/thepittreview/~/media/assets/www.cabinetoffice.gov.uk/flooding_review/flood_report_web%20pdf.ashx (accessed 20.06.08)

COMMUNICATING WITH THE PUBLIC DURING WATER CONTAMINATION EVENTS: ADDRESSING VULNERABLE POPULATIONS

P. A. Nsiah-Kumi, MD, MPH

Internal Medicine-Pediatrics, University of Nebraska Medical Center, 984085 Nebraska Medical Center, Omaha, Nebraska 68198-4085 USA Email: pnkumi@umc.edu

1 INTRODUCTION

The most effective health communication campaigns are centered around the audiences receiving them.[1] These messages must be relevant to the audience, appropriate to the audience's norms and expectations, and speak to their experiences. The messages must be delivered by a source perceived as credible by the audience as trust is essential for effective communication.[2,3] Thus, communicating public health messages during water contamination emergencies requires a thorough knowledge of the different populations being communicated to.[4]

In the marketing field, the importance of understanding the characteristics of different segments of the population is well understood.[5] The same core message may take on different forms to appeal most effectively to different audiences. The diamond industry has shown this very effectively over the past several decades with a series of advertisements appearing in various print media in the United States. The ads clearly appealing to different segments of the population. The primary message for all audiences (including single men, married men, married women, single women, African American women) is the same: "Buy a diamond ring." However, the content and images used in the advertisements targeting different audiences are drastically different.

One may wonder about the connection between diamond marketing and health-related messages in public health emergencies. However, targeted communication is an area in which public health, medical and water professionals can learn from those involved in the marketing of the diamond industry. Knowing and understanding the audience is essential to effective communication on any topic, and in a public health emergency, our communities' health depends on it.[6]

2 COMMUNICATING WITH THE PUBLIC BEFORE WATER CONTAMINATION EVENTS: EFFECTIVE RISK COMMUNICATION

Before and during a water contamination event, it is essential to communicate about risk with the public.[7] Risk communication is essential to prepare the public for potential future risks.[8] Risk communication is an interactive process of exchange of information and opinion among individuals, groups and institutions. It often involves multiple messages about the nature of risk or expressing concern, opinions or reactions to risk messages or to legal and institutional arrangements for risk management.[9] In essence, risk management is a dialogue about risk that involves multiple parties and multiple messages. It is important that effective lines of communication be created before a crisis occurs and that the nature of the communication process must be acceptable to the target audience.

3 COMMUNICATING WITH THE PUBLIC DURING WATER CONTAMINATION EVENTS: EFFECTIVE CRISIS COMMUNICATION

During a water contamination event, crisis communication is required. Its goals are to help the audience develop knowledge and understanding about the crisis, trust and credibility of the source, and cooperation and construction of dialogue between both parties.[7] Experts and laypeople do not always communicate using the same language and concepts. It is essential to bridge this gap.[10] Additionally, it is essential to consider that in a time of crisis, heightened public emotions, limited access to facts, and speculation lead to an unstable information environment.[11] Information is not necessarily received in the same way it would be in a non-crisis situation. While a number of guidelines exist for risk and crisis communication, only one set will be mentioned here. The National Center for Food Protection and Defense presents ten best practices for crisis and risk communication. These best practices appear in Table 1.

---

1. Risk and crisis communication is an ongoing process
2. Conduct pre-event (pre-crisis) planning
3. Foster partnerships with public
4. Listen to public's concern and understand audience
5. Demonstrate honesty, candor and openness
6. Collaborate and coordinate with credible sources
7. Meet the needs of the media and remain accessible
8. Communicate with compassion, concern and empathy
9. Accept uncertainty and ambiguity
10. Provide messages of self-efficacy

Table 1 *Ten Best Practices in Crisis & Risk Communication 2004 National Center for Food Protection and Defense*

4 DEFINING VULNERABLE POPULATIONS

Vulnerable populations are those who are vulnerable to being neglected by the health care system or have difficulties in comprehension. It is essential to reach every member of the community regardless of individual or collective communication barriers. To realistically reach that goal, one must first know who is in the community at any given time and how best to reach them with messages that can be understood and will motivate them to the desired action.

For the purpose of this discussion, we will consider the following populations as vulnerable populations: urban/rural poor, mentally ill and intellectually disabled, immunosuppressed, children, the elderly, racial/ethnic minorities, low literacy and limited English proficiency populations.

5 POTENTIAL COMMUNICATION DIFFICULTIES WITH VULNERABLE POPULATIONS

The urban/rural poor may have difficulties accessing information in venues that have cost associated with them. Additionally, they may have barriers to undertaking actions that have any more than minimal cost.[12, 13] Mentally ill or intellectually disabled individuals may have difficulty comprehending or processing complex health messages.[14-16] Individuals at high risk of waterborne illness, such as those who are immunosuppressed or on dialysis, are at risk of not receiving messages that target them specifically and address infection related risks.[17-19]

Children often have access to messages intended for adult audiences, and if these messages are not monitored, this can lead to undue psychological distress.[20, 21] The elderly, by reason of declining vision, hearing, and cognition may have difficulties accessing messages as well as carrying out recommended items.[22-25] Racial/ethnic minorities often perceive publicly available messages as not being pertinent to them. Differences in culture and communication styles place many racial/ethnic minority groups at risk for not receiving messages directed at the public in general.[26-28]

Literacy plays an important role in how people receive and use information. Community members with limited health literacy may have difficulty understanding and applying information given during water contamination events. In our very multicultural world, language diversity is a very real and growing issue. In English speaking countries, limited English proficiency populations may often not receive messages when they are presented only in English. It is important to present messages to populations speaking languages other than English.[29, 30]

6 RECOMMENDED STRATEGIES FOR EFFECTIVE COMMUNICATION WITH VULNERABLE POPULATIONS

Targeting communication to specific groups can be an important way in which to reach a variety of audiences with health messages.[31] Each group has specific needs and priorities that must be considered in the message design process. Targeting messages to receiving audiences is a practical and important way in which to make sure all segments of the population can understand the messages they receive.[31] It is essential to remember that there is no single audience. No single message or delivery method works for all populations.[10] Becoming acquainted with key community leaders and determining what venues are most appropriate are essential steps in effective communication. Assembling a group to be available for pretesting messages before releasing them to the community at large is also key.[32] Also, having spokespersons that are credible to a particular community are invaluable in establishing a dialogue with that community.[33]

6.1 Urban/Rural Poor

Individuals with limited resources may not have the same access to care or information as their counterparts with more income. When developing message for the urban or rural poor, it is essential to present low cost alternatives for all recommended actions.[34] The homeless, migrants and refugees should not be overlooked and strategies specifically for those audiences should be considered.[35-38]

6.2 Mentally Ill/Intellectually Disabled

Mental illness and intellectual disabilities can lead to difficulties with patient comprehension of and adherence to instructions in public health messages. Those who are mentally ill/intellectually disabled should have messages that are appropriate for them and have their caretakers and family members involved in delivering messages to them.[14-16]

6.3 Medically Vulnerable Populations

Individuals at high risk of waterborne illness, such as those who are immunosuppressed or on dialysis need messages that appropriately address the additional health related risks they face in a water contamination event.[19, 39] In addition to targeting patients, these messages should also be delivered to health care providers, as they are often a source of public health messages for their patients. Recent studies suggest that health care providers are often unprepared to act and communicate in public health crises. It is essential that resources be directed at providers so that they can better care for their patients. [39-42]

6.4 Children

No one message is appropriate for all children. Messages targeting children specifically need to take developmental stages into account. Messages being delivered between 3 and 6 pm, though they may not target children specifically, require special care. During these hours, children are often home from school while their parents are still working. The children are often exposed to television and radio without adult supervision. It is essential that messages be delivered in ways that do not cause undue stress for children. [20, 21]
Lastly, no matter what messages are delivered, parents and teachers need to be involved in the process.[21]

6.4 The Elderly

Older adults often face declining vision, hearing and cognition. Messages to the elderly need to consider potential vision, hearing, and cognitive impairment and should include their caregivers. Additionally, whether elders are community-dwelling or live in senior communities, assisted living or other types of facilities, is an important consideration as well. Communication challenges related to declining physical, cognitive, hearing and vision capabilities must be considered, Large print and illustrations for visual materials and media along with the use of clear language are essential. Additionally, volume is to be considered in audio materials or live meetings. [22-25]

6.5 Racial/Ethnic Minority Populations

Culture plays an integral role in how individuals perceive risk and receive and process information. With regards to racial/ethnic minority groups, it is important to understand cultural issues and make sure that they are addressed appropriately. Cultural competence is defined as "appropriate and effective communication which requires the willingness to listen to and learn from members of diverse cultures, and the provision of services and information in appropriate languages, at appropriate comprehension and literacy levels, and in the context of an individual's cultural health beliefs and practices.[43]

6.6 Low Literacy Populations

Up to 36% of the US population (or approximately 77 million Americans) read at a level below what is deemed to be intermediate or proficient.[44] However, information intended for the public is often written at a much higher level than the majority of people comprehend easily. This can pose significant challenges in the presentation of messages during or surrounding a water contamination event.

Health literacy is "the degree to which individuals have the capacity to obtain, process and understand the health information and services they need to make appropriate decisions about their health." This concept includes the skills of reading, writing, listening, speaking, arithmetic, understanding numerical data (such as percentages, ratios, and measurements) and understanding concepts.

A variety of factors are associated with increased likelihood of limited health literacy including having less than high school education or General Educational Development (GED), being a member of a racial/ethnic minority population, being an older adult, and being a non-native speaker of English. [45]

Addressing low literacy populations effectively requires making sure that messages are clear and use simple language. The use of plain language is one strategy by which to accomplish this.

Plain language is a strategy to present information in clear simple and coherent way. "Plain English is clear, straightforward expression, using only as many words as are necessary. It is language that avoids obscurity, inflated vocabulary and convoluted sentence construction. It is not baby talk, nor is it a simplified version of the English language. Writers of plain English let their audience concentrate on the message instead of being distracted by complicated language. They make sure that their audience understands the message easily." [46]

This method does not only apply to health information. It is increasingly being applied in a number of areas (such as government documents) to make information disseminated to the general public clear. [47]

An additional strategy for addressing literacy issues is testing messages with low literacy audiences. Through an iterative process of refining messages and taking them back to low literacy audiences for feedback, it is possible to design messages that are clear and attractive to this audience.[48] While this process can be time-consuming, it is essential to the health of low literacy populations.

6.7 Limited English Proficiency Groups

Addressing populations with limited English proficiency can be quite challenging. However, focusing on these groups is imperative for the health of these communities. It is also mandated by law. Title VI of the Civil Rights of 1964 states that "No person in the United States shall, on the ground of race, color, or national origin, be excluded from participation in, be denied the benefits of, or be subjected to discrimination under any program or activity receiving Federal financial assistance."[49]

To most effectively communicate with these groups, it is essential that public health messages be made available in a variety of languages common to the community.[50, 51] The use of professional translation services can aid significantly in this effort with regard to written documents. Additionally, the use of interpreters for verbal communication is essential, especially in public meetings. So, it is essential for those communicating public health messages to be aware of translation and interpreter services that are available in their communities.

7 Summary

In a water contamination event, all members of the community need to receive public health messages related to the pertinent risks, especially those in vulnerable groups that may be overlooked in the development of public health messages. While barriers to communication with vulnerable population exist, it is imperative that messages be designed intentionally targeting these segments of the population. The health of our nation depends on the health of our vulnerable populations, and developing effective strategies for communicating with these populations prior to a water contamination event will allow for effective communication in public health crises. Knowing the audience makes all the difference.

References
1. T.L. Albrecht, and C. Bryant. Advances in segmentation modeling for health communication and social marketing campaigns. *J Health Commun,* 1996, **1,** 65-80.
2. E.B. Arkin, and National Cancer Institute (U.S.). Office of Cancer Communications. *Making health communication programs work : a planner's guide,* U.S. Dept. of Health and Human Services Public Health Service National Institutes of Health Office of Cancer Communications National Cancer Institute, Bethesda, Md.; 1989, Page.
3. L.E. Boulware, L.A. Cooper, L.E. Ratner, T.A. LaVeist, and N.R. Powe. Race and trust in the health care system. *Public Health Rep,* 2003, **118,** 358-65.

4. Institute of Medicine (U.S.). Committee on Communication for Behavior Change in the 21st Century: Improving the Health of Diverse Populations. *Speaking of health : assessing health communication strategies for diverse populations*, National Academies Press, Washington, D.C.; 2002, Page.

5. G. Quinn, T. Albrecht, R. Marshall, Jr., and T.H. Akintobi. "Thinking Like a Marketer": training for a shift in the mindset of the public health workforce. *Health Promot Pract,* 2005, **6,** 157-63.

6. P. Nsiah-Kumi. Communicating effectively with vulnerable populations during water contamination events. *Journal of Water and Health,* 2008, **06,** 63-75.

7. V.T. Covello. Best practices in public health risk and crisis communication. *J. Health Commun.,* 2003, **8 Suppl 1,** 5-8; discussion 148-51.

8. E. Aakko. Risk communication, risk perception, and public health. *WMJ,* 2004, **103,** 25-7.

9. Center for Mental Health Services (U.S.). *Communicating in a crisis : risk communication guidelines for public officials*, U.S. Department of Health and Human Services Public Health Service, Rockville MD; 2002, Page.

10. R. Parkin, M. Embrey, and P. Hunter. Communicating water-related health risks: Lessons Learned and Emerging Issues. *Journal of the American Water Works Association,* 2003, **95,** 58-66.

11. U.S. Department of Health and Human Services. *Communicating in a Crisis: Risk Communication Guidelines for Public Officials.*, US Department of Health and Human Services,, Washington, D.C.; 2002, Page.

12. S.A. Haas. Health selection and the process of social stratification: the effect of childhood health on socioeconomic attainment. *J Health Soc Behav,* 2006, **47,** 339-54.

13. P. Jha, A. Mills, K. Hanson, L. Kumaranayake, L. Conteh, C. Kurowski, S.N. Nguyen, V.O. Cruz, K. Ranson, L.M. Vaz, S. Yu, O. Morton, and J.D. Sachs. Improving the health of the global poor. *Science,* 2002, **295,** 2036-9.

14. S.A. Cooper, J. Morrison, C. Melville, J. Finlayson, L. Allan, G. Martin, and N. Robinson. Improving the health of people with intellectual disabilities: outcomes of a health screening programme after 1 year. *J Intellect Disabil Res,* 2006, **50,** 667-77.

15. C.A. Melville, S.A. Cooper, J. Morrison, J. Finlayson, L. Allan, N. Robinson, E. Burns, and G. Martin. The outcomes of an intervention study to reduce the barriers experienced by people with intellectual disabilities accessing primary health care services. *J Intellect Disabil Res,* 2006, **50,** 11-7.

16. S. Trumble. Communicating with people who have intellectual disabilities. *Aust Fam Physician,* 1993, **22,** 1081-2.

17. S. Ahmad. Essentials of water treatment in hemodialysis. *Hemodial Int,* 2005, **9,** 127-34.

18. R.L. Amato. Water treatment for hemodialysis--updated to include the latest AAMI standards for dialysate (RD52: 2004) continuing. *Nephrol Nurs J,* 2005, **32,** 151-67; quiz 168-70.

19. D.M. Ward. Hemodialysis water: an update on safety issues, monitoring, and adverse clinical events. *Asaio J,* 2004, **50,** xiii-xviii.

20. J.F. Hagan, Jr. Psychosocial implications of disaster or terrorism on children: a guide for the pediatrician. *Pediatrics,* 2005, **116,** 787-95.

21. M.L. Wolraich, J. Aceves, H.M. Feldman, J.F. Hagan, Jr., B.J. Howard, A. Navarro, A.J. Richtsmeier, and H.C. Tolmas. How pediatricians can respond to the psychosocial

implications of disasters. American Academy of Pediatrics. Committee on Psychosocial Aspects of Child and Family Health, 1998-1999. *Pediatrics,* 1999, **103,** 521-3.

22. P. Kopp. Better communication with older patients. *Prof Nurse,* 2001, **16,** 1296-9.
23. C.D. Schewe, and H.E. Spotts, Jr. Principles for communicating with aging health-care consumers. *Clin Lab Manage Rev,* 1990, **4,** 352-7.
24. V. Cotrell, K. Wild, and T. Bader. Medication management and adherence among cognitively impaired older adults. *J Gerontol Soc Work,* 2006, **47,** 31-46.
25. D.M. Polk. Communication and family caregiving for Alzheimer's dementia: linking attributions and problematic integration. *Health Commun,* 2005, **18,** 257-73.
26. A.M. Apanovitch, D. McCarthy, and P. Salovey. Using message framing to motivate HIV testing among low-income, ethnic minority women. *Health Psychol,* 2003, **22,** 60-7.
27. M.W. Kreuter, and S.M. McClure. The role of culture in health communication. *Annu Rev Public Health,* 2004, **25,** 439-55.
28. D. Mulligan-Smith, S. Puranik, and S. Coffman. Parental perception of injury prevention practices in a multicultural metropolitan area. *Pediatr Emerg Care,* 1998, **14,** 10-4.
29. E. Wilson, A.H. Chen, K. Grumbach, F. Wang, and A. Fernandez. Effects of limited English proficiency and physician language on health care comprehension. *J Gen Intern Med,* 2005, **20,** 800-6.
30. C. Brach, I. Fraser, and K. Paez. Crossing the language chasm. *Health Aff (Millwood),* 2005, **24,** 424-34.
31. M.W. Kreuter, and R.J. Wray. Tailored and targeted health communication: strategies for enhancing information relevance. *Am J Health Behav,* 2003, **27 Suppl 3,** S227-32.
32. R.E. Rudd, J.P. Comings, and J.N. Hyde. Leave no one behind: improving health and risk communication through attention to literacy. *J Health Commun,* 2003, **8 Suppl 1,** 104-15.
33. National Cancer Institute (U.S.). Office of Cancer Communications. *Making health communication programs work a planner's guide,* U.S. Dept. of Health and Human Services Public Health Service National Institutes of Health, Office of Cancer Communications National Cancer Institute, Bethesda, Md.; 2002, Page.
34. L.B. Mauksch, W.J. Katon, J. Russo, S.M. Tucker, E. Walker, and J. Cameron. The content of a low-income, uninsured primary care population: including the patient agenda. *J Am Board Fam Pract,* 2003, **16,** 278-89.
35. J. Ensign, and A. Panke. Barriers and bridges to care: voices of homeless female adolescent youth in Seattle, Washington, USA. *J Adv Nurs,* 2002, **37,** 166-72.
36. T.D. Matte, and D.E. Jacobs. Housing and health--current issues and implications for research and programs. *J Urban Health,* 2000, **77,** 7-25.
37. A.R. Sutherland. Health care for the homeless. *Issues Sci Technol,* 1988, **5,** 79-87.
38. K. Fennelly. Listening to the experts: provider recommendations on the health needs of immigrants and refugees. *J Cult Divers,* 2006, **13,** 190-201.
39. P.L. Meinhardt. Recognizing waterborne disease and the health effects of water contamination: a review of the challenges facing the medical community in the United States. *J Water Health,* 2006, **4 Suppl 1,** 27-34.
40. C.E. Hsu, F.S. Mas, H. Jacobson, R. Papenfuss, E.T. Nkhoma, and J. Zoretic. Assessing the readiness and training needs of non-urban physicians in public health emergency and response. *Disaster Manag Response,* 2005, **3,** 106-11.

41. C.E. Hsu, F.S. Mas, H.E. Jacobson, A.M. Harris, V.I. Hunt, and E.T. Nkhoma. Public health preparedness of health providers: meeting the needs of diverse, rural communities. *J Natl Med Assoc,* 2006, **98,** 1784-91.

42. S.D. Martin, A.C. Bush, and J.A. Lynch. A national survey of terrorism preparedness training among pediatric, family practice, and emergency medicine programs. *Pediatrics,* 2006, **118,** e620-6.

43. T. Cross, Bazron, B., Dennis, K., Isaacs, M. . Towards a culturally competent system of care: A Monograph on Effective Services for Minority Children Who Are Severely Emotionally Disturbed: Volume I. Washington, DC: Georgetown University Child Development Center.; 1989.

44. M.A. Kutner, and National Center for Education Statistics. *The health literacy of America's adults*
results from the 2003 National Assessment of Adult Literacy, U.S. Dept. of Education National Center for Education Statistics, Washington, D.C.; 2006, Page.

45. L. Nielsen-Bohlman, and Institute of Medicine (U.S.). Committee on Health Literacy. *Health literacy : a prescription to end confusion*, National Academies Press, Washington D C; 2004, Page.

46. R.D. Eagleson. Adult literacy, plain English and official documents. In. Sydney, Australia: University of Sydney, Dept. of English; 1985.

47. R.E. Rudd, E.K. Zobel, C.H. Fanta, P. Surkan, J. Rodriguez-Louis, Y. Valderrama, and L.H. Daltroy. Asthma: in plain language. *Health Promot Pract,* 2004, **5,** 334-40.

48. C.C. Doak, L.G. Doak, and J.H. Root. *Teaching patients with low literacy skills,* 2nd,ed, J.B. Lippincott, Philadelphia; 1996, Page.

49. Guidance Memorandum. January 29, 1998. Title VI Prohibition Against National Origin Discrimination—Persons with Limited-English proficiency. In: US Department of Health and Human Services.; 1999.

50. G. Flores. Language barriers to health care in the United States. *N Engl J Med,* 2006, **355,** 229-31.

51. G. Flores, M. Abreu, and S.C. Tomany-Korman. Limited English proficiency, primary language at home, and disparities in children's health care: how language barriers are measured matters. *Public Health Rep,* 2005, **120,** 418-30.

MEDICAL PREPAREDNESS FOR WATER CONTAMINATION EVENTS

P. L. Meinhardt, MD, MPH, MA

Center for Occupational and Environmental Medicine, Arnot Ogden Medical Center, 600 Ivy Street, Elmira, New York, 14905 USA. E-mail: epidoc@twcny.rr.com

1 INTRODUCTION

Water supplies and water distribution systems represent potential targets for terrorist activity based upon the critical need for water in every sector of our industrialized and developing societies. Even short-term disruption of water service can significantly impact a community and intentional contamination of a municipal water system as part of a terrorist attack can lead to serious medical, public health, and economic consequences. The majority of practicing physicians and public health professionals has received limited training in the recognition and evaluation of waterborne disease from either natural or intentional contamination of water. Therefore, they may be poorly prepared to detect water-related disease resulting from intentional contamination and may not be adequately trained to respond appropriately to a terrorist assault on water.

The purpose of this review is to address this critical information gap and present relevant epidemiologic and clinical information for public health and medical practitioners who may be faced with addressing the recognition, management, and prevention water terrorism in their community. This summary will also discuss ten unique challenges facing the public health and medical community with regard to recognizing and managing water contamination emergencies resulting from accidental or intentional contamination of water. A series of recommendations for effective medical preparedness for water-related events will also be presented in this chapter as a guide for future reference for all responsible parties involved in emergency planning and threat management of water contamination events.

2 WATER CONTAMINATION AND PUBLIC HEALTH CHALLENGES

Deliberate or accidental contamination of water reserves with biological, chemical, and radiologic agents has the potential to affect the health of millions of citizens across the globe. Any successful strategy to ensure water quality and security must include a multi-disciplinary approach that embraces active coordination and cooperation among many different professionals typically not used to working together, which represents a significant challenge for any community. Preservation of water quality and prevention of the health effects of water contamination is a complicated task requiring a coordinated effort from many diverse disciplines including public health and medical practitioners, water utility and public infrastructure professionals, public safety and law enforcement officials, and emergency response and community disaster managers. Preparing for and responding to water contamination emergencies is a collective responsibility that requires collaboration on a universal scale in order to effectively address this potential public health challenge.

Another important public health challenge resulting from water contamination emergencies is the fact that water consumers are frequently unaware of the potential health risks associated with exposure to waterborne contaminants and often consult medical practitioners who are unfamiliar with water contamination hazards and their subsequent impact on human health. Mis-diagnosis and under-diagnosis of waterborne disease by the medical community is common and may result in morbidity for the general population and, potentially, mortality in vulnerable populations at increased risk of water-related disease.[1] Inadequate diagnosis and under-reporting of cases of waterborne disease by medical practitioners can confound waterborne disease surveillance efforts by the public health workforce. It can also delay implementation of health advisories initiated by government authorities and water treatment mitigation efforts by water utilities in the event of a breach in water security. Therefore, prepared medical and public health practitioners may make the difference between a controlled response to a water contamination event versus a public health crisis.

In addition, the events of September 11th in the United States (US) and other terrorist activities throughout the world emphasize the vulnerability of our societies and our public infrastructures to acts of terrorim.[2] These events have added new importance to the potential threat of water terrorism and the need for the public health and medical community to recognize unusual waterborne disease trends that may result from intentional contamination of water with biological, chemical or radiologic agents.[2] Public health and medical practitioners throughout the United States and across the world must be especially vigilant in light of the fact that they are likely to be the *first to observe* the early warning signs and changes in illness patterns that may result from intentional acts of water terrorism and must understand their critical role in protecting the public's health.[2] The medical and public health community are positioned to play a critical role in minimizing the impact of an act of water terrorism by practicing medicine with an increased index of suspicion that such an attack could occur in their community.

3 PUBLIC HEALTH CONSEQUENCES OF WATER CONTAMINATION

Disruption of water service can significantly impact a community and accidental contamination or intentional contamination of a municipal water system may lead to a myriad of negative medical, public health, and economic outcomes. These consequences are magnified by the fact that water plays an essential role in every segment of our industrialized and developing societies and access to potable water is a cornerstone of every community's public health.[3] Accurate diagnosis and timely detection of waterborne disease by the medical and public health community is critically important since the consequences of a waterborne disease outbreak are sobering, particularly if public drinking water is contaminated. In North America, contamination of water with biological, chemical, and radiologic agents has generally resulted from natural disasters or unintentional man-made accidents. A review of two important examples from this region of waterborne disease outbreaks resulting from accidental contamination of municipal drinking water systems illustrates the serious outcomes for both urban and rural communities from a water contamination event.

The massive outbreak of waterborne cryptosporidiosis in metropolitan Milwaukee, Wisconsin in 1993 illustrates how contaminated water distributed through a large municipal water system can lead to a major public health challenge for a metropolitan community.[4] An estimated 403,000 Milwaukee residents developed diarrhoea from the waterborne pathogen, *Cryptosporidium,* reflecting an attack rate of 52% of the population that was served by the contaminated municipal water system in Milwaukee.[5] In addition, more than 4,000 Milwaukee residents were hospitalized during the waterborne outbreak with cryptosporidiosis listed as the underlying or contributory cause of death in 54 residents subsequent to the outbreak.[6] It has been estimated that 725,000 productive days were lost as a result of the water contamination event at a cost in excess of $54 million in lost work time and additional expenses to local municipalities.[7]

In 2000, the municipal water supply of the small rural community of Walkerton, Ontario was contaminated with *E. coli* O157:H7 resulting in 2,300 symptomatic residents and seven deaths attributed to the waterborne disease outbreak.[1] More than $11 million was required to re-construct the rural community municipal water system and install temporary filtration after the *E. coli* O157:H7 water contamination event. One year after the outbreak, the estimated total cost of the Walkerton, Ontario waterborne disease outbreak and municipal water contamination event had already reached $155 million.[1]

4 POTENTIAL PUBLIC HEALTH THREAT OF WATER TERRORISM

Terrorist activity in the US has forced the public health and medical community, federal security and regulatory agencies, and state and local water authorities to consider the possibility of intentional contamination of US water supplies as part of an organized effort to disrupt and damage critical elements of the national infrastructure.[2,8-10] Water supplies and water distribution systems represent potential targets for terrorist activity based upon the critical need for water in industry, agriculture, food production, and maintenance of the public's health.[9]

The deliberate contamination of the wells, reservoirs, and other water sources with infectious pathogens has been employed as a method of attack by military forces throughout the history of war. Many armies have resorted to using this method of warfare including the Romans who frequently contaminated the drinking water of their enemies with diseased human cadavers and animal carcasses.[2] The mechanisms of dispersal of biological, chemical, and radiologic warfare agents have expanded considerably and currently include water as a delivery mechanism. Whether advanced scientific techniques or ancient warfare methods are used by terrorists, intentional contamination of water supplies remains a potential public health threat for the world's population.[2]

The plausibility of intentional contamination of water supplies as a potential terrorist scenario has been supported by recent US congressional testimony, a consensus statement by a US governmental review panel, and previous US Centers for Disease Control and Prevention (CDC) and US Environmental Protection Agency (EPA) water advisory health alert.[11,12] In 2002, the National Academy of Sciences concluded that water supply system contamination and disruption should be considered a possible terrorist threat in the US.[11] On February 7, 2003, the National Terrorism Threat Level in the US was elevated to a "high risk" threat level based upon information received and analyzed by federal intelligence agencies. Subsequent to this heightened alert, the CDC and the EPA issued a *Water Advisory in Response to the High Threat Level* detailing the enhanced vigilance required from the public health, medical, and water utility communities regarding the risk of a terrorist attack on the US water infrastructure (Figure 1). The water advisory emphasized the need for public health agencies and water utilities to work together, communicate effectively, and confirm operational plans for emergency response.[12]

This is an official
CDC HEALTH ADVISORY

Distributed via Health Alert Network
February 07, 2003, 20:56 EDT (8:56 PM EDT)
CDCHAN-000113-03-02-07-ADV-N

CDC and EPA Water Advisory in Response to High Threat Level

Today, the Department of Homeland Security upgraded the Homeland Security Advisory System from yellow level (elevated risk of terrorist attack) to orange level (high risk of terrorist attack).

While there are no data to indicate that water has been specifically targeted, our nation's water infrastructure remains at risk to terrorist attacks, or acts intended to substantially disrupt the ability of a water system to provide a reliable supply of water. Therefore, public health agencies and water utilities are encouraged to continue to work together, keep each other informed of any unusual activities, and confirm the proper operation of notification channels in emergency response plans.

Public health agencies should immediately notify local water utilities and the state's drinking water administrator in the event of an unusual number of cases of gastrointestinal illnesses or other indications of illness that may suggest water contamination by a biological, chemical or radiological agent.

Water utilities should immediately notify public health agencies 24/7 emergency operations number, and the state's drinking water administrator in the event of specific threats received at a water facility, customer complaints in water quality, or if circumstances lead the utility to believe that the water has been or will be contaminated with a biological, chemical or radiological agent.

The Centers for Disease Control and Prevention (CDC) and the U.S. Environmental Protection Agency (EPA) issue this advisory jointly.

Categories of Health Alert messages:
Health Alert: conveys the highest level of importance; warrants immediate action or attention.
Health Advisory: provides important information for a specific incident or situation; may not require immediate action.
Health Update: provides updated information regarding an incident or situation; unlikely to require immediate action.
===
You have received this message based upon the information contained within our emergency notification data base. If you have a different or additional e-mail or fax address that you would like us to use please notify us as soon as possible by e-mail at healthalert@cdc.gov.

Figure 1 *CDC and EPA Water Advisory in Response to High Threat Level distributed as part of the emergency notification Health Alert Network (HAN) on February 7, 2003.[12]*

5   MEDICAL CHALLENGES ASSOCIATED WITH WATERBORNE DISEASE

Recognizing and managing a waterborne disease outbreak and the health effects of exposure to contaminated water is a diagnostic challenge for public health and medical practitioners in the best of circumstances.[1] These challenges will be even more significant in an emergency situation resulting from waterborne exposure to potentially weaponized biological, chemical or radiologic agents as part of a terrorist attack.[2] The public health and medical challenges that are associated with water-related disease resulting from either accidental or intentional contamination of water include but are not limited to the following key issues:

5.1. Medical and public health practitioners may be the first to recognize a water contamination emergency and may need to act as "front-line responders" in detecting accidental or intentional water contamination and subsequent waterborne disease.

Although environmental detection methods for recognizing accidental or intentional contamination of a water reserves are improving, the most likely initial indication that a water contamination event has occurred will be a change in disease trends and illness patterns.[1,2,10] This initial indication could also present as a cluster of water-related cases of chemical or radiologic toxicity in the general population or in sensitive sub-populations at increased risk of disease. Local community healthcare providers are likely to provide the first warning to local public health authorities that an accidental or intentional water contamination event has occurred in their community. Early recognition, timely outbreak investigations, accurate diagnosis, and conscientious reporting by the medical and public health community of suspected water-related disease cases will be essential to maintaining water security and safety in the future.[1,2]

5.2. Prompt identification of waterborne disease and the health effects of water contamination resulting from a water-related emergency may be confounded by difficulties in early diagnosis.

The signs and symptoms of water-related disease are often non-specific and mimic more common medical conditions and disorders unrelated to water contaminant exposure.[1] In addition, many waterborne diseases resulting from exposure to weaponized or traditional biological, chemical or radiologic agents present with vague, non-specific symptoms in the early phases of illness and may be difficult to differentiate from other naturally occurring disease in a community.[13] A key factor in the accurate diagnosis and appropriate management of disease resulting from water-related contamination events is inclusion of water by public health and medical practitioners as one possible exposure pathway for the dissemination of biological, chemical, and radiologic agents at the time of initial case presentation.

5.3. Waterborne exposure to biological, chemical or radiologic contaminants may produce a broad spectrum of disease involving virtually every organ system and may enter the body through several portals or exposure scenarios.

Accidental or intentional contamination of water supplies with biological, chemical or radiologic agents may produce a broad spectrum of disease and involve virtually every organ system including but not limited to the gastrointestinal, respiratory, dermatologic, hematopoietic, immunologic, reproductive, and nervous system. In addition, waterborne agents may enter the body through various portals including: 1) ingestion and aspiration of contaminated water; 2) dermal absorption of contaminated water during bathing or recreational activities; 3) inoculation of skin lesions from direct contact with contaminated water; 4) consumption of food directly contaminated by water during food preparation; and 5) consumption of food indirectly contaminated by water through uptake in the food chain or through agricultural practices.[1,2]

5.4. Many traditional and weaponized biological, chemical and radiologic agents display a significantly different clinical picture when the route of exposure is ingestion through water consumption.

Whether water is contaminated accidentally or intentionally, if the route of exposure or mode of delivery is food or water, additional considerations are important for the public health and medical community to incorporate into their patient and outbreak evaluations.[13,14] When epidemiologic investigations and clinical assessments are restricted to evaluation of inhalation and cutaneous routes of exposure to biological, chemical or radiologic agents, exclusion of the waterborne route of exposure may confound diagnosis, delay treatment, and impede protective public health measures.[15-17]

5.5. Obtaining accurate exposure histories from exposed patients may be extremely challenging and often very confusing leading to delays in early detection of a water contamination emergency resulting from biological, chemical or radiologic agent exposure.

Most biological, chemical and radiologic compounds are not detectable by human senses since they are often colourless, odourless and tasteless.[2] Water may be contaminated without any change in appearance or physical characteristics obvious to patients or water consumers.[1] Obtaining accurate exposure histories from symptomatic patients may be difficult since they may be unaware of their water contaminant exposure prior to seeking medical evaluation and treatment. In addition, the majority of agents that may be the source of accidental or intentional contamination of water supplies are not unique to water. Most of these agents may be distributed through multiple routes of exposure and modes of delivery in addition to the waterborne route. Moreover, public drinking water may represent only one source of waterborne exposure with accidental or deliberate contamination of recreational waters, swimming pools, and bottled water also possible exposure routes.[1,2,15]

5.6. Water-related disease resulting from intentional or accidental contamination may present as benign symptoms or self-limited illness in a healthy population while the same waterborne exposure in a vulnerable patient population may result in significant morbidity and mortality.

The public health impact of a water contamination event depends upon not only the virulence and toxicity of the water contaminant but also upon the type of population exposed and their level of immunity or vulnerability.[15,18] Individual vulnerability to traditional or weaponized waterborne agents may vary widely and differences in host susceptibility factors may complicate recognition of an accidental or intentional water contamination event.[2] Susceptible or vulnerable sub-populations may experience significant medical sequelae from water-related disease at lower levels of exposure to waterborne contaminants than the general population. These factors reinforce the fact that the public health and medical community must address the special needs of susceptible or high-risk populations that may develop severe and fatal systemic disease from the same waterborne exposure that may present as an asymptomatic or mild illness in the general population.[1,2,19]

5.7. Water contamination emergencies resulting from accidental or intentional contamination may occur in any size community leading to serious water-related disease and major public health consequences for both rural and metropolitan communities.

Water systems in small rural communities are at equal risk for experiencing an accidental or intentional water-related contamination event as medium and large metropolitan systems. In the case of water terrorism, small community water systems may represent "testing grounds" for larger scale attacks in metropolitan municipal water systems.[1,2] Future terrorist activity may not follow an expected pattern of attack with respect to water requiring a high level of suspicion and ongoing diligence by the public health and medical community. An outbreak of terrorism-related waterborne disease in a small rural community may act as a warning of a more large scale attack. This potential scenario reinforces the need for incorporation of terrorism-related waterborne disease into the differential diagnosis of every public health and medical practitioner in practice no matter how small or large their community.[2,20,21]

5.8. A coordinated and effective response to water contamination events resulting from either accidental or intentional contamination depends upon cooperation among a multidisciplinary team of professionals.

In order for the public health and medical community to effectively respond to accidental or intentional water contamination events, they will need to develop and foster new partnerships and working relationships with water utility practitioners in order to protect the public's health and ensure water safety.[1,2] All medical preparedness and response activities targeting water-related events will require cooperation, coordination, and collaboration among a multidisciplinary team of healthcare providers, public health and water utility practitioners, law enforcement professionals, and community leaders in order to mitigate the negative public health consequences of water contamination from biological, chemical or radiologic agent exposure.

5.9. Medical and public health practitioners are frequently tasked with providing credible and timely risk communication and public notification of water-related disease risk and water contamination event management.

Healthcare providers are among the most trusted sources of information for the general public regarding drinking water quality and safety.[1,2] Community residents will immediately turn to their healthcare providers and public health leaders for advice regarding the safety of their drinking water during and after an accidental or intentional water contamination event. Therefore, the medical and public health community will be required to play a leading role in public risk communication following a water-related event in order to diminish: a) the public health impact of the water contamination event, b) the secondary disruption to potable water distribution and availability, and c) the psychological impact of the public's lack of confidence in water safety and quality following a contamination event.[1,2,9]

5.10. The medical consequences of intentional or accidental biological, chemical or radiologic contamination of water have the potential to be both short-term and long-term public health and community disasters.

Medical and public health practitioners will be tasked with developing medical preparedness strategies and countermeasures that address BOTH the short and long-term consequences of biological, chemical or radiologic contamination of water reserves and subsequent disease and illness in their patients and local community.[1] Long-term public health consequences of accidental or intentional water contamination may include delayed, prolonged, and environmentally mediated health effects for months to years after a water contamination event in an affected community.[2]

6 MEDICAL PREPAREDNESS AND RESPONSE FOR WATER EVENTS

Although the public health and medical community may not be able to prevent the first cases of illness resulting from a water contamination emergency, they are positioned to play a critical role in minimizing the impact of such an event in their communities.[1,2,14,17] The public health consequences of accidental or intentional contamination of water with biological, chemical or radiologic agents can be very serious as noted in this review. Prepared medical and public health professionals will make the difference between a controlled response to a water-related event instead of a public health disaster.[1,2,22]

Although environmental technology for monitoring agents in water is improving rapidly, real-time detection and agent identification is often not available for measuring many biological, chemical and radiologic compounds that may contaminate water.[1,2] Therefore, early detection and rapid response to water contamination events by medical and public health practitioners are critical elements to any effective natural disaster or terrorism response strategy. Several emergency response protocols and disaster tools have been prepared for the medical and public health community that are directly applicable to both accidental and intentional water contamination events.[23] However, the following medical preparedness and response strategies should also be incorporated into community emergency preparedness and planning activities that are specific to water contamination and provide additional guidance.

6.1. Medical and public health practitioners should have a working knowledge and basic understanding of the importance of water system vulnerabilities and potential water contamination pathways.

A major effort has been undertaken to improve and enhance the ability to detect and characterize accidental and deliberate contamination of water systems as part of a collaborative effort by both water utilities and public health agencies.[1,2] However, as is the case with all infrastructure protection and safety measures, several potential points in a water infrastructure remain vulnerable to either accidental or intentional contamination. Therefore, it is critically important for medical and public health practitioners to have a basic understanding of these water system vulnerabilities in order to be able to complete an accurate exposure history, thorough patient evaluation, and a comprehensive epidemiologic investigation of any water contamination event.

Table 1 provides a list of potential points of water contamination from both accidental and intentional contamination of water.[2] The information outlined in this table is presented as a preparedness resource for healthcare providers and public health professionals to keep in mind when evaluating an unusual symptoms complex or an atypical illness pattern that may represent a case of water-related disease from biological, chemical or radiologic contaminants.

Table 1 *Possible points of accidental or intentional contamination of water systems* [a]

Upstream of a community water supply system or collection point - Water supply systems are comprised of small streams and bodies of water, rivers, service reservoirs, aquifers, wells, and dams that may act as points for accidental or deliberate contamination of water.

Community water supply intake access point or at the water treatment plant – Many water supply systems are designed to receive water from source water reserves at a central intake point with this source water subsequently filtered and sanitized at the community water treatment facility for eventual distribution as potable water. Both water intake points and community water treatment plants may be targeted for deliberate contamination of water and may also be contaminated accidentally.

Selected points in the post-treatment water distribution system – Treated water is distributed to water consumers or end-users through transmission pipelines to homes and businesses. Selected portions of a water distribution system or water main are another potential point of water contamination that may be targeted by terrorists or contaminated accidentally and affect a sub-division, specific neighbourhood, school, medical centre or nursing home.

Private home or office building water supply connection, individual building water supply, water tanks, cisterns, or storage tanks – Treated water that is stored very close to the water consumer or end-user as well as individual house or building connections may serve as points of contamination of water during intentional or accidental contamination events.

Water used in food processing, bottled water production, or commercial water – Water used for food processing or preparation as well as bottled water production also represent points of potential water contamination from accidental or intentional contamination.

Accidental or deliberate contamination of recreational waters and receiving waters - Both treated and untreated recreational waters may serve as a point of potential contamination of water including swimming pools, water parks, and natural bodies of water (small lakes and ponds). Receiving waters such as rivers, estuaries, and lakes may be secondarily contaminated with wastewater from sanitary and storm sewer systems that may have been environmentally or intentionally contaminated by biological, chemical or radiologic agents.

---

[a]Modified and reprinted with permission from the author and Arnot Ogden Medical Center[2]

6.2. Healthcare and public health practitioners should rely upon the use of epidemiologic indicators and syndromic surveillance for early detection of water contamination events.

Healthcare practitioners will need to 'think like an epidemiologist' when evaluating any suspect case or unusual pattern of disease in their clinical practice in order for follow-up epidemiologic investigations to be initiated and appropriate remediation and prevention efforts to be instituted by the public health community following a water contamination event.[24] Several epidemiologic patterns and sentinel clues have been published and provide a valuable resource for both the medical and public health community facing the challenges of diagnosing water-related disease.[2,20,24-28] The medical and public health community will also need to embrace the process of using syndromic surveillance of disease trends as a valuable diagnostic tool in order to recognize water-related events resulting from accidental or intentional contamination of water.[2]

Table 2 summarizes several epidemiologic indicators and characteristic syndromes that may indicate the presence of a water contamination event and subsequent population disease. These epidemiologic indicators or sentinel clues may result from multiple exposure pathways including water to biological, chemical or radiologic agents and have universal application in a clinical and public health setting.

Table 2. *Epidemiologic indicators and sentinel clues indicating possible illness from accidental or intentional water-related bioevents* [a]

Point source illness patterns with record numbers of severely ill patients presenting within a short period of time

Severe and frequent disease manifestations in previously healthy adult populations

Increased and early presentation of immunocompromised patients and vulnerable population patients with debilitating disease since the dose of inoculum or toxin exposure required to cause disease may be less than for the general healthy population

Higher than normal numbers of patients presenting with gastrointestinal diagnoses for care

Localized areas of disease epidemics that may occur in a specific neighbourhood or sector possibly indicating contamination of a selected point in a post-treatment water distribution system

Multiple infections or cases of illness at a single location (school, hospital, nursing home) with a naturally occurring pathogen or toxin; atypical and rare biological pathogens; or chemical compound or radiologic agent not previously detected in that location

Unusual temporal or geographic clustering of cases with patients attending a common public event, gathering, or recreational venue

"Impossible epidemiology" with naturally occurring diseases diagnosed in geographic regions where the disease has not been encountered previously

Very high attack rates with 60-90 percent of potentially exposed patients displaying symptoms or disease from possible intentional contaminant exposure

Record number of fatal cases with few recognizable signs and symptoms indicating lethal doses near a point of dissemination or dispersal source of intentional contaminants

Lack of response or clinical improvement of presenting patients to traditional treatment modalities

Near simultaneous outbreaks of similar or different epidemics at the same or different locations indicating an organized pattern of intentional biological, chemical or radiologic agent release

Endemic disease presenting in a community during an unusual time of the year or found in a community where the normal vector of transmission is absent

Unusual or uncommon route of exposure of a disease such as illness resulting from a waterborne agent not normally found in the water environment

---

Note: - None of these indicators alone are pathognomonic for water-related disease but are presented as an educational tool for use by healthcare providers and public health practitioners as possible disease trends that may warrant further investigation.

[a]Modified and reprinted with permission from the author and Arnot Ogden Medical Center[2]

6.3 Active utilization of web-based preparedness resources and disaster response tools by the public health, medical, and disaster preparedness community should be encouraged for detection, management, and prevention of water contamination events.

The continuously changing nature of emerging water contaminant agents and the ongoing terrorist threat of an attack with an exotic or weaponized biological, chemical, or radiologic compound will continue to challenge medical and public health practitioners as they attempt to detect and diagnose waterborne disease.[1,2] Without quick access to constantly updated and credible information, most healthcare providers and public health practitioners will not be able to rapidly evaluate, manage, and prevent waterborne disease resulting from exposure to water that has been accidentally or deliberately contaminated with biological, chemical or radiologic agents.

Two free web-based preparedness resource guides and disaster response tools have been developed for the medical and public health community that address all aspects of the evaluation and management of water-related disease resulting from both accidental and intentional water contamination.[1,2] These two companion guides are posted at www.WaterHealthConnection.org and are highlighted in Figure 2 and Figure 3. The primary purpose and educational intent of these preparedness and response tools is to provide the medical and public health community with streamlined access to resources that will help guide them through the recognition, management, and prevention of water-related disease resulting from all types of water contamination.

This water preparedness website has been peer-reviewed by medical, public health, and military experts and provides comprehensive resources for healthcare providers, emergency responders, public health authorities, and water utility professionals faced with recognizing and managing water-related disease in their communities. In the past four years, this medical website has received more than 9 million hits for information from over

300,000 visitors located in 89 countries throughout the world. Sustained use of these types of disaster preparedness resources and response tools by the medical and public health community is a critical part of any strategy to protect and secure water supplies and prevent waterborne disease in the general public resulting from intentional or accidental contamination.

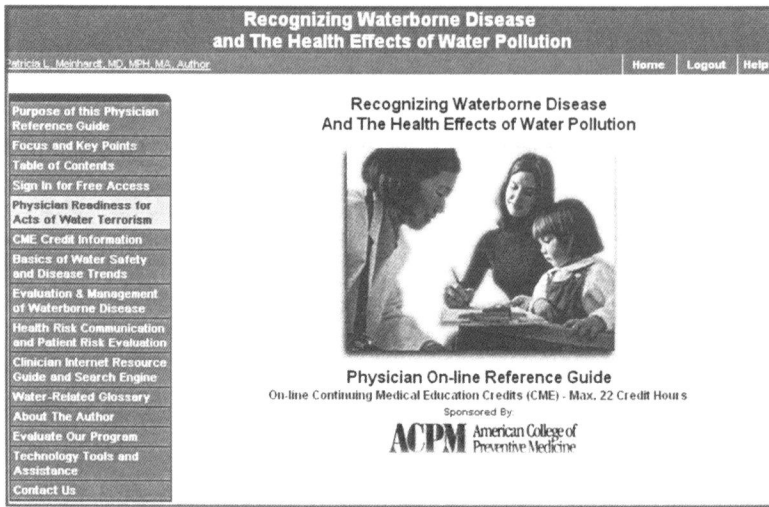

Figure 2. *Recognizing Waterborne Disease and the Health Effects of Water Pollution: Physician On-line Reference Guide accessible at www.WaterHealthConnection.org*[1]

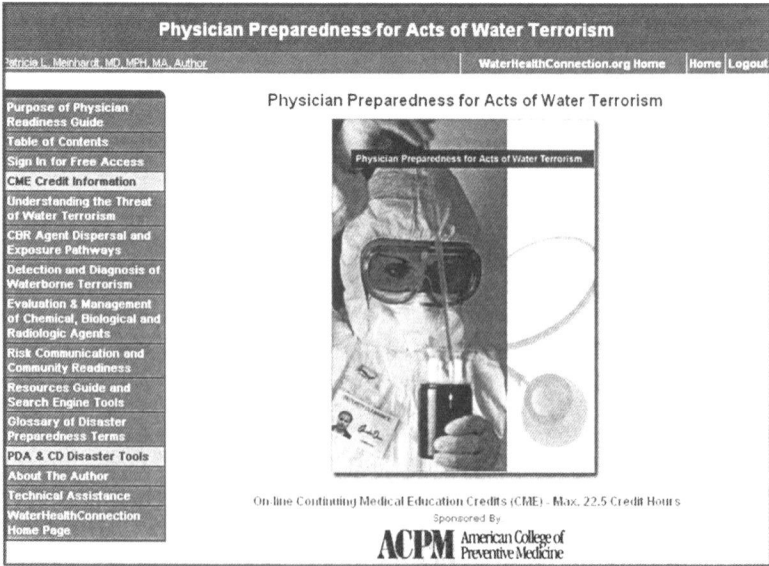

Figure 3. *Physician Preparedness for Acts of Water Terrorism: Physician On-line Readiness Guide accessible at www.WaterHealthConnection.org*[2]

References

1. PL Meinhardt, *Recognizing Waterborne Disease and the Health Effects of Water Pollution: Physician On-line Reference Guide,* American Water Works Association and Arnot Ogden Medical Center, New York, 2002; Accessed on April 1, 2008 at www.waterhealthconnection.org.
2. PL Meinhardt, *Physician Preparedness for Acts of Water Terrorism: Physician On-line Readiness Guide,* Environmental Protection Agency and Arnot Ogden Medical Center, New York, 2003; Accessed on April 2, 2008 at www.waterhealthconnection.org.
3. JM Last, *Public Health and Human Ecology,* Appleton and Lange. Stamford, CT, 1998, Chapter 1, p. 3.
4. TE Ford and WR MacKenzie, How safe is our drinking water? *Postgraduate Medicine,* 2000, **108**, 11-14.
5. WR Mackenzie, NJ Hoxie, ME Proctor, et al, A massive outbreak in Milwaukee of Cryptosporidium infection transmitted through the public water supply, *New England Journal of Medicine,* 1994, **331**, 161-7.
6. NJ Hoxie, JP Davis, JM Vergeront, et al. Cryptosporidiosis-associated mortality following a massive outbreak in Milwaukee, Wisconsin, *American Journal of Public Health*; 1997, **12**, 2032-5.
7. Cryptosporidium in water supplies: second report of the group of experts, Chairman, Sir John Badenoch. London, England: HMSO, 1995.
8. RM Clark and RA Deininger, Protecting the nation's critical infrastructure: the vulnerability of US water supply systems. *Journal of Contingencies and Crisis Management,* 2000, **8**, 73-80.
9. G Krieger, Water and food contamination. In: *Terrorism: Biological, Chemical and Nuclear from Clinics in Occupational and Environmental Medicine,* eds. KH Chase, MJ Upfal, GR Krieger, SD Phillips, TL Guidotti and D Weissman, WB Saunders Company, Philadelphia, PA, 2003, pp. 253-62.
10. S States, M Scheuring, J Kuchta, et al. Utility-based analytical methods to ensure public water supply security, *American Water Works Association Journal,* 2003, **95**, 103-115.
11. National Research Council. *Making the Nation Safer: The Role of Science and Technology in Countering Terrorism,* Committee on Science and Technology for Countering Terrorism, The National Academies Press, Washington, DC, 2002.
12. Centers for Disease Control and Prevention and Environmental Protection Agency, *Water advisory in response to high threat level,* CDC Health Advisory, Health Alert Network, Accessed on February 7, 2003 at http://www.phppo.cdc.gov/han/HealthAlerts/.
13. Departments of the Army, the Navy and the Air Force and Commandant Marine Corp, *Treatment of biological warfare agent casualties field manual,* No. FM 8-284/NAVMED P-5042/AFMAN (I) 44-156/MCRP 4-11.1C, Washington, DC, 2000.
14. PS Brachman, Bioterrorism: an update with a focus on anthrax. *American Journal of Epidemiology,* 2002, **155**; 981-8.
15. WD Burrows and SE Renner, Biological warfare agents as threats to potable water, *Environmental Health Perspectives,* 1999, **107**, 975-984.
16. DR Franz, Defense against toxin weapons, US Army Medical Research Institute of Infectious Diseases, Fort Detrick, Maryland, 2002, Accessed on September 28, 2002 at www.nbc-med.org.

17. DR Franz, PB Jahrling, DJ McClain, et al, Clinical recognition and management of patients exposed to biological warfare agents, *Clinics in Laboratory Medicine,* 2001, **21**, 435-473.
18. AF Kaufmann, MI Meltzer MI, GP Schmid, The economic impact of a bioterrorist attack: are prevention and postattack intervention programs justifiable? *Emerging Infectious Diseases,* 2003, Accessed on January 21, 2008 at http://www.cdc.gov/ncidod/EID/vol3no2/kaufman.htm.
19. ILSI Risk Science Institute Pathogen Risk Assessment Working Group, A conceptual framework to assess the risks of human disease following exposure to pathogens, *Risk Analysis,* 1996, **16**;841-848.
20. JA Pavlin, Epidemiology of bioterrorism. *Emerging Infectious Diseases,* 2003, Accessed on January 21, 2008 at www.cdc.gov/ncidod/EID/vol5no4/pavlin.htm.
21. C Smith, Biological warfare and bioterrorism. *Journal of the Medical Association of Georgia,* 2002, **91**, 12-15.
22. DA Henderson, Bioterrorism as a public health threat. *Emerging Infectious Diseases,* 2003, Accessed on January 21, 2003 at www.cdc.gov/ncidod/eid/vol4no3/hendrsn.htm.
23. Centers for Disease Control and Prevention, *Public Health Emergency Response Guide, Version 1,* Accessed on April 1, 2005 at http://www.bt.cdc.gov/planning/pdf/cdcresponseguide.pdf.
24. FM Burkle, Mass casualty management of a large-scale bioterrorist event: an epidemiological approach that shapes triage decision, *Emergency Medicine Clinics of North America,* 2002, **20,** 409-436.
25. Centers for Disease Control and Prevention, Recognition of illness associated with the intentional release of a biologic agent. *Morbidity and Mortality Weekly Report,* 2001, **50**; 893-897.
26. FM Henretig, TJ Cieslak, EM Eitzen, Biological and chemical terrorism. *Journal of Pediatrics,* 2002, **141**, 743-746.
27. MG Kortepeter, JR Rowe, EM Eitzen, Biological weapon agents, in *Disaster Medicine,* eds. DE Hogan, JL Burstein, Lippincott Williams and Wilkins, Philadelphia, PA, 2002, pp. 350-363.
28. TW McGovern, GW Christopher, EM Eitzen, Cutaneous manifestations of biological warfare and related threat agents, *Archives of Dermatology,* 1999, **135**, 311-322.

KEEPING THE PUBLIC ON-SIDE AND MAINTAINING REPUTATION

J .B. Shaw

Utilities Project Management Ltd
34, Nightingale Close, Droitwich Spa, Worcestershire. WR9 7HB

1 WHAT ARE WE TRYING TO DO?

Communications in this era are an absolutely essential ingredient and a yardstick of measurement in knowing how well you have performed, and as importantly perceived to have performed, by both external and internal stakeholders when your organisation has an emergency.

Before going into the "meat" of the presentation, I would just like to spend a few moments exploring the background and context of my later remarks

2 STOCK-TAKE

While good systems and procedures are an important part of any communications plan, I believe it is business critical to get the foundations (policies) and underlying principles established with customers and as importantly within the business.

A good reputation, which we all strive for both personally and organisationally, is derived from meeting customers' expectations and those who influence them, consistently in a structured, efficient and friendly manner.

In my view, we need to spend some time on the stock-take, as external and possibly internal stakeholders will ruthlessly expose not having done so during and following a major emergency.

3 REALITY CHECK

It is always worth remembering the Tom Peters famous statement "*It doesn't matter how good you think you are, the only thing that matters is how good your customers think you are.*" It is nearly always your record on how you have communicated to those affected by an emergency that will determine in the eyes of others success or failure.

4 GETTING TO THE PROMISED LAND

Reputation, capability and organisation are key customer service "needs & wants" because I think this encapsulates exactly what we are trying to achieve.

Customer service needs and wants at the very heart of our thinking, underpinned by capability, and organisation will deliver the reputation we are seeking.

Reputation isn't just about being a "Rolls Royce" or Mercedes! We need to aim consistently for the reputation that best fits with our customers' needs and wants, and most importantly our capability and organisation to deliver. There is nothing wrong with a Morris Minor.

With a good reputation comes a store of good-will and you need a huge amount of that in a major emergency, and also during the recovery and restoration phase to normal supplies.

5 ORGANIZING FOR CUSTOMER CARE

I think it is vital for any plan to work that the organisation has an agreed vision, and is backed up by the organisation and capability to deliver that vision consistently.

Good reputations are derived not just from performing well once, but consistently both in day-to day transactions and emergencies.

Focusing on things from the customer's perspective is very important: -

What do we do for customers....   to make life easy? ...to make life difficult?

Service we give to our customers is determined by ..... the service we give to each other?

 Service excellence is about ..... doing it continuously?

Table 1 illustrates this key aspect.

6 TAKING THE TEMPERATURE

Unless we have the structure and organisation in place to measure before an emergency occurs, we will certainly not do it effectively in the white-heat of "battle".

Everything is put to the test-policies, strategy, procedures and decisions, and actions. The test moreover is almost certainly going to be a public one, with media spotlights seeking error, fault and of course blame.

7 MAJOR SUPPLY EMERGENCIES ARE RARE EVENTS

Table 2 highlights two such events where the companies, while doing many things right, were "pilloried" by the media and regulatory organisations for not having been seen to get the communications organised effectively. These organisations that have had "experience" have learnt the lessons and are probably best prepared for the next-time.

Both events highlighted particularly the need to establish the organisation and resources quickly to cover both inbound telephone/written communications and inform customers proactively through letters/cards and other means.

BUSINESS ISSUES	PUBLIC SEES e.g.	REQUIREMENTS e.g.
VISION What we aim for	• "Reputation" (perception)	• Continuous debate • Definition/focus/integration • Strategic objectives • Commitment • Consistency
ORGANISATION Policy, Procedures and Systems	• Problem/end results • Response/timing • Telephone interaction • Web site • Mailings	• Structure • Goals set/KPI's • Responsibilities identified and agreed • Standards agreed/SLA's • Contingency planning • Measurement • Monitoring &evaluation & action
CAPABILITY: **TEAM WORK** What we show the world	• Timely and up to date accurate information • Appreciation of inconvenience caused • Speed of work • Response to Problems/questions/ complaints	• Awareness of likely problems/questions • Co-ordination • Management style • Plans & instructions carried out. Including changes • Media relations • Informed releases (written & verbal)
CAPABILITY: PERSONAL What we say and do as individuals	• Language/politeness • Accurate information/ problems /questions /complaints • Working Methods • What we say we do	• Standards expected • Competency • Supervision • Encouragement • Recognition • Training/coaching/ corrective action

Table 1 *Organising for customer care*

I see as the key issues coming out of the official post incident reports the need to provide customers with effective communications, also the internal structure and staff briefing mechanisms also need to be in place.

Other organisations and agencies can also have a role to play in delivering communications. Police, Health Trusts, Social Services, elected representatives, amongst others.

If customers and the general public are having difficulty in obtaining information from you they will go elsewhere doing whatever it takes to get satisfaction.

While the media are very important in helping get messages out quickly, in my view we shouldn't put over reliance on this channel of communication. Keeping as much control as possible of the message in our own hands is vital.

Remember the only one who cares about your reputation is you and your organisation!

Case Study 1
Burncrooks diesel spillage 1997 – West of Scotland Water
"Water utilities should review all the methods available to them for informing their customers in an emergency, and incorporate the most appropriate methods in various circumstances into their emergency plans" – **The Scottish Office**

Case Study 2
Kemsing/Plaxtol Supply interruptions 1999 – Mid Kent Water
" Review all arrangements for communicating with Customers including manning the company's telephones as well as taking active steps to deliver sound information and advice to customers" **OFWAT**

Table 2 *Two examples of major supply emergencies*

8 WHAT IS POSSIBLE?

I come back to the point about agreeing the vision. It is important that as a minimum, organisations are at one with the vision and the plans, and resources available to deliver it. If we can get customer "buy-in" so much the better.

You will recall that the government a few years ago sent out a leaflet to every residential and business address in the land called "Preparing for Emergencies"- What you need to know?

Perhaps there is now another opportunity for the companies to start educating and have on-going dialogue with customers in how both parties can help each other in an emergency. Having pre-determined contingency plans with major accounts would also be helpful.

The benefits to your reputation, while hard won, would be significant even if you did not have a major event, as you would be differentiated positively from other organisations in the minds of your customers and other stakeholders.

It is important to remind ourselves that it is not always the water company that directly induces a water supply emergency. The floods of last year present a long-lasting reminder.

It is also important that all those who play a part in an emergency, including support staff, have a prior awareness of their role and access to guidance checklists within an organized structure and framework.

9 KEY ISSUES 1

We should remember that the Water Industry has one of the widest diversification of customers of any industry delivering products and services in UK. Getting the message and delivery mechanisms right applicable to the area affected should not be left to chance.

Vulnerable Customers in an emergency fulfil different criteria to those utilized by Customer Services in normal day-to-day situations. While we have self-elective registers by customers this is only the "tip of the iceberg" and we need to build a database of contacts that will help us deliver the additional services required, at the time of any emergency, to the people that need them most.

We also need to ensure that we know where we have significant immigrant populations, - European and others- that we take account of their requirements in all our external communications.

The blind, and the deaf are also vulnerable in different ways and we need to consider how the message(s) is effectively transmitted to them.

Finally we have those visiting the area, - such as tourists on caravan and camping sites that need to be catered for.

I believe there is a duty to prioritise, as it is impossible to reach everybody immediately. Policies, procedures and databases should be in place to help this process.

10 KEY ISSUES 2

Consistency of message is important when dealing with all inbound and outbound communications and to achieve that, the internal communication system needs to be very slick and precise.

I just want to touch on briefly the issue of delivering card/letter communications in an emergency. Whether it is an outside agency including Royal Mail, or undertaken in-house there is a significant amount of thought and preparation required for the task to be undertaken rapidly, efficiently and attending to the priorities.

GIS has become a valuable tool in helping to identify properties that could be in the area affected. However, unless along with other systems it can produce address points and usable maps of the affected area quickly and reliably, we can lose valuable hours.

If you are contemplating delivery by your own staff or other agencies outside the Royal Mail you will need usable street address lists- no's of customers in each road so as to provide an effective split of walk-orders. Also consider typography of the area, including availability of pathways and potential no-go areas, as these will have a significant effect on delivery and other resources required. Robust Health and Safety assessments and regimes should not be forgotten!

Printing and production lead-times of communications, especially where there are large volumes should also be considered and tested, and dovetailed into your target delivery plan.

It is also perhaps worth considering whether you should have "all your eggs in one basket" with the Royal Mail if that is what you currently have?

As part of the process of effective monitoring, I believe the setting and agreeing of realistic targets should be in place including the service provided by Royal Mail. I also think it is worth pointing out that once the message is in the Royal Mail system you have lost effective control if circumstances rapidly change.

11 PRIORITIZED & PRACTICAL COMMUNICATION PLANS

As companies have slimmed down their own labour force, it may well be that you need support from your contractors and other suppliers to help with the communication process. To contribute to a successful outcome they need to have an understanding and awareness of their role and the level of resources that they might be required to provide.

In some organisations the reliance on contractors may well become a Business Continuity issue.

I believe that it is wise to have your own intelligence network monitoring what is going on, not just from a technical viewpoint. As well as planning and deploying management, you can use other agency contacts such as Neighbourhood Watch to help you monitor performance.

I come back to the point made at the start that if we are not undertaking effective monitoring, other potentially hostile individuals and organisations will!!!

Tidy up the site- when we finish an excavation we do the reinstatement and take barrier boards etc away. In communication terms make sure that customers and the public, plus all other relationships formed with other stakeholders during the emergency are thanked for their help, know the emergency is over and any follow-on actions specified. Unless these messages get through, there will be ongoing confusion and disappointment, as well as potentially high costs in dealing with people subsequently.

12 SUMMARY

- Understand the expectations
- Organise for and communicate what is possible
- Test out your structures, policies, procedures and facilities
- Monitor and regularly review your plans
- Turn a potential negative into a positive experience for customers and supplying organisation
- Don't walk away when the emergency is over. – Use and maintain key relationships. Lightning can and often does strike twice for the unprepared

The company's reputation is on the line!

SOCIOLOGICAL AND PSYCHOLOGICAL CONSTRAINTS TO LEARNING FROM FAILURE

B.H. MacGillivray

Department of Management Science, Lancaster University Management School, Bailrigg, Lancaster, LA1 4YX, UK

"Experience is a revelation in the light of which we renounce our errors of youth for those of age."

Ambrose Bierce

1 INTRODUCTION

Despite recent advances in the methods, technologies and resources for the provision of safe drinking water, contamination events remain depressingly common within developed nations. Overlooking for a moment the human and financial toll that such incidents exact, perhaps the most disturbing insight to be drawn is that, beyond their specific technical or physical characteristics, they are uncomfortably reminiscent of one another.[1] Moreover, such incidents do not merely happen; they are typically preceded by warning signals or near-misses, the lessons of which are all too rarely heeded. In this paper, I argue that this tendency to repeat the failures of the past cannot be attributed solely to complacency, nor blamed on isolated glitches such as poor communications or resource limitations. These explanations certainly have their merit in individual cases, but it seems plausible to suggest that the ubiquitous nature of this dysfunctional learning, in the sense of neglecting to pay due regard to past errors or even learning the wrong lessons from them, is deeply rooted in social organisation and human psychology. Drawing upon inquiries into near-misses and outbreaks within the water sector, along with the sociological and psychological literatures, I set out an argument that near-miss and incident investigations are all too often contaminated by political interests, social forces, and psychological biases. Combined, these constraints encourage the suppression of error, the denial of responsibility, the distortion of evidence, and a systematic tendency to explain away contamination events as narrow anomalies ("acts of God") or simple matters of "human error." I'll end on a more positive note, by speculating how these factors can be attenuated, such that organisations are not prisoners of their pasts.

2 LEARNING FROM FAILURE

By learning from failure, I simply mean improving system safety through a trial and error approach. This trial and error process begins with a contamination event or near-miss, following which incident analysis is undertaken to determine its cause(s), and concludes with a technical, operational or administrative solution designed to resolve any identified weaknesses (*e.g.* adapting design standards or operating procedures). Ideally, these investigations should address not just the immediate, proximate causes (*e.g.* operator error), but also the deep and often latent organisational failings which generate those causes (*e.g.* an emphasis on efficiency over reliability) and render them problematic. In other words, they should accept that humans are, well, human, and therefore are fundamentally error-prone, and that these errors are often encouraged or amplified by actions, decisions or strategies made "upstream" in the organisational sense.[2] Similarly, investigations should acknowledge that technology will inevitably fail at times, and not view this as the end of the story; instead questioning, for example, whether the design assumptions about reliability were realistic, or whether appropriate redundancy was built into the system.

The catalyst for learning, then, is the recognition that something has "failed," in the sense that there is an observed discrepancy between our expectations of how the system should work and the reality, and the mechanism for learning is causal attribution and resolution (detect the problem, find the cause, identify the cure). In the following sections, I shall set out the difficulties pervading each of these three stages. Although my discussion will focus on constraints to explicit approaches to learning (*e.g.* incident and near-miss investigations), it applies equally to learning occurring outside of such formal processes.

3 THE AMBIGUITY OF THE PAST

The first and most fundamental constraint to learning is that the truth or meaning of historic events is rarely self-evident; instead, the past and the lessons that should be drawn from it are characterised by ambiguity.[3] Water quality incidents and near-misses often leave scope for a broad range of plausible explanations as to why they occurred, and an equally wide array of fixes to prevent their recurrence. This is because any given failure has a plethora of pre-conditions, some immediate and visible, others remote and latent.[2] For example, the pre-conditions for a particular contamination event may include high-precipitation, poor land management practices, operator error, and an organisational culture which implicitly promotes efficiency over reliability. Moreover, such pre-conditions do not exist in a vacuum, they each have their antecedents. This would be problematic enough, even without the dearth of knowledge underpinning this process. Incident analysts are not all-seeing all-knowing beings; they are more like amateur historians seeking to reconstruct and interpret the past from the incomplete and often conflicting accounts of others. This ambiguity means that conclusions as to what has gone wrong and what should be done about it are inherently malleable. This in turn provides the space for political, social, and psychological influences to intrude upon and distort the learning process. This wouldn't be so interesting in its own right if these distortions were random; but to the contrary, I shall argue that they tend to be systematically biased towards the suppression of errors and the protection of the status quo.

4 SOCIOLOGICAL CONSTRAINTS TO LEARNING FROM FAILURE

There are a variety of social pressures found within all organisations which can significantly impede efforts to learn from the past, some self-evident, others more subtle. In this section I will focus on how organisational norms (*e.g.* failure, whistle-blowing as taboos; management resistance to bad news) and politics may combine to suppress meaningful learning through encouraging cover-ups and distorting causal attributions.

4.1 Fear of Blame and Organisational Intolerance of Dissent

Perhaps the most obvious constraint to error reporting is the fear of blame and subsequent sanction, whether to the individual reporting or to his colleagues.[4] This stems from Western cultural norms,[5] which: place primary importance on individual rather than collective responsibility; view the practice of whistle-blowing as tantamount to apostasy; and frame failures as indicative of incompetence rather than as resources for learning. To show that this is not mere theory, consider the Koebels' evasive and dishonest responses to regional health and environmental units as the Walkerton incident was unfolding: they failed to disclose the findings of *E. coli* and total coliforms; lied about chlorine residual results; and ultimately falsified operating records.[6] Is there a plausible explanation for this behaviour beyond that they didn't want to reveal their unsafe operating practices and receive the inevitable sanction?

However, it is not merely a fear of being blamed which encourages the widespread withholding of information about potential problems by employees. Organisations are generally intolerant of dissent, and so employees often fear the negative repercussions of speaking up, and may not believe that doing so would make any difference (thus view it as both dangerous and futile; see[7]). This is amplified in that management often feel threatened by negative feedback, and so try to avoid receiving it, and when they cannot, they my try to ignore it, dismiss it as mistaken, or attack the credibility of the source.[8] For example, testimony of one operator to the inquiry into the Battleford outbreak revealed that he was aware of a public health risk when a settling problem occurred in the solids contact unit in March 2001 (later identified as the proximal cause of the outbreak).[9] However, because he was applying for a promotion, he did not press the issue with his superior.[9] The strength of these tendencies vary strongly according to organisational structure and the environments in which they operate; for example, research has found that organisations which place a primacy on predictability, control and efficiency, and which operate in mature and stable industries, tend to be less open to dissenting opinions, and often place a primacy on unity of views in the belief that it will improve performance.[7] The former characteristics, at least, bear more than a passing resemblance to the water sector.

4.2 Organisational Politics: Power Structures and Conflicts of Interest

The desire to avoid blame and sanction is but one crucial interest which can conflict with the broader organisational need to learn from failure.[3] There are myriad other interests, some legitimate and some self-serving, which can strongly influence the selection, construction and interpretation of past failures (*e.g.* desires to protect existing power hierarchies and thereby maintain organisational stability, to avoid legal liability, to maintain public and regulatory trust, to reduce the cost of corrective measures, *etc.*). These produce powerful incentives to apportion blame in the light of an incident or near-miss to low-level or even deviant members of the organisation, rather than to those in power or, worse still, to the organisation itself (*e.g.* to find a scapegoat and so resolve the issue without threatening existing power structures or challenging deeply held assumptions about system design and operation). Such politicisation of the past is achieved through a range of mechanisms and at various stages in the learning process. The most basic but perhaps most important mechanism is the implicit pressure placed on investigators, who are often sharply aware of the political implications of their findings, and so may shy away from finding fault in the actions or decisions of higher authorities. More subtly, those in power have the ability to shape the architecture within which incident investigations are conducted, for example, through: establishing criteria for what passes as an incident or near-miss; defining the scope and depth of investigations; establishing criteria for what passes as plausible accounts; controlling which cues are highlighted or suppressed, and so forth.[10] If the history of warfare is written by the victors, then the history of organisations tends to be written by their elites, and at times in a self-serving fashion.

This partly explains why so many incidents and near-misses are blamed on the most proximate cause – human error by operators – rather than on more subtle, latent problems such as poor design or mismanagement by higher authorities.[11, 12, 13] This is not to suggest that water utilities are hotbeds of Machiavellian manoeuvring, but merely to highlight that creating sense out of the past does not occur in a political vacuum: people have motivations beyond simply the search for the truth, and at times they can subsume the latter. Power struggles, politics and conflict are normal realities of organisational life, and not unique to dysfunctional organisations.[14]

5 PSYCHOLOGICAL CONSTRAINTS TO LEARNING FROM FAILURE

Having covered the social and political problems that organisations face when seeking to learn from the past, I now turn to summarise some key psychological phenomena which pose similar challenges.

5.1 Counterfactual Thinking and Heuristic Judgements

In an ideal world, incident investigators would be the modern incarnation of Pascal, applying formal logic and probability theory to uncover and evaluate causal structures of failure. In the real world, limited knowledge, cognitive capacities and time make this an unobtainable norm. And so investigators are left with a form of reasoning known to psychologists as counterfactual thought.[21] Here, the chain of occurrences leading to a past undesirable event is reconstructed so as to produce a desirable outcome, and through this process the cause(s) of failure are isolated (*e.g. if only* we had issued a boil water notice on receipt of the *E. coli* monitoring results, the outbreak wouldn't have happened). Given the often broad range of causes which can be isolated, people subconsciously use various rules of thumb, or heuristics, to structure and simplify their counterfactual thinking.

Foremost amongst these is the availability heuristic,[22] which leads investigators to conflate the ease with which an explanation comes to mind with its explanatory power. This ease correlates highly with saliency; thus, the most familiar and striking explanations for an incident will be viewed as the most plausible. Of course this is a fairly useful way of thinking and generally leads to legitimate judgements; however in certain situations it can generate predictable and systematic error. For example, it is problematic when there have been recent changes to the water supply system, whether technical, organisational or in the surrounding environment, such that past experience is no longer a reliable guide to the present (*i.e.* familiarity is no longer a useful cue). Experimental findings have revealed other rules of thumb in counterfactual thinking which bias causal attribution; people focus on: errors of commission rather than omission; unusual pre-conditions rather than normality; and aspects perceived to be under the direct control of individuals.[21] Again, the point is not that the biases inherent to counterfactual thinking will lead to erroneous judgements *per se*, but that they encourage the neglect of the broader causal structures of events, discounting the influence of latent, systemic factors and privileging the narrative of narrow anomalies and isolated human errors. Of course, the human mind is not captive to its intuitive, subconscious system; when people are made aware of the biases governing their thought processes, whether through self-reflection or the help of others, their deliberative, rational system is quite capable of correcting for them.

5.2 Confirmation Bias and Groupthink

Above I shed light on some distortions in causal attribution by comparing how counterfactual thinking departs from the norms of logic and probability theory. However, counterfactual thinking is but one, cognitively oriented, way of framing the process of incident investigation. Another, complimentary, frame is to compare it to the scientific method; in other words, to view investigators as forming and testing hypotheses which speak to the causal structure underpinning an event. This is useful as psychological research has shown that, in contrast to the tenets of the scientific method, people more commonly seek to confirm rather than reject their hypotheses, and have a strong tendency to seek out information in support of them, whilst ignoring or neglecting disconfirming evidence.[23] This is known as confirmation bias. It means that preconceived norms, values and beliefs about system safety are somewhat insulated from critical inquiry, radically restricting the scope for what can be learnt from a crisis.

A similar phenomenon occurs at the social psychological level, known as groupthink.[24] It's best viewed as a type of thought exhibited by group members who try to minimize conflict and reach consensus whilst avoiding the critical testing and evaluation of ideas. Here, people have a preference for consensus over truth, and often mistake the former for the latter. This may stem from the desire to avoid ridicule, embarrassing colleagues, or threatening the status quo or existing power structures. This can lead investigations to freeze on an articulated conventional wisdom in an excessively hasty manner. For example, in an enquiry into the Battleford outbreak, one water plant operator indicated that he was aware of cryptosporidium and the health risk it potentially held, and moreover was aware of misconceptions held by others: that chlorine could kill cryptosporidium.[9] Yet whenever he tried to correct these misconceptions, he was, in his words, "ridiculed." This neatly illustrates not just the malign influence of groupthink, but also that its distorting effect extends beyond simply incident investigations to suppressing knowledge of hazards before they are realised.

5.3 Hindsight Bias

A further crucial psychological constraint is hindsight bias, which refers to the tendency to view past events as more predictable and inevitable than they in fact were.[19] At root, it stems from a deterministic view of the world, and a subconscious tendency to be wise after the event. There are two key ways in which it makes learning history's lessons problematic:[20] it distorts our perceptions of other people's responsibility for the outcomes of their decisions or actions by making them seem more foreseeable than they were; a more general consequence is that it limits the incentives for organizations to make fundamental changes following an incident, because if those reflecting on events consider that "they knew it all along", and that the outcome was in a sense inevitable, then they have little reason to change their assumptions, beliefs, and patterns of thinking.

In summary, by masking the uncertainty characterizing the tasks and situations of practitioners at the sharp-end of operations, hindsight bias encourages investigators to overestimate the former's knowledge and culpability, thus amplifying the pre-existing tendency to infer and focus upon human error to the detriment of more subtle and chronic weaknesses.

5.4 The Ego and its Defence Mechanisms

I shall end this section by discussing a rather Freudian concept, which is that organisations and their members seek to protect their emotional states and senses of identity in the face of crises, and that this desire can conflict with the goal of learning from failure.[15] Water quality incidents and near-misses, which are the triggers for organisational learning, tend to generate intense, negative emotions in individuals, such as fear, stress, anxiety, anger and guilt. Such emotional states elicit defensive responses that seek to ameliorate them and to return the mind to a state of equilibrium (*e.g.* denial, idealisation, *etc.*). There is a growing consensus that this applies not only at the individual level; similar defence mechanisms are activated to protect organisational self-esteem and self-identity in the face of crises. In other words, the need to protect the organisational ego encourages the re-affirming of core values, beliefs and norms about system operation in the event of an incident: those central, enduring and distinctive features that collectively define the organisation.[16]

This ego-defence becomes pathological when it occurs at the expense of dealing with the underlying problem: *i.e.* the cause of the failure. Let's turn to some concrete examples of these mechanisms. Most prominent amongst them is *denial* that there is a problem in the first place, for example, consider the Sydney "incident," where the first response of the CEO was to query the validity of the monitoring test results rather than to investigate the integrity of the system or to issue an immediate boil water notice, despite the high level of independent confirmation of results.[17] Further defence mechanisms include *rationalization* (*e.g.* focussing on the unique aspects of an incident so as to view it as idiosyncratic, or as an aberration beyond internal control), and *idealisation*, where one seeks to portray failure in a positive light (looking beyond the water sector, the Russian authorities sought to portray the Chernobyl disaster as a triumph of technology and human heroism, prompting one Ukrainian cynic to ask "Is there no room for improvement? Let's blow up one more unit to make the situation really splendid!"[18]). These, in concert with the other social and psychological constraints, can create in organisations a perverse resistance to meaningful learning.

6 CONCLUSIONS

I began by showing how the stigmatisation of failure and the strong desire for allocating blame and punishment in the aftermath of incidents encourages the suppression of error and cover-ups. I then argued that staff seek to frame and influence incident analyses in ways that protect their power, interests and authority, rather than being solely concerned with a detached search for the truth. I then summarised various cognitive biases which further distort attributions of causality. Combined, these constraints encourage the suppression of error, the denial of responsibility, the distortion of evidence, and a systematic tendency to explain away contamination events as narrow anomalies ("acts of God") or simple matters of "human error."

Given the pessimistic picture painted, it may be tempting to view learning from failure as beyond the bounds of possibility, and to see organisations as condemned, like Prometheus, to relive the pain from their pasts. Yet a mere glance at modern trends in mortality rates from waterborne diseases gives the lie to this position; the provision of an increasingly safe and reliable drinking water supply is perhaps the greatest public health achievement of recent times, and is a powerful testament to the fact that utilities can and frequently do learn meaningful lessons from near-misses and contamination events. Learning may be constrained by the influences discussed above, but it is not prevented. In the spirit of this more optimistic note, I conclude with a few practical principles which can help mitigate those social and psychological constraints:

- If reports are used to punish mistakes, employees will conceal or distort more information than when it is viewed as a sign of solidarity when one admits to mistakes that others can learn from.[25] Thus, incentivise, rather than penalise, error reporting;

- Protect the identity of reporters and whistleblowers, and shield them from unwarranted reprisals;[5]

- Maintain sufficient distance between those involved in investigations and those responsible for the processes being examined (so that they do not fear retribution or upsetting close colleagues, *etc.*), and ensure that those with overt conflicts of interest are screened from investigation teams;

- Ensure that investigators focus on uncovering the full causal schemas behind failures, rather than seek to identify an often convenient root cause (*i.e.* emphasise that human error or technical failure is rarely the end of the story);

- Designate a member of each investigation team to act as the devil's advocate[24], whose role would be to: directly tackle any issues that may normally be considered taboo; continually challenge the legitimacy of the rest of the team's assumptions, data and findings; and seek to re-interpret the evidence in support of alternative conclusions. In particular, this can provide a strong corrective to confirmation bias and groupthink.

- Make investigators aware of the common rules of thumb that people tend to rely on when determining causality. Revealing to them their likely sources of bias should reduce their inclination to believe that their judgements are purely fact-based, and so increase their openness to alternative hypotheses;[26]

- Seek out and provide staff with feedback that highlights the practical value of problem reporting and investigation, thus emphasising that their concerns are taken seriously and that changes are made in response where appropriate. In other words, show employees that the incident learning system is not merely another exercise in bureaucracy;

- Enhance the focus on identifying and ameliorating potential causes of failure *before* they arise, for example *via* simulation exercises, risk analyses, and so forth. This is valuable because the exploration of hypothetical failure scenarios gives rise to less conflict and power struggle than does interpreting actual failures from the past, because issues of blame and punishment are not relevant. Moreover, these approaches are ideally suited to imaginative thinking, which encourages the challenging of deeply ingrained organisational beliefs and assumptions about system safety which may otherwise remain sacrosanct under purely retrospective approaches to managing risk.

References

1   S.E. Hrudey, P. Payment, P.M. Huck, R.W. Gillham and E.J. Hrudey, *Water Science and Technology*, 2003, **47**, 7.
2   J. Reason, *Human Error*, Cambridge University Press, United Kingdom, 1990.
3   S.D. Sagan, *The Limits of Safety: Organizations, Accidents, and Nuclear Weapons*, Princeton University Press, New Jersey, 1993
4   C. Argyris, *Reasoning, Learning and Action: Individual and Organizational*, Jossey-Bass, San Francisco, 1982.
5   J. Reason, *Managing the Risks of Organizational Accidents*, Ashgate Publishing, United Kingdom, 1997.
6   N. Leveson, M. Daouk, N. Dulac and K. Marais, *A Systems Theoretic Approach to Safety Engineering*, 2003, available at:
    http://sunnyday.mit.edu/accidents/external2.pdf
7   E. W. Morrison and F. J. Milliken, *Academy of Management Review*, 2000, **25**, 706.
8   D.R. Ilgen, C.D. Fisher and M.S. Taylor, *Journal of Applied Psychology*, 1979, **64**, 349.
9   SERM, *Submission of the Province of Saskatchewan to the North Battleford Water Inquiry*, 2002, available at:
    http://www.northbattlefordwaterinquiry.ca/pdf/finalsubmission-GOVERNMENTOFSASK.pdf
10 K.E. Weick and K.M. Sutcliffe, *Organization Science*, 2005, **16**, 409.
11 C. Perrow, *Normal accidents: Living With High Risk Systems*, Basic Books, New York, 1984.
12 S.D. Sagan, *Journal of Contingencies and Crisis Management*, 1994, **2**, 228.
13 D. Vaughan, *The Challenger Launch Decision: Risky Technology, Culture, and Deviance at NASA*, University of Chicago Press, Illinois, 1996.
14 M. Easterby-Smith, *Human Relations*, 1997, **50**, 1085.
15 A. D. Brown and K. Starkey, *Academy of Management Review*, 2000, **25**, 102.
16 D.A. Gioia and J.B. Thomas, *Administrative Science Quarterly*, 1996, **41**, 370.
17 P. McClellan, *Sydney Water Inquiry - Final Report*, 1998.
18 D. R. Marples, *The Social Impact of the Chernobyl Disaster*, St. Martin's Press, New York, 1988.
19 B. Fischhoff, *Journal of Experimental Psychology: Human Perception and Performance*, 1975, **1**, 288.
20 H. Blank, J. Musch and R.F. Pohl, *Social Cognition*, 2007, **25**, 1.
21 N.J. Roese, *Psychological Bulletin*, 1997, **121**, 133.
22 A. Tversky and D. Kahneman, *Science*, 1974, **185**, 1124.
23 B. Brehmer, *Acta Psychologica*, 1980, **45**, 223.
24 I.L. Janus, *Groupthink*, Houghton Mifflin, Boston, 1982.
25 P. Mascini, *Journal of Contingencies and Crisis Management*, 1998, **6**, 35.
26 U. Bar-Joseph and A.W. Kruglanski, *Political Psychology*, 2003, **24**, 75.

LESSONS LEARNED FROM MAJOR CONTAMINATION INCIDENTS – A DISCUSSION

M. Furness

Consultant-Water Science and Public Health

1 INTRODUCTION

This paper was presented to stimulate debate and discussion from the delegates and highlight any areas where improvements in collective preparedness for water contamination incidents would be appropriate , or were needed.

2 THE PLANNING CYCLE

Preparedness is the key to demonstrating effective, fully integrated and timely responses to any operational incident. The typical Plan, Respond, Review cycle, Figure 1, is appropriate to all incidents, including water contamination incidents. The response stage contains those critical elements that decide success or otherwise; risk management and decisions, operational solutions, communication and remediation. The Review stage can then proceed to identify asset and procedural improvements and share the critical lessons learned. The findings, along with emerging issues, then aid the future planning and security scenarios.

Major water contamination incidents demand a multi-agency approach. Integrated decisions, responses and communication are very evident. There has to be a collective responsibility to act, based on sound reasoning, to protect public concern and health. Any subsequent reviews need to be internal (to the company affected) and external (other agencies involved) but also to non contributing organisations who have a direct interest in any findings- the industry, national or inter-national agenda. There appear to be few opportunities to undertake the latter and ensure all relevant practitioners receive the messages. The benefits to best practice adoption and collaborative research of this are self-evident and should also help clarify training needs and define 'exercise ' opportunities.

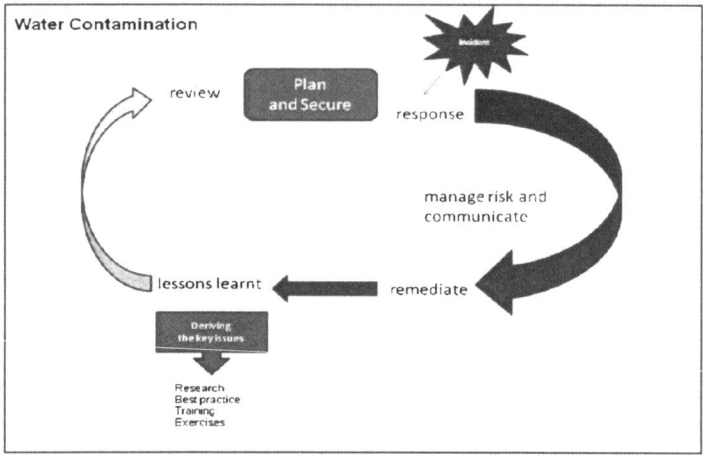

Figure 1 *The planning cycle*

3 LEARNING FROM INCIDENTS

A quick glance at recent major water contamination incidents, Table 1, shows 1) there quite a few in number, 2) there are a variety of causes but some common themes and 3) all have something to offer as learning points.

Table 1 *Major water contamination incidents*

1998	Glasgow, Scotland	Chemical
1998	Sydney, Australia	Cryptosporidium
2000	Taranto, Italy	Microbiological
2000	Clitheroe, England	Cryptosporidium
2000	Belfast, NI	Cryptosporidium
2000	Walkerton, Canada	Microbiological
2001	Asikkala, Finland	Microbiological
2001	N. Battleford, Canada	Cryptosporidium
2001	Hawkes Bay,	Microbiological
2005	Hurricane Katrina, USA	Floods
2005	NW Wales	Cryptosporidium
2005	Portsmouth, England	Cryptosporidium
2006	Cornwall, England	Chemical
2007	Galway, Ireland	Cryptosporidium
2007	Gloucestershire, England	Floods

All the incidents presented major challenges to the water company and involved agencies and some had a significant impact on public health. The causes and subsequent contamination of water were mostly well understood and could be classed as conventional. None would be classed as wilful contamination and thankfully this action is a rarity. The thwarted attempt to poison Rome's water supply in 2002 showed how real such an attack could be. In recent years more countries have stepped up their security measures for fear of terrorist activity. Water security has become more prominent in this third WCEC conference.

4   KEY LESSONS

The key lessons from most water contamination incidents are often quoted as:
- Catchment management
- Robust multiple barriers
- Capable and responsive operators
- Change often precedes incidents
- Hygienic operations
- Security vigilance
- Effective inter-agency co-operation
- Sound hazard and risk assessments
- Judgement based decisions

So, are we learning from past water contamination incidents? Are we refining and improving response procedures, where necessary? Where would we go for such information? The Chief Inspector's Report, DWI in the UK is one such source. It is now increasingly drawing on aspects of concern which could threaten water quality and referring these to water companies to note. Where an individual incident is particularly noteworthy then the specific circumstances are made available. There are text books which describe incidents and key learning points. Safe Drinking Water by Steve Hrudey is one such source. Accessing the internet can reveal a wealth of information on water quality but little in terms of collation of lessons to be learnt.

Within the UK Water Industry, there have been past calls for more sharing between water companies and the relevant agencies. At the first WCEC conference in 2003 two of the outputs were:

> "......*a forum to be sought whereby Water Industry Incidents may be reviewed and best practice shared on operational responsiveness, communication, cascades and water quality monitoring.*"

> "........*a public health network to be established.*"

At the second WCEC conference in 2005 there was again a call for:

> ".... "*a lessons learnt process and regular exercises, both nationally and internationally.*"

5   BEST PRACTICE

So, if better preparedness can come from learning lessons, sharing of best practice, training and emergency exercises, what is available nationally and internationally? Do the networks exist to promote such activity? This conference has shown us that there excellent examples of benefit to gained by the water industry, individual water companies and relevant agencies.

Some networks are well established, like the Water UK CBRN group, Public Health Network and Laboratory Users (Mutual Aid) group. The DWI has established formal liaison links with the HPA. The WSMART (Water Security, Management, Research and Technology) association is an alliance of international water utilities and related organisations to promote information sharing and collaboration.

In terms of best practice, we have over the past two days received presentations on Consequence Management Plans and the Response Protocol Toolbox from the EPA. Threat Management Plans have been utilised in the Victorian State Government and Scottish Water has developed its Waterborne Hazard Plan.

Few examples of training and exercises have surfaced at the conference. Most exercises that are organised tend to be tailored to localised circumstances. More sharing of outcomes would be beneficial. Water contamination training days are delivered by the HPA.

With respect to learning lessons, there are few examples of structured cascades and this conference continues to re-enforce past calls for such a programme.

The challenge then to delegates was should we do more? Drawing on the conference presentations were there initiatives, ideas, technologies that could take collective preparedness forward? More specifically and to stimulate comment:

- Do we ensure that effective networks with key players operate at the policy and technical levels?
- Is there a structured approach to learning lessons that cascades to the practitioners?
- Do we share best practice nationally and internationally?
- Are we collaborating on training and exercises?

6   CONCLUSION

An orchestrated audience discussion led to a general appreciation that sharing and learning experiences were invaluable and more could probably be done. The position in the UK regarding preparedness for water contamination incidents was felt to be well managed. Contributors from the USA and Australia were particularly keen to share their own approaches on enhancing and systemizing their water security and emergency procedures. It was concluded that more formalised networks should be encouraged and progress made to ensure tangible outputs from learning lessons and sharing best practice.

REVIEW OF CONFERENCE

J. Gray

Consultant – Water safety and security. 1 Faraday Ride, Tonbridge, Kent, TN10 4RL, UK

1 INTRODUCTION

This third international conference on water contamination, has attracted some 167 attendees from regulatory agencies, water companies, the health community, research organisations, academia and treatment and monitoring suppliers. There are delegates from 15 different countries who heard 47 presentations supported by 14 posters.

I hope in this short presentation to summarise the main points arising out of the various presentations and pose a few challenges to the water professionals. Time does not permit me to mention all contributors and to those I omit, I apologise.

2 OVERVIEW

Gordon Nichols set the scene with a consideration of past incidents, especially microbiological incidents, and what they should have taught us. In 28% of water contamination incidents, adverse weather conditions were a significant factor, and included drought as well as floods. Planning and preparedness, security and initial responses, HPA incident management, the summer floods, and the importance of regular contact between water companies and Health Protection Units were considered. He considered also the need to share lessons learned and to improve further the linkages between organisations. These themes recurred throughout the conference.

The Operations stream considered how different countries approached security of assets, assessment and management of risk, monitoring of the supply chain, provision of bottled water and decontamination after incidents.

The Data/Information stream considered the assessment of water quality in live systems, including rapid sensing technologies to detect contaminants and the management of radiological incidents.

The Communications stream received presentations on Emergency Planning approaches, water security initiatives, liaison with the HPA and the medical community, how to communicate key messages and the legal implications of a major incident.

The final sessions in the conference sought to bring together best practice and learning from lessons to stimulate delegate discussion.

Keith Weston gave an informative insight into what actions might take place should a terrorist linked incidence occur including an overview of command and control structures, media strategy and multi-agency working. There was also a timely reminder of the availability of published literature.

Martin Furness, Independent Consultant, gave an historical background on key water contamination incidents and resultant lessons and posed a number of questions to the audience as a challenge to stimulate discussion.

3 PRESENTATIONS

Gill presented an Australian view of water industry resilience and identified similarities between Australian and UK industries. He considered a risk based method based on All-Hazards approach and a multi-disciplinary build of business resilience and protection of critical infrastructure. Common issues of concern included e-security, business resilience, international cooperation and resilience models.

Haught considered on-line toxicity monitors to address some of the vulnerabilities of sources and distribution systems. He emphasised that early detection permitted early response and that there was a need for real-time quality information to permit informed decisions.

Pickard considered consequence management and described the Water Security Initiative and contaminant warning systems and their on-site testing. He emphasised the key role of consequence management planning to minimise response and recovery times and considered it vital to train and exercise and to develop and maintain good communications between key players.

Evans and Rink looked at the legislative and regulatory framework applicable to managing drinking water emergencies and identified areas of risk and future challenges. They considered the need for on-line early warning system, the vulnerability of distribution and SCADA systems.

Mills present the Phase 1 report prepared by Water UK's review group on the summer floods. Some 18 recommendations were made dealing with, among other issues, command and control structures.

Bradshaw considered diverse causes of incidents and interdependency of effects and identified the need for a competent approach to risk management, supported by a strong organisational structure. Consistent risk assessments were needed for robust management decisions. There was a need for continuous learning and training.

Danneels presented the risk assessment methodology (RAM-W) and confirmed that to identify critical assets may take prolonged discussion or even argument. There was a need to evaluate the threat spectrum. Understanding distribution systems leads to an understanding of contamination events. There was a need to develop and test contingency plans, to recognise cyber threat and to subsequently question and validate responses to audit.

A number of papers presented latest developments in analytical techniques.

Clay-Chapman of Three Valleys Water took the audience through the experience of the Buncefield fire and the resulting environmental impact that demanded intensive inter-agency working.

Fenton presented an overview of UK Emergency Planning including a new system where one company would act as coordinator for minor incidents. He confirmed that a national exercise was planned for 2010.

McGuinness presented an overview of the Scottish Waterborne Hazard Plan which included standard templates for press releases, letters, agendas. She confirmed that administrative staff are key assets and that accurate and timely information is critical. There was a need to review and refine external communications and to train.

Woolloff considered inter-agency communications and discussed challenges faced in inter-agency incidents with regard to control and public communication. He confirmed the need to develop a response plan which was subject to review and refinement.

Meinhardt considered medical preparedness and confirmed that, as water supplies are potential targets, early detection and response was critical. Since a probable early indicator will be change in disease and illness patterns, she considered that health professionals were key in minimising effects by being (suspiciously) aware of potential for attack. She concluded that preparation makes the difference between controlled response and a public health crisis.

Scoggins considered liability and blame when it does go wrong. He identified that there were other potential concerns for water companies than being sued, including an internal enquiry, the DWI report, the public enquiry, a possible inquest, prosecution, attention of the media, central and local government, employees and disciplinary hearings. If there are no records covering the basic questions of who, what, when, how and why, then there is no proof. Evidence may be needed for events before the day and during the event. Key questions include are roles defined, is there adequate training and are there appropriate plans, procedures, support and resources? Could it have been done better?

Shaw considered reputation and the need for complex messages to be communicated clearly despite competing expectations of the board, shareholders, customers, the public, and regulators etc. There must be realistic targets which are monitored regularly to manage stakeholder expectations. Audiences should be researched before communicating with them with consistent messages.

McGillvray emphasised the need to learn from failure and reflected on the background of uncomfortably similar preventable events and near misses. He asked why do we not learn from the past? There is no single co-ordinated database and a skewing of data and selective recollection arising from a need to protect organisational egos. This can lead to rationalisation and denial of failure. He confirmed that investigations must be non-adversarial to permit learning.

4 CONCLUSIONS

Despite a generally excellent record as regards drinking water quality, there have been significant challenges arising from specific water quality incidents, both accidental and intentional. Even an unfulfilled threat to contaminate drinking water requires significant resources to respond effectively, appropriately assess risk and then protect consumers. In many countries, maintaining water security requires significant effort with concerns that water supplies could be a possible terrorist target. Indeed, it has been said that an intentional attack of drinking water supplies must be regarded as inevitable[1].

This conference has shown clearly how we all plan for, manage risks associated with and respond to water contamination incidents, including the development of communication strategies. However, after the incident, how do we share, learn and develop? It seems to me that whilst the water industries in the UK and other countries can be justifiably proud of their drinking water quality, there remains no formal system to take forward some important issues, most notably the sharing of lessons.

The aim of this conference was to promote the value of effective networks to share best practice. Examples have been presented and potential gaps identified. Could we in the UK learn from the Consequence Management Plans used in Australia? Should the UK have an anglicised version of the EPA's Response Protocol Tool Box? Where is the UK's structured lessons learned programme available to all practitioners? Should there be more links with the Health Protection Agency and the water industry at both policy and technical level? The international community, notably from the USA and Australia are keen to promote further learning and sharing.

When considering collective responsibility when faced with an emergency, there is an essential requirement for all participants to understand how the organisation and the individual plans, responds, communicates and manages risk. I question whether, following an incident, organisations and individuals learn, share best practice, have effective networks, collaborate and target research and share in exercises. I remain unconvinced that there is a common set of values for evaluating risk and prioritisation, nor an enthusiasm and desire to make a common approach work. Organisations appear reluctant to modify their own systems where necessary to achieve these aims.

Collective responsibility is all about the need to understand dynamics and psychology of teams, where an individual's role in a team needs to be clarified and understood by all. Team roles are difficult to maintain when structures of organisations frequently change and individuals move on. There is a need to recognise the requirements of the task, the team and the individual. Good teams are characterised by smooth professionalism. When it "hits the fan" there will be no time to discuss who does what on site. All should be task focussed. It can be invaluable to exchange roles beforehand to increase mutual understanding. Exercises provide a perfect opportunity for this in a risk-free and non-confrontational environment. There must be open and honest debriefing afterwards to identify and share lessons learned.

Consumers (and indeed shareholders and regulators) need to be reassured that water companies and the relevant agencies are constantly reviewing practices, procedures and structures to keep water supplies safe and secure. At the end of the day it is incumbent on everyone involved in emergency response to ask themselves the question: "Am I that good that others want to work with me?"

References

1   M.S. Field, Assessing the risks to drinking-water supplies from terrorist attacks in Water Supply Systems Security, L.W. Mays, McGraw-Hill, 2004. 0 07 142531 4

LIST OF POSTERS

Biodetection in drinking water
S. Cunningham[1] and B Dowling[2].
[1]Dstl Porton Down, Salisbury, Wiltshire SP4 0JQ, UK. [2]Health Protection Agency Newcastle Laboratory, Institute of Pathology, Newcastle General Hospital, Westgate Road, Newcastle upon Tyne. NE4 6BE, UK

Making rapid point-of-need immunoassays work for natural and synthetic toxins and pollutants
C.H. Self.
Selective Antibodies Ltd, CELS at Newcastle, 1st Floor William Leech Building, Framlington Place, Newcastle-upon-Tyne NE2 4HH, UK

Online monitoring of radioactivity
C.E. Engeler, J.van Steenwijk and O.J. Epema.
Netherlands Center for Water Management (RWS WD), Directorate General of Public Works & Water Management (RWS), Ministry of Transport, Public Works & Water Management, Zuiderwagen-plein 2, NL 8224 AD, P.O. Box 17, NL 8200 AA, Lelystad, The Netherlands.

Toxscreen3 - a rapid and sensitive bioluminescence-based test for early warning of drinking water contamination
N. Ulitzur.
CheckLight Ltd, P.O.Box 72, Qiryat Tivon 36000, Israel

Laboratory proficiency testing for potable water contamination emergencies
B. May.
LEAP, Central Science Laboratory, Sand Hutton, York YO41 1LZ, UK

Iron sulphide – a broad spectrum decontaminant for aqueous matrices
M.J. Ashworth, I.W. Croudace, J.H. P Watson and C.W. Keevil.
School of Biological Science, University of Southampton, Bldg. 62, Boldrewood Campus, Southampton, SO16 7PX, UK

Requirement for molecular methods to detect pathogens in potable water systems
S.A. Wilks and C.W. Keevil.
School of Biological Science, University of Southampton, Bldg. 62, Boldrewood Campus, Southampton, SO16 7PX, UK

Combination of a biological sensor using light emitting bacteria and a UV spectrometer probe for water quality monitoring and drinking water security monitoring
J. Appels.
microLAN On-line, Biomonitoring Systems, P.O. Box 640, NL5140 AP Waalwijk, The Netherlands

Increased sensitivity with combination of on-line solid phase extraction and bioluminescent toxicity monitor for water quality and drinking water security monitoring
J. Appels.
microLAN On-line, Biomonitoring Systems, P.O. Box 640, NL5140 AP Waalwijk, The Netherlands

Mutual aid radioactivity sub-group intercomparison exercises for gross alpha and beta analysis of potable water samples
P. Frewin[1] and K.C. Thompson2.
[1]South West Water, Exeter Laboratory, Bridge Rd, Countess Wear, Exeter EX2 7AA, UK.
[2]ALcontrol Laboratories, Templeborough House, Mill Close, Rotherham S60 1BZ, UK

Can chlorophyll fluorescence emitted by the natural flora within water supplies be used to detect contamination events?
K Oxborough.
Chelsea Technologies Group Ltd, 55 Central Avenue, West Molesey, Surrey, KT8 2QZ, UK

Recent improvements in detection limits of trace elements in water by ICP-OES spectrometry using radial plasma viewing
M. Haigh and O. Schultz.
Spectro Analytical Instruments, Boschstrasse 15, Kleve, Germany

The use of fast gas chromatography-time of flight-mass spectrometry (GC-ToF-MS) for the rapid analysis of water for the presence of environmental contamination
R. Brown and S. Kippin.
LGC Ltd, Queens Road, Teddington TW11 0LY, UK

RSC Publications
Royal Society of Chemistry
Thomas Graham House, Science Park, Milton Road, Cambridge, CB4 0WF, UK

Subject Index

2,4-D, 248, 249
2,6-Dibromophenol, 248

aflatoxin, 234
aldicarb, 125, 126, 127, 234, 248
algae, 25, 219, 309
algae Toximeter, 25
alpha, 239, 285, 286, 287, 292, 300, 305
Am-241, 264
American Water Works Research Foundation, 126
amitrole, 248, 279
ammonia, 28, 74, 75, 76, 219, 221, 276, 281
aniline, 248
anions, 221, 269, 270, 281
antimony, 269
aromatic hydrocarbons, 222
arsenic, 198, 199, 200, 202, 203, 204, 205, 269
asulam, 273, 274
atrazine, 21, 274
AwwaRF, 126, 192, 193, 194
Azinphos – methyl, 234

B. cepacia, 344, 345, 346, 347, 351
Bairnsdale Sewerage Treatment Plant, 16
Battleford, 391, 393, 396, 398
beta, 281, 285, 286, 287, 288, 290, 292, 304, 305, 406
biofilms, 155, 165
biosecurity, 243, 244
bottled water, 4, 6, 41, 107, 139, 140, 141, 142, 143, 146, 147, 148, 149, 152, 153, 168, 173, 180, 181, 184, 250, 303, 374, 377, 401
Bovine Viral Diarrhoea Virus, 243
bowser, 6, 139, 142, 143, 149, 183
British Standards Institution, 207
BS 6068, 211, 212
Buncefield, 57, 252, 253, 256, 258, 259, 354, 402
Burkholderia cepacia, 343

cadmium, 269
Campylobacter, 4
carbon, 7, 22, 121, 236, 241, 243, 249, 258, 275, 276, 278, 280, 309, 310, 355, 356
Castlemeads, 179
CBRN, 2, 128, 129, 130, 131, 132, 134, 135, 136, 137, 277, 279, 400
Ce-144, 263
centrifugal filtration, 343, 347
Chemical Hazards and Poisons Division, 255, 325, 326, 330, 333, 335, 336, 353
chemiluminescence, 220, 221, 222, 224, 282, 291
chemiluminescent, 282, 283
Chernobyl, 262, 264, 266, 394, 396
chlorine, 25, 28, 44, 58, 69, 72, 73, 74, 75, 76, 77, 79, 118, 119, 121, 122, 125, 133, 159, 160, 163, 164, 222, 234, 268, 278, 282, 306, 308, 391, 393
chlorpyriphos-ethyl, 245
chlorpyriphos-methyl, 245
clarification, 299, 356
climate change, 166, 171, 172, 174, 187
Clostridium novyi, 4
Clostridium perfringens, 4, 222, 224
Co-60, 264
cobalt-60, 261, 265
coliphage, 201
colony count, 222, 224
communication, 1, 2, 14, 40, 47, 50, 52, 54, 82, 84, 102, 103, 104, 109, 168, 180, 181, 182, 184, 209, 218, 227, 317, 336, 339, 340, 360, 361, 362, 364, 365, 366, 367, 375, 386, 387, 388, 397, 399, 403
communications, 2, 4, 14, 40, 48, 52, 54, 56, 60, 84, 175, 177, 181, 182, 184, 208, 290, 314, 318, 337, 339, 383, 384, 385, 387, 389, 402, 403
conductivity, 26, 31, 44, 69, 71, 72, 73, 76, 219, 220, 233, 234, 269, 276, 281, 306, 310
consequence management plan, 45

Consumer Council for Water, 175, 181, 190, 191

contaminant, 1, 21, 26, 28, 44, 69, 95, 117, 118, 119, 122, 123, 124, 125, 131, 156, 157, 158, 159, 160, 161, 163, 164, 194, 199, 220, 229, 231, 233, 237, 247, 249, 250, 270, 279, 284, 291, 306, 307, 308, 309, 321, 322, 323, 373, 374, 378, 379, 402

contaminant warning system, 125

contaminants, 22, 23, 28, 29, 44, 45, 59, 69, 90, 117, 125, 135, 155, 156, 158, 160, 164, 194, 195, 196, 199, 219, 229, 231, 232, 233, 234, 245, 246, 247, 248, 249, 250, 257, 270, 278, 279, 282, 290, 291, 294, 308, 321, 322, 323, 324, 342, 370, 373, 374, 377, 379, 401

contamination, 1, 2, 4, 5, 6, 7, 21, 22, 25, 28, 34, 44, 45, 47, 48, 49, 50, 52, 53, 54, 57, 59, 63, 77, 87, 100, 117, 118, 119, 121, 122, 123, 124, 125, 127, 129, 130, 131, 132, 133, 134, 135, 136, 137, 140, 144, 147, 155, 156, 157, 158, 160, 164, 165, 194, 196, 198, 199, 200, 201, 203, 206, 213, 218, 219, 220, 221, 222, 225, 226, 227, 228, 229, 237, 243, 244, 245, 250, 252, 253, 254, 258, 267, 268, 275, 276, 277, 282, 287, 289, 291, 298, 300, 301, 302, 303, 304, 305, 309, 314, 316, 317, 321, 322, 323, 325, 326, 328, 337, 338, 349, 355, 360, 361, 362, 363, 364, 365, 366, 367, 369, 370, 371, 372, 373, 374, 375, 376, 377, 378, 379, 380, 381, 389, 390, 394, 395, 397, 398, 399, 400, 401, 402, 403, 405, 406

contamination warning system, 44, 45, 53

contingency plans, 195

cost benefit analysis, 109

Council Food Intervention Levels, 285

Critical Infrastructure Modelling and Analysis Program, 11

Critical Infrastructure Protection, 11, 12, 20

cryptosporidiosis, 4, 5, 8, 371

Cryptosporidium, 4, 5, 8, 47, 201, 202, 227, 371, 381, 398

Cryptosporidum parvum, 201

Cs-137, 262, 264, 265

cyanide, 29, 221, 276, 281

Daphnia, 21, 24, 283

Daphnia magna, 21, 24, 283

decontamination, 155, 156, 160, 161, 163, 164, 226, 320, 323, 324, 326, 401

derivatisation, 246, 270, 271

dicamba, 245, 248, 249, 279

dichlobenil, 245

diesel, 58, 163, 275, 276, 280, 281, 386

diode array detection, 245

diode array ultraviolet spectrometry, 273

diquat, 274, 276

dissolved air flotation, 299

dissolved oxygen, 28, 30, 219, 220, 282

distribution system, 21, 23, 25, 29, 38, 44, 45, 47, 53, 58, 68, 69, 71, 74, 76, 77, 81, 88, 91, 101, 117, 118, 122, 125, 127, 156, 158, 159, 160, 161, 194, 195, 196, 201, 202, 220, 228, 229, 277, 288, 306, 307, 309, 322, 323, 377, 378

DOC, 230, 231, 233, 235, 236

drainage, 160, 164, 171, 178, 185, 186, 187, 188

Drinking Water Inspectorate, 8, 55, 56, 57, 60, 67, 94, 115, 144, 266, 314, 331, 333, 351, 355

drought, 4, 5, 9, 15, 140, 146, 401

due diligence, 139, 140, 142, 143, 145, 146, 147, 148, 153, 154, 313

DWI, 8, 56, 58, 61, 62, 139, 144, 145, 148, 184, 255, 258, 266, 286, 288, 314, 333, 351, 399, 400, 403

E. coli, 6, 163, 201, 371, 391, 392

E.coli O157, 201, 202

early warning system, 21, 34, 126

ECLOX, 220, 221, 222, 224, 282

ecotoxicity, 267, 268, 277, 282, 283, 290, 292

emergency, 1, 2, 6, 7, 10, 15, 17, 18, 26, 39, 42, 45, 47, 51, 52, 54, 55, 56, 58, 60, 82, 85, 103, 107, 140, 142, 144, 145, 146, 147, 149, 152, 166, 167, 168, 169, 170, 172, 174, 175, 176, 177, 178, 179, 180, 181, 182, 183, 184, 185, 189, 195, 212, 218, 220, 221, 245, 246, 247, 249, 250, 252, 257, 259, 267, 268, 269, 270, 271, 272, 277, 279, 280, 281, 282, 283, 285, 286, 287, 288, 289, 292, 300, 304, 305, 313, 314, 315, 320, 321, 323,

325, 328, 333, 334, 338, 339, 340, 354, 355, 356, 360, 367, 369, 370, 372, 373, 374, 376, 379, 383, 384, 386, 387, 388, 399, 400, 404

Emergency Management Australia, 10, 19

Event Monitor, 71, 72

exercises, 41, 42, 53, 84, 145, 195, 227, 246, 279, 286, 287, 288, 292, 395, 399, 400, 404, 406

fenamiphos, 234
fenchlorphos, 245
fenitrothion, 245
Flame Ionization Detector Gas Chromatograph, 22
flocculation, 299
flood prevention, 139, 170, 190
flooding, 4, 6, 7, 16, 60, 141, 144, 160, 167, 169, 170, 171, 172, 173, 174, 175, 176, 178, 179, 181, 185, 186, 187, 188, 189, 190, 229, 259, 354, 355, 356
fluorescent reporter molecules, 238

gas chromatography, 222, 245, 270, 406
Gippsland, 15, 16, 17, 18, 20
glyphosate, 248
gross alpha, 281, 285, 286, 287, 288, 290, 292, 406
groundwater, 5, 57, 188, 200, 206, 253, 258, 263, 265, 266

HACCP, 38, 39, 86
Hach, 68, 69, 71, 77, 126
health protection, 6, 8, 356
Health Protection Agency, 4, 6, 8, 170, 177, 180, 184, 191, 255, 285, 296, 305, 325, 331, 335, 336, 338, 342, 348, 353, 354, 355, 356, 400, 404, 405
herbicides, 222, 270, 272, 273
high performance management, 82
Homeland Security Presidential Directive, 44, 192, 196, 320
HPLC, 21, 223, 224, 245, 270, 271, 273
HPLC-DAD, 21

IAEA, 264, 265, 266
ICP-MS, 269
immunocompromised, 143, 378

incident management, 2, 4, 60, 86, 96, 100, 101, 103, 104, 105, 106, 107, 108, 109, 114, 258, 259, 333, 338, 401
incidents, 1, 4, 6, 7, 8, 10, 14, 45, 53, 58, 71, 80, 82, 84, 86, 88, 89, 90, 91, 92, 93, 94, 96, 97, 98, 99, 100, 101, 102, 103, 104, 105, 106, 107, 108, 109, 110, 112, 114, 129, 155, 167, 170, 172, 177, 178, 181, 188, 193, 259, 262, 263, 264, 266, 267, 268, 269, 272, 277, 279, 280, 285, 287, 289, 292, 302, 305, 314, 315, 316, 318, 321, 325, 326, 327, 328, 329, 330, 331, 332, 333, 334, 356, 389, 390, 392, 394, 397, 398, 399, 400, 401, 402, 403
Intelligent Aquatic Biomonitoring System, 24
ISO TC224, 39
Isoproturon, 234

Kr, 265
Kyshtym, 263

Laboratory Environmental Analysis Proficiency, 245
LD50, 223, 233
lead, 41, 55, 85, 108, 143, 144, 147, 148, 158, 175, 180, 188, 198, 200, 211, 219, 232, 245, 269, 296, 302, 314, 315, 320, 324, 337, 338, 340, 361, 362, 363, 369, 370, 371, 372, 387, 393, 403
LEAP, 245, 246, 247, 248, 249, 271, 279, 288, 292, 405
learning, 2, 7, 14, 17, 82, 84, 101, 102, 112, 231, 288, 389, 390, 391, 392, 393, 394, 395, 398, 399, 400, 402, 403, 404
Legionella, 212, 227
Lepomis macrochirus, 24
linuron, 234
Little Miami River, 31
LSD, 234

mass spectrometer, 222
mass spectrometry, 237, 245, 269, 270, 279, 406
MCPA, 248, 249
membranes, 233, 343, 344, 345, 346, 347, 348, 349, 351
mercury, 269, 323

metamidophos, 234
methyl tertiary butyl ether, 57
methylene Chloride, 22
mevinphos, 234, 248, 279
microsporidiosis, 4
Milwaukee, 371, 381
Mississippi River, 30, 34, 35
MS, 224, 227, 245, 269, 270, 271, 272,
    273, 275, 278, 281, 293, 294, 309, 406
MS/MS, 245, 270, 271, 272
Musselmonitor, 219
mutual aid, 42, 140, 142, 153, 166, 167,
    168, 169, 172, 176, 178, 179, 183, 184,
    267, 272, 286, 287, 314, 315, 400
Mythe, 6, 7, 140, 141, 142, 145, 146,
    147, 149, 153, 173, 175, 178, 184, 188

National Homeland Security Research
    Center, 54, 126, 312, 320, 321, 324,
National Poisons Information Service,
    326, 335
National Resilience Working Group, 10
nitrate, 28, 219, 230, 237
nmr, 245, 246, 247, 248, 249, 250, 278,
    279, 291, 293, 294
NPIS, 326
nuclear   magnetic   resonance
    spectroscopy, 245

Office of Civil Nuclear Security, 263
Ohio River, 22, 30
Olympic Games, 218, 227
online toxicity monitors, 21
organisational reliability, 82
organisational resilience model, 11, 12
oxamyl, 234

paraquat, 274
parathion-ethyl, 245
parathion-methyl, 246
paratyphoid, 4
PAS, 115, 210, 212, 213, 214, 215
pathogen, 117, 201, 202, 243, 371, 378
PCR, 227, 238, 239, 240, 241, 242, 243,
    244, 276, 342, 343, 344, 345, 346, 347,
    348, 349, 350, 351
pentachlorophenol, 246, 249
pentachlorophenol, 248, 249
perfluoro octanoic acid, 257
perfluorooctane sulphonate, 57

pesticides, 57, 155, 219, 222, 233, 270,
    272, 273, 290, 323, 330
PFOA, 257
PFOS, 57, 257, 258
pH, 26, 28, 29, 30, 69, 71, 72, 73, 74, 75,
    76, 118, 133, 219, 220, 221, 222, 224,
    233, 247, 276, 306, 308, 309, 310
phenol, 274
phosphamidon, 248
photoionisation, 275
picric Acid, 248, 279
Pitt Review, 8, 140, 170, 175, 259, 356
plutonium, 262, 263
polyaromatic hydrocarbons, 57
polymerase chain reaction, 238, 342, 351
polynuclear aromatic hydrocarbons, 222
Pr-144, 263
protozoa, 203, 221, 225, 351
Pseudomonas aeruginosa folliculitis, 4
Pu-238, 264, 265
public health, 4, 6, 7, 26, 27, 50, 56, 59,
    101, 110, 114, 117, 139, 144, 154, 167,
    169, 177, 178, 181, 184, 198, 199, 202,
    204, 205, 229, 256, 257, 266, 303, 308,
    318, 320, 321, 323, 325, 326, 327, 333,
    334, 337, 338, 352, 353, 355, 360, 363,
    365, 366, 367, 369, 370, 371, 372, 373,
    374, 375, 376, 378, 379, 382, 391, 395,
    398, 399, 403
public health, 28, 45, 166, 168, 169, 184,
    228, 367, 370, 372
Publicly Available Specifications, 210
pyridine, 248, 279

Q-TOF Mass Spectroscopy, 21

radioactivity, 219, 228, 265, 267, 268,
    285, 286, 287, 288, 289, 290, 292, 297,
    298, 305, 405, 406
radiological incident, 296, 297, 298, 304,
    305
radionuclides, 44, 219, 221, 225, 261,
    262, 264, 286, 296, 297, 298, 300, 303
radon, 285
RAM-W™, 192, 193, 194, 197
rapid methods, 267, 288, 342
redox potential, 219
regulations, 5, 36, 68, 143, 144, 147, 148,
    184, 209, 342

remediation, 1, 2, 26, 47, 48, 50, 53, 170, 190, 260, 378, 397
resilience, 8, 9, 10, 11, 12, 13, 14, 15, 17, 18, 19, 20, 40, 83, 100, 109, 114, 142, 147, 152, 154, 166, 169, 176, 177, 183, 189, 338, 402
resilience, 10, 11, 12, 13, 14, 18, 19, 20, 166, 170, 177, 184, 188, 190, 314
Response Protocol Toolbox, 54, 228, 400
reverse osmosis, 258
ricin, 163, 227, 234
Risk Assessment Methodology for Water Utilities, 197
risk management, 12, 13, 14, 37, 39, 58, 62, 67, 108, 115, 130, 136, 212, 215, 317, 361, 397, 402
River Meuse, 21
rotavirus, 201, 202

Sale Water Treatment Plant, 16
Salmonella, 4, 202, 224
Salmonella Enteritidis, 4
Sandia, 126, 192, 193, 194, 196, 197
Sandia Corporation, 126, 196
Sandia National Laboratories, 192, 197
saxitoxin, 234
SCADA, 57, 58, 59, 60, 312, 321, 402
scientific and technical advisory cell, 6, 353
security, 1, 2, 4, 10, 13, 29, 36, 40, 41, 42, 43, 44, 47, 55, 56, 59, 60, 62, 63, 64, 66, 68, 71, 80, 81, 128, 151, 152, 153, 169, 183, 192, 193, 194, 195, 196, 212, 213, 214, 215, 216, 218, 219, 220, 221, 227, 228, 232, 237, 263, 264, 265, 290, 305, 307, 310, 313, 314, 320, 321, 369, 370, 371, 373, 381, 397, 399, 400, 401, 402, 403, 405
Security and Emergency Measures Direction, 55, 56, 141, 143, 145, 148, 170, 175, 180, 183, 185, 313
Security Vulnerability Risk Assessment Guideline, 11, 12
selenium, 269
Sellafield, 262, 263, 266
sensors, 28, 29, 69, 76, 77, 79, 117, 118, 119, 121, 123, 125, 129, 131, 132, 133, 134, 136, 137, 219, 234, 239, 308, 322
Shigella dysenteriae, 202
Shigella sonnei, 4

simulation, 117, 121, 122, 124, 125, 129, 132, 136, 156, 395
sodium hypochlorite, 58
solid phase extraction, 405
solid phase microextraction, 222
solvents, 155, 222, 249, 275, 276, 281
spatially resolved delta spectroscopy, 236
SPE, 223, 224, 246, 249, 272, 281, 351
spores, 155, 158, 160, 163, 242, 351
Sr-90, 263, 264, 265
STAC, 6, 177, 353, 354, 355, 356
SUDS, 187, 188
Supervisory Control and Data Acquisition, 57, 59, 102, 113
Sustainable Urban Drainage System initiative, 187
syndromic surveillance, 378
Synechococcus, 17

tankers, 139, 141, 142, 143, 144, 145, 150, 151, 152, 153, 166, 168, 176, 178, 179, 180, 181, 183, 184, 303
telemetry, 22, 25, 30, 31
temperature, 26, 30, 147, 219, 220, 233, 241, 242, 247, 268, 279, 294, 310, 343
terrorism, 37, 40, 140, 229, 366, 367, 369, 370, 375, 376, 382
thallium, 269
Thamnocephalus platyurus, 269, 283, 284
Three Mile Island, 262, 266
TOC, 28, 29, 69, 72, 75, 118, 122, 123, 126, 219, 221, 222, 224, 230, 232, 233, 234, 276, 280, 306, 308, 309, 310, 311, 312
total organic carbon, 28, 44, 69, 118, 119, 121, 125, 133, 276, 306
Town & Country Planning Act 1990, 186
ToxControl system, 25
toxicity, 21, 22, 25, 28, 29, 30, 31, 35, 125, 219, 220, 221, 223, 269, 272, 276, 280, 282, 283, 291, 292, 373, 374, 402, 405
toxins, 24, 44, 155, 221, 225, 227, 233, 250, 282, 378, 381, 405
ToxProtect, 25, 29
training, 8, 41, 47, 52, 53, 54, 58, 60, 82, 84, 101, 102, 112, 113, 114, 129, 167, 177, 178, 207, 216, 239, 243, 244, 287, 296, 298, 302, 304, 305, 312, 334, 340,

354, 356, 366, 367, 369, 397, 399, 400, 402, 403
tritium, 261, 263
tubewell, 203
tubewells, 198, 202, 203, 204
turbidity, 5, 16, 26, 28, 29, 30, 69, 72, 73, 75, 76, 77, 79, 118, 133, 219, 220, 230, 231, 232, 233, 235, 237, 274, 276, 308
Turin, 218, 219, 220
typhoid, 4

U.S. Environmental Protection Agency, 21, 44, 54, 126, 155, 165, 320, 324, 372
UK Recovery Handbook, 292, 296, 297, 301, 302, 303, 304, 305
uranium, 260, 261, 263, 269, 292
USEPA, 35, 42, 54, 228, 290, 306, 307, 308, 310, 312, 324
UV/VIS spectrometry, 231
UV/Vis spectroscopy, 229, 236
UV-Vis spectra, 219

vandalism, 40, 56, 61, 141, 152, 229, 321
Vibrio fischeri, 25, 30, 282, 283, 291
Vienna Waterworks, 232, 233
vulnerability, 5, 13, 14, 57, 58, 59, 60, 62, 95, 168, 171, 179, 192, 193, 194, 196, 219, 321, 370, 374, 381, 402
vulnerability assessment, 193, 219
vulnerable populations, 360, 362, 365, 366, 370
VX, 234

Walham, 7, 179

Walkerton, 371, 391, 398
Water Industry Act 1991, 55, 56, 115, 185, 341
Water Infrastructure Protection Division, 320
water quality data, 21, 235
water quality monitoring, 21, 22, 23, 29, 30, 33, 44, 70, 236, 306, 308, 399, 405
Water Safety Plan, 38, 39, 57, 67
Water Security, Management, Research and Technology, 400
Water UK, 139, 140, 141, 142, 143, 144, 145, 147, 166, 167, 169, 170, 173, 176, 178, 181, 183, 184, 187, 189, 191, 272, 315, 400, 402
Water Undertakers (Information) Direction 2004, 94, 115, 341
waterborne disease, 4, 367, 369, 370, 371, 372, 373, 375, 379, 380
waterborne outbreak, 371
WHO, 36, 38, 39, 67, 198, 199, 200, 201, 202, 203, 204, 205, 219, 228, 290, 327, 337
WHO Guidelines for Drinking-Water Quality, 201, 205
Windscale, 262, 264, 266
WIPD, 320
World Health Organisation, 38
W-Smart, 39, 42
WSMART, 400

Xn, 265

Zr-95/Nb-95, 263